RENEWABLE
ENERGY IN CITIES

RENEWABLE ENERGY IN CITIES

Center for Renewable Resources

VNR VAN NOSTRAND REINHOLD COMPANY

Manufactured in the United States of America

Published by Van Nostrand Reinhold Company Inc.
135 West 50th Street
New York, New York 10020

Van Nostrand Reinhold Company Limited
Molly Millars Lane
Wokingham, Berkshire RG11 2PY, England

Van Nostrand Reinhold
480 Latrobe Street
Melbourne, Victoria 3000, Australia

Macmillan of Canada
Division of Gage Publishing Limited
164 Commander Boulevard
Agincourt, Ontario MIS 3C7, Canada

15 14 13 12 11 10 9 8 7 6 5 4 3 2 1

Library of Congress Cataloging in Publication Data

Main entry under title:

Renewable energy in cities.

Includes bibliographical references and index.
    1. Renewable energy souces—United States. 2. Cities
and towns—United States—Energy consumption. I. Center
for Renewable Resources (U.S.)
HD9502.U52R45 1984   333.79'3   83-23455
ISBN 0-442-21654-8

PROJECT STAFF

Project Manager: Beth McPherson

Principal Author and Research Director: Gerald Mara

Editor: Kathleen Courrier

Research Associate and Contributing Author: Margaret Hilton

Contributing Authors and Principal Consultants: Kathleen Courrier, Kevin Finneran, Jack Gleason, William Snyder

Research Assistants: Karl Golovin, Daniel Brasuk, Rick Beckwith, Tom Giller

Administrative Manager: Linda Doherty Marsh

Production Assistants: Paula Volk, Page DeMello

DISCLAIMER

NOTICE

Preface

Since research for this study began in 1979, the renewable energy scene has changed in several ways, some of which have hastened solar development whereas others have set it back. The most visible change occurred in the Federal solar research and development budget, which dropped from almost $600 million in 1981 to about $175 million in 1984. Efforts at demonstration and commercialization were virtually eliminated during this period as federal priorities shifted to long-range, high-risk research. Moreover, the Solar Bank, which subsidizes loans to low- and middle-income people for solar and conservation measures, is now receiving only $25 million a year instead of the $300 million originally authorized. The economic recession of 1981 and 1982 also hurt solar development. The demand for new housing dropped dramatically, and builders were reluctant to adopt conservation measures, solar systems, or anything else that would boost the price of a new home. Likewise, individuals and city governments had less money to spend on renewable energy investments. When oil prices finally stabilized in 1983 and the housing slump ended, energy seemed a less pressing issue than it had been during the "energy crisis."

Other variables, meanwhile, were encouraging solar development. The Public Utility Regulatory Policies Act (PURPA), passed in 1978 to promote small, nonutility electric generation from renewable energy sources and cogeneration, was ineffective in its early years because of court challenges from the utility industry. These delays ended, however, when the Supreme Court upheld all PURPA's major provisions in a June 1983 ruling, and utilities are now required to interconnect with small power producers and pay them for their electricity at a rate equal to "full avoided cost," that is, what it would cost the utility to produce the electricity from other sources. The Supreme Court ruling stimulated a spurt in cogeneration, solar photovoltaic, and windfarm development. Cogeneration capacity now stands at 11,000 megawatts, and market forecasters project that capacity will reach 35,000 megawatts by 1995. California alone added 250 megawatts of capacity from 3,000 wind turbines in 1983. Photovoltaic cells are also entering the utility market faster than expected. Several multimegawatt facilities will go on line in 1984, and peak watt prices should drop below $5. The Sacramento Municipal Utility District is one of the leaders in photovoltaic development, with construction already begun on a 100-megawatt facility.

Innovative financing has helped spur this growth in solar electric generation. Smart investors saw a promising opportunity in a technology backed by tax credits, investment credits, and accelerated depreciation. Large solar thermal installations also attracted investor interest. Creative financing opened the commercial water-heating market for such applications as hospitals and food-processing plants. Several multimillion-dollar projects of this kind were installed in 1983, including a $30-million water-heating system at the Packerland meatpacking plant in Green Bay, Wisconsin. Third-party investors are also taking over energy management for apartment houses and office buildings, guaranteeing energy savings for the owners and making a profit for themselves.

Technical progress has not been dramatic, but it has been steady. Not only have single-crystal silicon photovoltaic cells dropped in price, but polycrystalline and amorphous silicon cells have entered the market. Large-scale photovoltaic facilities became feasible in 1983. Wind developers appear most satisfied with machines of 100- to 200-kilowatt capacity, since the multi-megawatt machines have been disappointing. Although solar water heaters made exclusively of plastic continue to be heralded, the traditional glass and metal systems still dominate the market. Thermosyphon collectors, less expensive than other types because they do not require pumps, are gaining a growing share of the California market. It is now possible to boost the efficiency and cleanliness of some wood stoves with a catalytic converter. One can also buy a refrigerator, freezer, air conditioner, electric water heater (or furnace), or gas water heater (or furnace) that uses less than half as much energy as the average model for sale in 1980.

In spite of the setbacks and triumphs of the early 1980s, the data base used for this study of renewable energy is aging remarkably well. Conventional fuel prices have remained relatively stable, and so too have the problems faced by cities. Public attention may have wandered, but city officials are still aware of the importance of energy to a local economy. High energy prices continue to influence cities' ability to attract and hold industry, keep housing affordable, and maintain basic services. Indeed, cities are becoming more involved in energy considerations as the focus of concern shifts from oil to electricity. The sudden crisis of nuclear power, reflected in rapidly rising costs and widespread plant cancellations, is creating financial chaos for the many municipal utilities with a share in these projects.

Whether the energy problem concerns uncertain oil, gas, or electricity prices, however, *Putting Renewable Energy to Work* can provide an answer because it is based on stable supplies of renewable energy and permanent energy-efficiency improvements. Its cost equations are as useful in this period of stable oil prices as they were during the tumultuous 1979 oil crisis. Throughout all the shifts in energy costs and fads, the need to improve energy efficiency and increase the use of inexhaustible and inexpensive renewable energy continues to make itself felt.

KEVIN FINNERAN

Acknowledgments

More people contributed to this book than can be easily named. Mary Proctor of the Office of Technology Assessment reviewed the bulk of the manuscript, Frank Weinstein helped develop the economic analyses of active solar and space and water heating systems, and Bruce Green reviewed the heating and cooling chapters. Eric Leber of the American Public Power Association, Scott Noll of Resources for the Future, and Arthur Reiger of the U.S. Department of Housing and Urban Development also gave worthy counsel. David Engel of HUD, Richard Mounts, Leonard Rodberg, David Morris, and Travis Price helped shape the research. Special thanks go to U.S. Department of Energy staff: James Quinn, William Hochheiser, and Mary Glass. Martin Jaffee of the American Planning Association offered invaluable comments on the urban energy context. Most of all, individuals in cities and communities across the country provided examples, insights, encouragement, and inspiration.

Contents

RENEWABLE
ENERGY IN CITIES

I. THE RIGHT KIND OF ENERGY PLANNING

1. Making the Best Use of Scarce Resources

This book supplies some of the information needed to develop local energy plans and policies based on the use of energy conservation and renewable energy technologies. Although the analysis is objective, it reflects and supports the belief that these options together represent the country's best energy choices if weighed in terms of long-range economy, environmental benevolence, increased community control, and strengthened national security.

Three basic assumptions undergird this book. The first is that city energy officials need technical information to discriminate between appropriate and inappropriate technologies and options. Here, both the strengths and the weaknesses of various technologies are discussed to indicate which technologies would work best in which types of cities. For example, to show how conservation or solar approaches can be combined, the presentations are organized according to end-use (such as water heating or space heating) rather than technology.

The second assumption is that urban energy managers need to understand existing constraints and opportunities: land-use patterns, building stocks, energy supply and distribution networks, and social and economic conditions.

The third assumption is that no urban energy program can afford to overlook the links between energy and housing, employment, and economic development.

Cities themselves must come up with detailed plans for implementing renewable resources programs. But this book summarizes what is known now. It outlines not distant proposals, but immediate possibilities.

ENERGY AND THE CITY

When the U.S. Department of Energy surveyed local government energy activities in 1978, most local governments had not yet linked energy to high unemployment or a declining tax base.[1] But all that has changed. The National League of Cities reported in mid-1981 that 76 percent of city officials it polled identified energy as a major factor in national economic stagnation, and 67 percent felt energy costs were undermining the local economy as well.[2]

3

Rising energy costs do account for ever larger portions of municipal budgets. Testifying before Congress in 1977 during the relative calm before the most precipitous oil price rises, former Hartford City Councilman Nick Carbone put things this way:

> [W]e must understand how rising energy costs over the past few years have affected local budgets and the delivery of services. Hartford's experience with inflation is relatively typical. Since fiscal year 1971–2, Hartford's budget has increased from $84 million to $130 million . . .
>
> [E]ven though our budget has increased by 28 percent, we have been forced to cut back on services by 31.6 percent . . . at the heart of these inflationary price increases is the rising cost of energy.[3]

Of course, more recent gasoline and fuel price hikes mean that still more is being spent to deliver even less.

A Drain on the Local Economy

Spiralling energy costs have meant more to cities than simply higher figures on municipal budget sheets. Cities' overall economic health, measured by the size of the tax base, the availability of decent housing, and the number of jobs provided by business and industry, is also being threatened. Energy problems exacerbrate other economic woes.

Energy expenditures can drain the local economy, especially in such solar-energy-poor regions as the Northeast and Midwest.[4] In Washington, D.C., the Institute for Local Self Reliance found in 1979 that for every dollar that city residents spent on energy, only 13 cents stayed in the city.[5] In New Orleans, say city planners there, the monetary drain caused by energy price increases means "less local capital for investment, less purchasing power for consumers, and, consequently, slower economic growth than could be possible in the absence of such a cash outflow. Additionally . . . less tax revenue is generated for the city."[6]

Some urban groups and interests are even worse off than the big picture suggests. Struggling industries must endure energy price rises even as they face the threat of energy cutoffs or curtailments. In Philadelphia, industries use relatively little natural gas, and yet the instability of gas supplies has been a driving force behind the city's recent energy-management efforts. Minneapolis, St. Paul, and other cities located at the far end of regional gas pipelines are commensurately more vulnerable to shortfalls.

When energy shortages or curtailments are severe, industries are forced to fold or relocate, costing workers jobs. Indeed, hard on the heels of the energy price rises and shortages of 1977 came the loss of 11,000 jobs in Philadelphia, twice the previous year's loss.[7]

Substandard housing is also hard hit by energy price inflation. Energy price hikes contribute to the premature demise of the marginal units that constitute much of inner city housing. In Rochester, for example, the number of abandoned buildings jumped from under 400 in 1970 to 1,900 by late 1978. During this same period, fuel oil prices rose by 153 percent, natural gas by 105 percent, and electricity by 108 percent.

The Plight of the Urban Poor

Housing abandonment and growing unemployment are only the most dramatic examples of the steady and severe pressure that energy costs exert upon the urban poor. Families with annual incomes under $5,000 devote three to five times more (on a percentage basis) to energy than higher-income families do. In colder areas with high-priced fuels, the poor are in even worse straits.[8] In Hartford, families with an average yearly income of $3,800 or less spend 30 to 46 percent of their total income on energy alone.[9] What's more, people who cannot afford to fill their oil tanks at each delivery pay a 10 to 15 percent premium as a delivery charge. Since the same people who cannot spare another dollar for energy or who cannot afford to buy heating oil by the tankful will ·most likely be the first to lose their jobs when businesses or industries cut back, price increases are a matter of double jeopardy.[10]

THE BEGINNINGS OF LOCAL ACTION

Many local governments have responded to this challenge by developing policies to conserve energy and to tap renewable energy technologies. In a landmark study, Portland, Oregon, discovered it could reduce energy consumption to 34 percent below predicted levels by 1995 by adopting a broad range of conservation policies.[11] Davis, California, has cut residential, commercial, and transportation energy use radically by adopting energy-conserving building codes and zoning ordinances, developing land-use controls to encourage the use of solar energy, and officially discouraging automobile traffic and encouraging bicycling. A nine-week campaign in Fitchburg, Massachusetts, paid off handsomely by motivating over 3,000 homeowners to take low-cost or no-cost conservation measures that reduced their heating bills an average of 14 percent.[12]

More cities have launched conservation programs than have developed incentives to encourage alternative energy technologies. But urban solar projects have proven successful. Solar waterheating systems in multifamily structures dot cities from Boston to Honolulu. New infill housing using passive solar energy is planned for neighborhoods in Denver and Buffalo. St. Paul may use a community-scale solar energy system to provide heat and hot water in a developing industrial park that will house mostly energy firms.[13]

Just as energy problems interlock and take their toll in many economic sectors, the benefits of developing conservation and renewable energy in cities would reverberate throughout the urban economy. Far fewer dollars would leave the local economy if money now spent for foreign oil or out-of-state natural gas were instead spent on manufacturing and installing conservation or solar energy materials and equipment. Besides assisting businesses and creating jobs, investments in local energy would indirectly improve housing and reduce energy costs.

Local energy development based on conservation and renewable resources will be neither automatic nor simple. While the pioneering efforts of a few cities provide precedents and inspiration, few local governments have any experience in community energy management. To develop renewable energy resources, cities will need integrated energy plans. How else can they expect to combine energy-supply and conservation options economically and to make sure that beneficial, long-term energy options are selected over "quick fixes" that ultimately backfire?

Strategies for Developing Local Energy

One strategy open to cities is to *implement federal and state conservation programs* vigorously. The Portland study just cited made this option central to its energy-conservation projections: fully 27 percent of the city's projected 34 percent energy savings in 1995 are to be derived from state and federally sponsored conservation programs.[14] But to get such results, a city would have to lobby legislatures and agencies to adopt energy conservation policies and devise local initiatives to turn federal or state "paper policies" into real programs. State weatherization requirements, for instance, mean little if cities don't have the capacity to conduct energy audits, install energy equipment, and provide financing. With the federal government retrenching on its energy conservation and renewable energy initiatives, support for state and local programs has been diluted in state block grants and reduced. This is bad news, but it may jolt cities to take more responsibility and be more creative. Cities can work more speedily than the federal bureaucracy, and municipal regulatory or incentive programs can be designed to fit each city's energy needs precisely.

Another approach favored by cities is contingency planning. Good contingency plans do force several city departments to coordinate their activities. But while planning for energy emergencies is necessary, it does little to alleviate the underlying problems that caused the energy shortage in the first place.

(One exception is Los Angeles, where contingency planning is part of a general plan to reduce future energy consumption.[15])

The limits of meeting energy emergencies by stepping up lobbying for guaranteed, ongoing conventional energy supplies are especially severe in cities like those in New England that must import nearly all their energy supplies. In contrast, New Orleans has oil and gas deposits within its boundaries, so it need only reduce the legal and environmental obstacles that threaten fuel production to guarantee continuing supplies.[16] But in the end, securing additional oil and gas does nothing to reduce vulnerability to price escalations, and it may delay the search for energy sources even as they grow more scarce.

Still another energy policy available to cities is the use of market barriers and incentives. Rather than rely on mandatory powers to compel energy efficiency, the city in effect changes the pattern or mix of opportunities and constraints that the private energy user faces by, for example, reserving certain roadways exclusively for carpools, providing density bonuses for developers and renovators who include energy conservation features, and providing tax breaks for users of solar energy equipment.[17] Even more common is the adoption of such energy-saving pricing policies as increasing local gasoline taxes and charges for public parking while holding mass transit fares down.

Despite its significant strengths, the barriers-and-incentives approach lacks breadth. As John Alschuler, City Manager of Santa Monica, California, notes:

> [The] use of municipal regulatory power and fiscal resources is not . . . an alternative to the market approach. Successful local programs depend on the present price structure and on the anticipation of continued price increases. However, market strategies without local government support will not yield the anticipated or necessary benefits . . . equal amounts of savings can be achieved with less price-induced hardship if local government action is coupled with price signals.[18]

Moreover, simply changing the constraints on private decisions will not necessarily have the desired public effects. To the city, any policy that preserves its tax base, reduces the flow of dollars from the local economy, and minimizes consumer hardships without costing more than it saves is economical. But the private consumer is more interested in his own cash flow or his rate of return on investment. Thus, what is most economical for the city over time may not appeal to the consumer in the short term. The degree to which the barriers-and-incentives approach can make some publicly beneficial choices more privately appealing without some form of regulatory support is questionable, and the most rational incentive program may be short-circuited if one market factor can simply shift the costs of energy consumption to others.

A final limitation of this approach is the difficulty of ensuring that its costs and benefits are distributed equitably. Barriers or constraints, including taxing energy use or installing individual utility meters in rental units, may be too weak. They may also hurt most those whose ability to pay is lowest and whose demand for energy is least elastic.

The drawbacks to approaches that rely on voluntary compliance suggest that *mandatory programs* are the best solution. Used alone, however, mandatory energy policies have some serious deficiencies. First, political battles over mandatory requirements will be difficult. No matter who wins, protracted political struggles may stifle private initiatives until they are resolved. For example, in Wisconsin several owners of multifamily buildings decided against installing energy conservation devices because the equipment might not have met mandatory standards then being considered by the legislature.

A second serious deficiency is that the compromise measures resulting from political squabbles tend to be the weakest standards. In Minnesota, energy-conservation practices mandated for multi-family buildings do not extend to heating systems, arguably the key to conservation in larger buildings. In many places, "light" standards have become a maximum, rather than a minimum. And a third deficiency is that mandatory requirements will merely limp along without the support mechanisms and broad sweep of a comprehensive community energy-management effort. Why require efficiency improvements in lower-income homes or apartments without providing some means of helping lower-income people finance them? Then too, though some end-use efficiency improvements may occur outside a broad program, larger improvements in thermal or electrical distribution cannot be expected. Finally, the overall efforts of mandatory standards on other urban problem areas are poorly understood. Will mandatory improvements in marginal apartment buildings result in better and less costly housing? Or more abandonments and condominium conversions?

Recognizing the limits of each strategy or policy option described thus far, some cities are gambling on still a fifth strategy—*a "shotgun" approach* designed to attack energy problems from many vantages. The Portland study, for example, identifies 90 separate policy options. Numerous other policies are spelled out in the reports produced by 16 communities whose planning efforts have been supported by the U.S. Department of Energy's Comprehensive Community Energy Management Planning (CCEMP) program.[19] In St. Paul, five citizen subcommittees devised lists of options.[20] Some frequently endorsed include the following:

- Encouraging end-use conservation in buildings through incentives and regulation.

- Developing more efficient energy-distribution systems, including cogeneration facilities and district heating systems.
- Managing municipally controlled energy use more efficiently.
- Using land-use controls to encourage energy efficiency in buildings and transportation.
- Requiring energy impact statements for large new urban projects.
- Encouraging the use of renewable resources.

While a list of such actions can help a city formulate an energy plan, turning lists into plans and programs requires making choices about priorities, timetables, and resource commitments. Lacking these choices, community energy efforts will continue to reflect only the energies and initiatives of committed individuals, some of them at cross purposes.

INTEGRATED URBAN ENERGY PLANNING

The final and best option for cities is integrated energy planning. Because the right strategy for a particular community will depend on its own problems, resources, and political situation, this approach cannot be boiled down to a few specifics. But some general guidelines and perspectives are invaluable to cities beginning to grapple with energy policy development.

Couple Conservation and Production Planning

End-use conservation and improvements in energy distribution efficiencies are part of larger efforts to secure more reliable, less expensive energy supplies. Installing simple conservation measures in buildings is, dollar for dollar, the cheapest way to reduce conventional energy consumption. But efforts to reduce consumption can often be markedly more effective if teamed with the use of on-site renewable energy or low-temperature, community-scale thermal distribution systems. A good example is supplementing conservation measures with solar water heating. On a grander scale, the Swedes now combine strict end-use conservation in buildings with small, centralized community or district-heating systems that draw on coal-fired boilers or waste heat to meet the reduced load.[21]

Once the wisdom of wedding energy conservation to energy production becomes plain, so do the hazards of focusing too narrowly on end-use conservation. Portland's energy-use study illustrates this point. Among the many end-use efficiency improvements expected by 1995 in Portland is the wide-scale conversion of gas- and oil-fired heating units to electrical units.[22] But although electric heat is more efficient within a building than gas or oil furances and uses fewer Btu's of energy, the energy loss that occurs in electrical generation is staggering. If hydropower were available, the error would not be grievous. But if costly new nuclear generating facilities are the only available power source, stressing building-line efficiency is a mistake.[23]

The failure to relate conservation and production planning can also cause missteps in the improvements of energy-distribution techniques. Efforts to conserve electricity are often hampered because initial investments in centralized power-generation plants cannot be recouped unless energy demand is sustained at predicted levels. If demand drops significantly, rates must be increased to compensate for lost revenues. By the same token, if large-scale community thermal systems like district heating are sized and financed on the assumption that end-use conservation will be minimal, any subsequent energy improvements to buildings could reduce demand enough to undermine the community system's financial viability.[24] The lesson is clear: plan building-conservation measures before planning community energy-distribution systems.

Make Realistic Projections

Mistakes in anticipating the future feasibility of different energy supply options can lead to oversizing systems and foregoing cost-effective energy-conservation measures. For example, rapid growth in electrical demand will not occur again, even in fast-growing areas, because prices are increasing and the market for more efficient applicances is growing. Nor will large-scale nuclear or coal-fired plants be built any faster or at lower cost than is possible right now. Their construction will in all likelihood become even more time-consuming and costly.[25]

Be Politically Realistic

Successfully carrying out energy programs will necessarily involve cooperation among various local individuals and groups. The key here is gauging how the cooperation or opposition of affected groups will influence the programs' success. A less stringent or comprehensive program relying on voluntary compliance may be widely acceptable where a stricter program could succumb to heavy opposition. The more controversial regulatory proposals are bound to run up against political opposition.

Compare and Evaluate Multiple Options

To assess the feasibility of a large-scale project, cities should compare the economics of the project not only to a "business as usual" scenario, but also to alternative actions whose energy-saving benefits could be in the same range. For example, energy savings from using a large-scale, high-temperature district-heating system in leaky buildings and inefficient heating systems should be compared with the savings achievable through extensive building and heating-systems improvements. Similarly, the costs and benefits of passive solar retrofits requiring substantial building alterations should be compared with the costs and benefits of more conventional energy-conservation approaches.

Choose a Flexible Mix

Good energy planning involves finding the right mix of complementary technologies for the city. Pinning all hopes on one technology seems foolhardy, regardless of its promise, since it might mean passing up attractive energy options that come along later. Decentralized sources of heat and power tend to be compatible with this goal of flexibility. In contrast, the substantial economies of scale once enjoyed by centralized facilities disappeared as the capital costs of construction and equipment spiraled upward, the price and availability of conventional fuels became less certain, and the demand for energy began growing less rapidly.

Consider Links to Other Issues

Finally, planners should acknowledge the obvious links between energy and problems in service-delivery, housing, employment, economic development, and public welfare. The appropriateness of different energy solutions depends largely on their potential impacts on all such consumers and producers. Accordingly, urban energy planning and program efforts should be opened to the scrutiny of all affected groups.

ASKING THE RIGHT QUESTIONS

Given cities' financial and staff limitations, energy difficulties that exacerbate existing problems should receive top priority. However, even with only scant resources, cities can make headway against longer-term energy problems by delving into both the context and the impacts of narrowly defined projects. Formal "rules" are of little use in this process. What is needed is a broad understanding of how specific energy applications and programs are influenced by a few basic factors that vary from city to city.[26] An older, densely populated city, for instance, is likely to have a considerable stock of multifamily housing and a well-developed mass-transit system. Just as likely, many of its buildings will be poorly insulated. But a sprawling newer city with better-insulated buildings could still require more energy per dwelling and more transportation energy per person. Obviously, the conservation and renewable energy strategies best suited to one city would ill suit the other. The older city would have relatively little leeway for making development more compact, but would have much to gain from promoting energy conservation in its standing building stock. The newer city, on the other hand, might gain most from revising land-use controls to encourage denser development and protect access to sunlight.

This lack of fixed rules makes facts about a particular city's energy usage all the more important. Since information needs are determined by the city's problems, the logical first step is identifying problems and relating them to some of the city's broader housing, employment, and economic development needs. Yet, extensive, across-the-board data collection could yield an ocean of details of dubious validity and use. Data become useful information only if collected in the right form, quantity, and level of detail. Is energy consumption data needed in Btu or in some other measure? Does the city need to know how many hospitals and food stores lie within city limits? How important is exact information on materials used in buildings' exteriors? Such questions are best answered soon after energy problems have been delineated.

Perhaps the best known of the energy planning methodologies that concentrate on extensive *a priori* data collection is the one developed for DOE's 16 "CCEMP" communities. But that approach overemphasizes data collection and rests on several other flaws. For instance, it advises extrapolating current energy consumption to anticipate energy shortfalls, which can lock a local energy-management program into an energy future predicated upon today's mistakes. It also focuses exclusively on end-use consumption—thus slighting opportunities to distribute conventional energy supplies more efficiently or to substitute renewable energy for conventional fuels—and ignores socio-economic characteristics.

By subjecting proposed activities and programs to a careful review of their potential city-wide impacts, cities can avoid these pitfalls and integrate many nontechnical factors into their energy policy. If this fluid and impressionistic approach is perhaps too tolerant of tangents and false starts, it is also self-correcting: as cities gain experience and citizen involvement broadens, relationships among separate projects can be clarified. And if the approach lacks academic rigor, its compatibility with practical problem-solving more than compensates.

There are at least four other distinct advantages to an adaptable, incremental planning approach. First is economy. This kind of planning is not likely to lead to a financial commitment to oversized and overpriced energy projects that the city will be stuck with for years. In fact, Seattle's recent decision to rely less on nuclear power plants and to pursue a broad public energy-conservation program grew out of such planning. Second is sensitivity to other urban concerns. This approach calls for integrating energy programs with other urban programs—the only way to gauge the full range of the program's potential impacts. For instance, some large-scale energy supply or distribution systems earn local government endorsements simply because they would create new jobs—no matter that the jobs might be temporary or go to out-of-city workers, or that other strategies for creating urban jobs could be more cost-effective or socially beneficial. Third is accountability. Nothing would be more foolish than to exclude energy suppliers from community energy-management efforts. But city energy policy cannot be developed by energy supply firms alone. A balanced approach would involve energy suppliers but make them accountable to the public.

The final advantage is political stability. The right kind of community energy management enables the city to exert more control over energy problems and solutions. In St. Paul, Mayor George Latimer mobilized more than one hundred citizen representatives to help develop a comprehensive community energy plan. The letter he and Councilwoman Joann Showalter sent to each participant said:

The key to our becoming an energy-conserving, self-reliant city is community technology—all of us bringing our ideas together to find the solutions. By working together we can help one another to avoid obstacles which would waste many years of precious time that we no longer have. Our society is vastly complex and energy is the common denominator.[27]

APPLYING RESOURCES TO END-USES

Resources

The first resource to consider is conservation. In heating, cooling, lighting, and the operation of appliances, the potential for increasing the amount of "energy out" while reducing the total "energy in" is staggering because so much of the heat and electricity now used in poorly insulated buildings, overdesigned heating systems, and inefficient appliances is simply wasted. Reducing the amount of energy needed to supply heat and power or tapping supplies of waste heat can be viewed as creating new energy resources and extending the nonrenewable energy sources now used.

The next resource is "renewables." All renewable energy does come directly or indirectly from the sun's light and heat. But renewables include energy produced in the movement of winds, tides, and fresh water flows, too. Because they are continually renewed, all kinds of natural resources cannot be depleted. By this same token, biomass or plant matter, organic wastes, and geothermal energy are usually considered renewable since they will last indefinitely. Overall, then, three types of energy resources are at issue here: renewable resources (sunlight, wind energy, and water flows), abundant finite resources (wood, organic wastes, and geothermal energy), and resources supplied by energy conservation.

End-Uses

Thermal, mechanical, and electrical energy can be used in various applications or "end-uses." Thermal energy can heat water and buildings and provide heat for industrial processes. Electrical energy can power home appliances, lighting, and public transit systems. In some applications, different kinds of energy can be combined. For example, heat produced by burning garbage or wood chips can run a turbine to help generate electricity, which can then be used to heat a building.

How successful conservation and renewable energy are in each instance depends upon how well energy resources are matched to end-uses. Making this match is a matter of assessing available energy sources, suitable applications, and current technologies. Mismatches lead to energy waste. For example, the temperatures reached at an electrical generating plant far exceed the operating range of on-site boilers, furnaces, or solar collectors. Thus, the latter are clearly much better suited than electricity to the end-use of home heating. When the true energy demands imposed by various end-uses are recognized, the advantages of many smaller-scale, lower-temperature energy sources become more apparent.

Technologies

Table 1.1 suggests the range of technologies available to capture or control various energy resources for various end-uses. Strictly speaking, a few are not "technologies." Passive solar design elements, for instance, are simply windows, walls, floors, and overhangs modified to increase the building's natural heat gain or to reduce its natural heat loss. Solar thermal-electric systems, on the other hand, are highly sophisticated technologies.

Size varies, as does sophistication. Renewable energy technologies can be decentralized—matched "one to one" with individual buildings. But, they can also be centralized to provide power to thousands of people. In some cases, the same technology can be used in applications at both ends of the scale. A few flat-plate solar collectors on the roof of a small house can provide most of its occupants' hot water, while the same kind of solar collectors multiplied over

Table 1.1. Matching Resources and End-Uses by Technology.

RENEWABLE RESOURCES	TECHNOLOGIES	END-USES
Sunlight	flat-plate solar collectors passive solar design solar-assisted heat pumps ground-coupled heat pumps	low-temperature heat
	concentrating solar collectors	medium-, high-temperature heat
	flat-plate solar collectors, concentrating solar collectors passive solar design	cooling
	photovoltaics solar thermal-electric systems	electricity
Water	solar ponds	low medium-temperature heat
	passive solar design (roof ponds)	cooling
	low-head hydro production solar ponds ocean thermal-electric production	electricity
Wind	mechanical friction	low-temperature heat
	wind energy conversion systems	electricity
Wood/Agricultural Crops	wood stoves	low-temperature heat
	central wood combustion	low-to-high temperature heat
	wood-fired electrical generation	electricity
	methanol, ethanol production	liquid fuels
Organic Wastes	central waste combustion	low-to-high-temperature heat
	waste fired electrical generation	electricity
	methane, biogas production	gaseous fuels
	pyrolysis bioconversion	liquid fuels
	refuse-derived fuel production	solid fuels

several thousand square feet as part of a community thermal energy system can heat a cluster of office buildings. Similarly, photovoltaic (solar electric) systems can be sized for use on a single residence, a shopping center, or in a public utility grid. Overall, size flexibility affords energy planners great latitude in matching resources to tasks.

The renewable energy technologies listed so far also vary by levels of technical and economic feasibility. The performance of solar water-heating systems is a matter of record. High-temperature solar collectors, on the other hand, need more research and development. Active solar water heating and passive solar design can be economical choices in many situations, while photovoltaics and solar ponds are still prohibitively expensive.

PRACTICAL BENEFITS OF CONSERVATION AND RENEWABLE ENERGY TECHNOLOGIES

Energy conservation investments are, dollar for dollar, the most effective of energy purchases. But there are other reasons why they make sense from a community energy-management perspective. Since they require "appropriate" matching to end-uses, conservation and renewable energy technologies are particularly responsive to the specific energy needs of different cities and neighborhoods. This kind of matching means that it is less likely that "high-level" energy (e.g., electricity generated in nuclear power plants) will be used for "low-level" jobs (e.g., space heating). These technologies and techniques also afford long-range flexibility. One step at a time, a city can retrofit housing with basic conservation measures, add passive solar energy features where practicable, and ensure that new infill housing is energy-conserving and properly oriented to make use of solar heat. This approach allows the city substantial freedom of choice as it further reduces thermal energy needs in the future.

Building an urban energy system "from the ground up," allowing for incremental growth, also enables cities to replace outdated technologies rationally. Adding energy-conservation and passive solar features to a house today does not make it harder to add a renewable electric system three or four years from now. In contrast, a city loses flexibility when it becomes dependent on an electric power plant to supply waste heat to a large district heating grid. It is forced to ensure the plant's continued viability, even though electricity use may taper off or opportunities to generate electricity from renewable energy sources arise.

Decentralized conservation and renewable energy technologies are also more open to public inspection than most larger-scale energy technologies. This is not because renewable energy technologies are less complicated (most are not) or because they need not interrelate with conventional energy sources (all must). But their development coincides with greater public interest in community energy planning, and their use may be subject to a greater degree of local control—a key factor given the increasing public demand for greater self-reliance.

THE ENERGY PLANNING CONTEXT

Before cities can find the right technologies to match available energy resources to end-uses, they must answer two important questions.

First, what kinds of conservation opportunities and renewable energy resources are available locally? Ann Cline, an energy planner in Richmond, Indiana, waxes enthusiastic about "discovering" renewable resources there. Says Cline:

> A search for local energy resources can be one of the most fascinating projects any city can undertake. In general, it is a new look at the city which produces maps the like of which no one has heretofore seen. Visually representing the geography of the city's energy uses and potentialities has opened our imaginations.[28]

Of course, it doesn't require genius to identify sunshine in Phoenix or wind in Chicago. But for most cities even a crude inventory of local renewable energy supplies means collecting new information, some of it with new or unfamiliar techniques.

Second, how is energy now being used? Consumption patterns and needs for energy in specific forms vary enormously from city to city. But the variations are closely linked to a few basic factors: climate, land use, building characteristics, energy supply and distribution systems, and socio-economic factors—key considerations in any energy technology assessment.

2. The Natural Context: Climate and Resources

Climate and geographical characteristics influence both urban energy demands and the amount of renewable energy available. Temperature is the most obvious climatic influence on energy demand since heating and cooling can account for 60 to 80 percent of the energy used in a city's homes and commercial buildings.[1] (See Table 2.1.) Humidity and wind within or near city limits (where "heat islands" form) are also important influences on demand, while climate and topographical features partially determine the availability of renewable energy. To help cities assess their energy production potential, this chapter specifies different types of renewable resources and indicates in broad terms where they can be found and how they can be measured.

SUNSHINE

How much sunlight is available in any spot depends upon latitude, topography, microclimatic influences, and seasonable differences. Although most U.S. cities are sunniest in summer, substantial amounts of winter sunshine fall nearly everywhere in the country. From November through April, for example, Cleveland averages about 675 Btu of sunshine per square foot of horizontal space per day, almost six times as much energy as a single-family home there needs for space heating, cooling, and water heating. Even in a dwelling unit with five times more living space than roof area, the sunshine falling on a horizontal roof is more than enough to meet those same needs. As for collecting that energy efficiently, the potential is enormous and prospects will improve as the costs of photovoltaic cells drop and interseasonal thermal storage techniques are perfected.

Differences in solar intensity, which occur because the sun strikes the northern latitudes at steeper angles than it does the more southerly ones, can be a key determinant of solar potential. (See Table 2.2.) For example, while Philadelphia and Houston receive approximately the same number of hours of sunshine yearly, Houston gets about 20 percent more Btu per square foot per year. How much sunshine an area receives also depends upon how rainy and cloudy it is. Cleveland gets sunshine only 52 percent of daytime hours, while

Table 2.1. Heating and Cooling Degree Days in Selected Cities.

CITY	HEATING DEGREE DAYS[1]	COOLING DEGREE DAYS[2]
Boston	5621	668
Providence	5972	610
New York	4848	919
Rochester	6719	531
Buffalo	6927	540
Newark	5034	1024
Pittsburgh	5278	836
Philadelphia	4865	1247
Baltimore	4729	1274
Washington, D.C.	4211	1732
Norfolk	3488	1535
Charlotte	3218	1696
Atlanta	3095	1773
Jacksonville	1327	2559
Miami	206	4183
Birmingham	2510	1975
New Orleans	1465	3059
Memphis	3227	2357
Louisville	4640	1539
Cincinnati	4844	1070
Cleveland	6154	896
Detroit	6228	894
Fort Wayne	6209	748
Indianapolis	5577	1300
Chicago	6127	982
Milwaukee	7444	548
Minneapolis/St. Paul	8159	811
St. Louis	4750	1519
Kansas City	4711	1535
Omaha	6049	1321
Dallas	2382	2965
Austin	1737	2908
Houston	1434	2866
San Antonio	1570	2994
Oklahoma City	3695	2418
Denver	6016	748
Salt Lake City	5983	1018
Albuquerque	4292	1398
Phoenix	1552	4343
Los Angeles	1819	827
San Diego	1507	722
San Francisco	3080	144
Oakland	2909	128
Sacramento	2843	1132
Portland	4792	343
Seattle	4727	161
Anchorage	10911	0

1. A heating degree day measures how much the average daily outside temperature falls below 65°F. Figures here are annual (sums of heating degree days for each day) averaged over 30 years.
2. Measures how much the average daily temperature exceeds 65°F. Annual data averaged over thirty years.
Source: U.S. National Oceanic and Atmospheric Administration, Comparative Climatic Data.

Solar insolation map

SOURCE: SOLAR ENERGY RESEARCH INSTITUTE

Table 2.2. Availability of Sunshine—Selected Cities.

CITY	SUNNY HOURS OF ANNUAL DAYLIGHT HOURS	INTENSITY OF SUNSHINE ON HORIZONTAL SURFACE, NOV-APRIL ($BTU/FT^2/DAY$)
Boston	60	738.9
Providence	57	767.7
New York	59	759.9
Rochester	45	641.9
Buffalo	52	630.9
Newark	n/a	825.4
Pittsburgh	49	693.4
Philadelphia	58	830.3
Baltimore	57	872.8
Washington, D.C.	58	850.5
Norfolk	63	1000.5
Charlotte	66	1040.1
Atlanta	61	1038.9
Jacksonville	61	1209.2
Miami	67	1161.9
Birmingham	58	1027.1
New Orleans	59	1148.9
Memphis	65	998.3
Louisville	57	840.7
Cincinnati	57	781.0
Cleveland	52	674.2
Detroit	54	719.71
Fort Wayne	59	730.2
Indianapolis	58	779.0
Chicago	57	799.9
Milwaukee	56	775.0
Minneapolis	58	767.8
St. Louis	59	921.8
Kansas City	67	936.6
Omaha	62	910.4
Dallas	n/a	1114.6
Austin	61	1139.2
Houston	57	1046.7
San Antonio	61	1161.3
Oklahoma City	67	1101.8
Denver	70	1165.4
Salt Lake City	70	1055.7
Albuquerque	77	1402.7
Phoenix	86	1441.1
Los Angeles	73	1260.4
San Diego	67	1296.1
San Francisco	67	1092.6
Oakland	n/a	1095.5
Sacramento	79	1053.1
Portland	48	619.0
Seattle	49	574.7
Anchorage	56	n/a

Source: U.S. National Oceanic and Atmospheric Administration, *Comparative Climatic Data.*

Table 2.3. Average Wind Speeds for Selected Cities.

CITY	AVERAGE ANNUAL WIND SPEED (MILES PER HOUR)
Boston	12.6
Providence	10.7
New York	12.2
Rochester	9.8
Buffalo	12.3
Newark	10.1
Pittsburgh	9.4
Philadelphia	9.7
Baltimore	9.4
Washington, D.C.	9.3
Norfolk	10.6
Charlotte	7.5
Atlanta	9.0
Jacksonville	8.4
Miami	9.2
Birmingham	7.3
New Orleans	8.3
Memphis	9.1
Louisville	8.4
Cincinnati	9.1
Cleveland	10.8
Detroit	10.2
Fort Wayne	10.3
Indianapolis	9.7
Chicago	10.4
Milwaukee	11.8
Minneapolis	10.5
St. Louis	9.5
Kansas City	10.3
Omaha	10.8
Dallas	10.9
Austin	9.3
Houston	7.6
San Antonio	9.4
Oklahoma City	12.8
Denver	9.0
Salt Lake City	8.8
Albuquerque	9.0
Phoenix	6.3
Los Angeles	6.2
San Diego	6.7
San Francisco	10.5
Oakland	8.2
Sacramento	8.2
Portland	7.8
Seattle	9.2
Anchorage	6.7

Source: U.S. National Oceanic and Atmospheric Administration, *Comparative Climatic Data.*

70 percent of Denver's daylight hours are sunny. Still, diffuse solar radiation reaches the earth's surface through cloud cover, and many solar collection techniques can use diffuse radiation efficiently.

How much available sunshine is usuable also depends on an area's topography and the built environment. On south-sloping land, incident sunlight is more intense than on adjacent flatlands. On north slopes, it is less. Where hills, vegetation, and buildings are densely clustered, shading occurs, and shading problems are exacerbated by seasonal changes. (See Chapter 3.)

Air quality also influences the amount and intensity of solar radiation available in cities. Fog, smog, and pollution absorb or scatter the sun's rays, diffusing solar radiation. These impacts upon solar availability vary over the day. In San Francisco, for instance, sunlight becomes more intense and direct once morning fog dissipates. By the same token, it becomes more diffuse as industrial pollution and auto exhausts increase.

WIND

The winds flow because the earth's surface is not heated evenly. Solar radiation striking the earth at various angles is absorbed and reradiated at different rates, depending on latitude, vegetation, topography, and geology. As a result, the air near the earth's surface warms up at different rates. These warm and cool air masses interact as high and low pressure systems. Dense air from low-pressure systems tends to flow under the "lighter" air of high-pressure systems, producing winds as a result of convection. The rotation of the earth combines with these factors to create prevailing winds, which in North America are northwesterly.

Winds of different speeds embody different amounts of potential kinetic energy. Ordinary windmills can convert this energy into mechanical energy, while more elaborate wind energy conversion systems (WECS) can transform it into electricity.

Within densely settled cities or neighborhoods, electricity generation from wind energy will be minimal, however. Spatial and safety considerations interfere, and large buildings tend to create troublesome turbulence. But many cities can import wind power produced outside the city, and some cities do have unpopulated and unobstructed tracts of land suitable for wind energy collection.[2] In general, wind energy is available to cities within regions with average wind speeds of 10 miles per hour or more. (See Table 2.3.)

If the region is windy enough, the availability of wind energy depends heavily on natural topography. But the built environment is even more important since turbulence can damage a wind machine's blades. Given these potential problems, thorough windpower assessments are needed. In general, though, the prime sites for larger wind systems are lake fronts, land abutting water, and open areas. For small wind energy conversion systems, the flat roofs of relatively low apartment or commercial buildings are suitable if neighborhood building heights are more or less uniform. (See Chapter 11.)

WATER RESOURCES

Both ground water and surface water can be tapped as energy resources. Ground water is most commonly used to supply and store thermal energy. Surface water can do both and also generate electricity in hydropower facilities or solar ponds.

Few large cities will be able to use natural water resources within city limits to meet much of their energy demand, though they can buy hydroelectricity produced outside the city. But in some areas within such cities, water resources may be able to supply relatively large amounts of energy. Smaller cities have more opportunity to use in-city water resources on a larger scale. For example, many northeastern cities with populations under 100,000 are finding that existing dams on adjacent rivers can be used as hydroelectric facilities.[3]

Ground water, which stays between 40 and 60°F year-round in most of the United States, makes a good heat source in winter and heat store (or heat sink) in summer.[4] Such technologies as ground-coupled heat pumps can be used to transfer heat to and from the water found in confined aquifers, underground reservoirs, and subterranean channels.

Whether it is technically feasible to use ground water to supply or store heat depends primarily on where the ground water is found. Two rules of thumb are that aquifers are larger than subterranean channels and that underground reservoirs experience greater temperature swings than confined aquifers. However, confined aquifers are relatively inaccessible and can be thermally polluted if ground water is used for heat storage.

Most large cities lie above ground water resources. The questions are whether those resources are worth developing, whether ground water reserves are readily available, and whether local environmental constraints limit their use for supplying and storing heat. In general, though, useful aquifers are far more prevalent under large cities than hunch would allow.[5] Surface water resources can also be used to obtain or store heat, though doing so may prove tricky where surface water temperatures vary more by season than ground water temperatures do.

Artificial bodies of water can also be used to collect and store heat or to store cold.[6] Solar and ice ponds can be constructed to supply heating or cooling to many smaller buildings, a few bigger ones, or a single large structure. However, they require relatively large areas of land.[7]

Natural or man-made bodies of still water in sunny areas can produce enough heat to generate electricity if they contain natural or implanted layers of salt that increase in density with the pond's depth. As the salt absorbs heat, higher temperatures concentrate toward the bottom while the less dense salt solution moves upward, preventing the convection of concentrated heat throughout the pond. Even more than solar thermal ponds, ponds producing electricity have water, land, and climatic requirements that are almost impossible to meet in urban areas. But cities could obtain electricity produced by solar ponds in outlying areas.

Using flowing surface water to rotate turbines that generate hydroelectricity is perhaps the most common way to derive energy from water. How much electricity can be produced this way depends largely on the size of the vertical drop (the "head") and the consistency and rate of stream flow. All but the smallest of hydroelectric projects (less than 100 kw) require one or two types of dams to direct the flow of water. Storage dams are best suited to large-scale (or "high-head") sites, where flow rates usually vary. They work best in deep valleys, where they can form large, deep reservoirs. In contrast, run-of-the-river dams are used within city boundaries. These dams control the flow of a river rather than create a reservoir. They abound in many cities in the Northeast, upper Midwest, and the Northwest. (See Table 2.4.) Lowell, Massachusetts,

Table 2.4. New Hydropower Potential by Region (1978–2000).

	SITES	CAPACITY MEGAWATTS	ANNUAL ENERGY PRODUCTION GIGAWATT HOURS
Totals, Nation	2,100	57,000	149,000
Northeast Power Coordinating Council (NPCC)	32%	8%	11%
Western Systems Coordinating Council (WSCC)	30%	55%	50%
Southeastern Electric Reliability Council (SERC)	9%	12%	9%
East Central Area Reliability Coordinating Agreement (ECAR)	9%	8%	9%
Mid-American Interpool Network (MAIN)	4%	2%	2%
Alaska	3%	6%	10%
Electric Reliability Council of Texas (ERCOT)	2%	1%	1%
Mid-Atlantic Area Council (MAAC)	2%	2%	1%
Mid-Continent Area Reliability Coordinating Agreement (MARCA)	2%	2%	1%
Puerto Rico and Hawaii	2%	1%	1%
Southwest Power Pool (SWPP)	5%	4%	5%

Source: U.S. Army Corps of Engineers.

for example, is renovating the Pautucket Dam in the Lowell Canal System off the Merrimack River to produce about 80 million kilowatt-hours (kwh) per year.

Since it is cheaper to renovate than to build run-of-the-river dams, existing dams make the most economical hydropower facilities. In addition, the economic feasibility of developing a low-head hydro site depends on the condition of the existing dam, the area's electricity demand, and the local utility's policies. (See Chapter 1.)

BIOMASS RESOURCES

The solar energy stored in biological material—primarily wood, crops, and organic wastes—can be turned into heat or steam (for electricity generation) or into solid, liquid, or gaseous fuels. Wood and crop "biomass" resources include harvest residues, other waste by-products, and resources grown specifically as energy sources. Most organic wastes are found in the community's solid waste stream.

Large cities have easier access to municipal solid wastes than to agricultural or forest resources. But liquid or solid fuels or electricity from wood-fired power plants can be "imported," and many smaller or middle-sized cities are close enough to forests, wastewood industries, or crop-growing areas to use biofuels to generate electricity in intra-city power plants or to economically convert wood or crops into fuel in urban facilities.

Any use of biomass for energy must be evaluated in light of environmental constraints. In municipal solid-waste operations, cities must distinguish between wastes that have no other uses (and thus pose disposal problems) and wastes than can be recycled.

WOOD

Wood can be used to heat buildings or to fuel power plants with up to 50 Mw of capacity.[8] Alternatively, wood can be converted into liquid or gaseous fuels for space heating or transportation uses.

Wood resources are available from several sources. In New England, lumber and paper mills and furniture manufacturing plants all produce wood and bark wastes that these industries then convert into energy.[9] Logging residues also make good fuel. A third wood source for biofuels is wood grown or harvested specifically for energy. For large-scale harvesting, look to overgrown woodlots that need thinning to improve yields and wood-energy plantations on which trees could be harvested every four to seven years.

The expense of hauling wood will probably limit its intra-city use for electricity generation or wood-derived fuels production. For example, fueling a 50-Mw electric generating plant in Burlington, Vermont, will require the equivalent of 105 truck trips a day to haul 1500 tons of wood chips from within a 50-mile radius of the city.[10] Distances much in excess of this would be uneconomical.

Environmental factors limit the number of trees that can be harvested for energy. Erosion results when trees on steep land or close to stream beds are harvested, and using wood for energy could sap nutrients from the soil.[11] "Whole tree chipping" is quite profitable, but removes nitrogen-rich branches and leaves.[12]

CROPS

Crops, crop surpluses, and crop residues can be put to many of the same energy uses as wood. In the Midwest, numerous plants are already turning crops into liquid fuels, and permits for other conversion facilities of all sizes are being sought. Dried and treated crop residues can be used alone to fuel small-scale, on-farm electric generators or used with coal in larger power plants.[13]

Surplus crops—those stocks not used for food, animal feed, or export—can be channeled to energy uses. Which surpluses are used depends partly on such natural or climatic factors as those determined by the Department of Agriculture's acreage set-aside and grain-reserve programs.[14] In addition, the long-term impact on food supplies of growing energy crops on a large scale remains to be established.[15] Harvest residues are most plentiful where corn, soybeans, wheat, and other grains are produced.[16] But between one third and three-fourths of all crop residues must be left on the soil to prevent nutrient depletion.[17]

In general, cities will be hard-pressed to make direct use of agricultural energy resources if large crop-growing areas are more than 50 miles away. Such cities may be able to purchase crop-based liquid fuels from rural producers, but they may also have to compete with farmers for these fuels.

MUNICIPAL SOLID WASTE

All big cities have only too ready access to municipal solid wastes. Indeed, disposing of mountains of solid waste can pose severe environmental and land-use problems. "Clean air" statutes limit burning, and land for waste disposal is growing scarcer. On average, cities process 3 to 5 lbs. of waste per person per day.[18] To help offset the costs and headaches of waste control, some now produce salable energy as a by-product of waste disposal, either by using solid waste as fuel in "garbage-to-energy" operations or by obtaining methane gas from landfills.[19] While getting energy from garbage is no justification for abandoning environmental or land-use controls, it can reduce the amount of trash that must be burned or else buried in landfills.[20]

The size and composition of the solid waste stream vary by day and season. Yet, both must be known before any resource-recovery system can be properly sized. Particularly important is accurately estimating the amount of available recyclables (paper, ferrous metals, aluminum, and glass) since building oversized plants may actually stifle some recycling programs by providing an incentive to keep the volume of waste high. Equally important is finding reliable markets for the plant's products.[21]

GEOTHERMAL RESOURCES

Heat stored within the earth can be used to provide thermal energy to run electric generators. In some geological formations, large quantities of the earth's heat are concentrated close enough to the surface to be extracted economically. While geothermal wealth is not equally distributed, usable stores of heat are available in a surprising number of locations throughout the country. The West is better endowed geothermally, but the Atlantic coast and other parts of the East have potential, too.[22] In fact, the Department of Energy confirmed the presence of geothermal hot water under the Atlantic Coastal Plain when it drilled an exploratory well in peninsular Maryland.[23] Also, the geothermal potential of the eastern U.S. may be underestimated since most geothermal "prospecting" has been for high-temperature reserves only.

Most of the higher-temperature (over 200 °F) reserves are in the West. Cities there may be able to produce electricity and feed it into local utility grids. In contrast, the moderate- and low-temperature reserves found in Eastern cities are suitable primarily for heating clusters of buildings.

Geotheral energy is stored in hydrothermal (hot water) resources, geopressured resources (water and dissolved methane stored at high temperatures and high pressure), and dry hot rock.[24] It can be tapped by drilling and then used to provide heat or to run electric generators. Hydrothermal resources are easiest to recover, and hydrothermal technologies are closest to commercial readiness.

Hydrothermal resources are now being used to heat clusters of buildings (as in Boise, Idaho) and to provide electricity (as in the Geysers in Sonoma, California). Technologies for using the methane in geopressurized reserves are expected to become operational in the mid-1980s. The technologies for drilling, fracturing, and recovering heat from hot dry rocks are still under development.[25]

Cities interested in geothermal resources also have institutional difficulties and negative environmental impacts to consider. Some geothermal recovery systems produce large quantities of environmentally menacing brine. (Also, since most geothermal reserves are in earthquake belts, the withdrawal of hydrothermal fluids needs to be monitored carefully.) Questions over the definition of ownership and state laws precluding the development of geothermal resources may also hamper development of geothermal resources.[26]

WHERE TO FIND LOCAL INFORMATION

Local energy potential depends first of all on resource availability. The need for progressively more detailed and site-specific analyses will depend on the resource, its likely applications, and the projected city role. For larger-scale technologies that require capital expenditures by the city, resource assessments and feasibility analyses are absolutely essential. Unfortunately, most sources of information on climate and renewable energy resources are too sketchy to do more than help city energy offices decide whether more detailed studies would be fruitful.

Before undertaking any analysis, a city should check with the state energy office to see if such assessments have been done already. Also, many utilities collect data on sunshine, wind, and other resources for their own projects. Other possible sources of resource analyses are state colleges and universities and state solar energy associations.

Primary sources of information on sunshine and wind are the nearest weather station and airport. To interpret such information as intensity of sunshine, the average annual percent of sunny hours, and average wind speeds, find out how the weather station's microclimate differs from that of the city and then correct for the effects of air pollution on direct solar radiation. Topographical variations can create fairly substantial differences in wind speeds for sites that are close geographically. (Airport sites tend to be selected for their moderate winds, so wind speeds are likely to be greater elsewhere.) Information more detailed than weather-station data will be required to evaluate different neighborhoods' potentials for using solar energy systems or wind energy conversion systems. For any system using sunshine as a resource, information on shading patterns is needed. (See Chapter 3.) For wind energy systems, information on average wind speeds and wind-distribution patterns is essential. A wind data compiler (which records wind speeds each second so that average wind speeds per hour can be calculated) can in two months or so provide the data needed to calculate wind speed distribution.[27] Where weather-station data cannot be used to predict wind speed distributions at the site, site data may have to be collected for up to a year. A few states (Alabama, California, Connecticut, and New Jersey) have established wind energy offices, and 91 utilities around the country had some kind of wind program by 1980.[28]

General information on an area's surface and ground water resources will be available from state or regional departments of environmental or natural resources, bureaus of water or land management, or public interest or education groups that focus on water issues. To gauge the potential of a body of surface water to supply or store heat, consider the size of local heating and/or cooling needs, the size of the water resource, average seasonal temperatures, the proximity of surface water to end-uses, and such potential environmental problems as thermal pollution. To assess hydropower production potential, find out the size of the average stream heads and the rates and consistencies of flow at existing dams. (Ask the state agency responsible for water resources of the U.S. Geological Survey first.) Topographical maps may indicate an area's general potential, but engineers are needed to measure the maximum and minimum flow rates of a stream—the determinants of its maximum and minimum electrical output potential.[29] Similarly, water-resource assessment experts will be needed to determine a ground water reserve's capacity and to assess the environmental consequences of heat removal or heat storage on both the ground water source and connected water resources. Cities interested in recommissioning an abandoned hydro-electric plant should check the plant's old records for stream flow data.

General information on how much and which type of wood resources are available can be obtained from state offices of natural resources, environmental resources, or forestry services, and from university forestry departments. With this site-specific information, the city official can determine whether to commission experts on energy, forestry, and environmental issues to determine the energy potential of available wood stocks, the transportation costs of hauling wood, and the optimal level of tree harvesting.

General data on the availability of crops and crop surpluses for energy use can be obtained from state or regional agricultural offices, and the Stanford Research Institute's *An Evaluation of the Use of Agricultural Residues as an Energy Feedstock* estimates the county-by-county availability of agricultural wastes. However, more detailed studies will be needed to assess local possibilities for converting crops to biofuels. The location of existing or proposed processing plants, transportation costs, and the effects of crops-to-energy projects on food supplies and soil quality must all be taken into account. In the case of energy plantations, consider whether land used for energy farms would be used more profitably for food production.

Before a city commits itself to a major resource-recovery study, it should survey the city's goals (for example, to use resource recovery to generate revenue or to solve a waste-disposal problem, an energy problem, or both) together with analyses of its existing disposal options, the cost of resource recovery versus other options, and markets for electricity, steam, and recycled materials. To estimate the size and composition of a community's municipal solid waste stream, examine the weight records of the refuse disposed. Alternatively, estimate total weight by recording the weights of a seasonally and geographically representative sample of disposal trucks and multiplying the average by the number of truckloads of waste generated per day, and estimating the recycling potential of the waste stream.[30] Where preliminary assessments are favorable, commission a qualified firm to determine the precise composition of the waste stream on a seasonal and annual basis. To make preliminary feasibility assessments of resource-recovery plants, turn to the Environmental Protection Agency's Office of Waste Management guidebooks. Also, examine the U.S. Conference of Mayors' case study reports on cities using resource-recovery plants, the International City Management Association's up-to-date files on resource-recovery projects around the country, and the American Consulting Engineering Council's directory of engineering firms engaged in resource recovery.

General information on geothermal resources is available from the U.S. Geological Survey. "An Assessment of Geothermal Resources of the U.S.— 1978" (U.S. Geological Circular 790) tells the location of known geothermal resources, gives estimated average temperatures, and estimates the amount of energy that can be extracted. Also, many states are compiling geothermal data under DOE sponsorship through the state bureau of geology, water resources department, bureau of mines, or energy office. Where geothermal potential is known to exist, a more detailed study of possible sites will be needed. Since different applications require different temperatures, the flow rate, water temperature, and the water quality must be ascertained early. An engineering firm should be hired to determine the expected duration of the recoverable hydrothermal reserves under different usage patterns, probable losses of heat during transmission, and the environmental effects of geothermal recovery and transport, brine removal, and brine disposal.

3. Land Use and Energy

Undertaking even modest energy-conservation or renewable-energy programs raises land-use issues that cannot be resolved without accurate information and a keen awareness of building density, the nature of the industrial/commercial mix, and each sector's consumption patterns and conservation potential.

BUILDING DENSITY

Typical densely populated areas have many different types of built environments: high- and mid-rise apartment buildings and shopping complexes, and lower-rise garden apartments or attached townhouses interspersed with commercial strips. In reducing energy use, building density can be a plus. Multi-family buildings tend to use less energy per unit than single-family houses do because the common walls and ceilings in apartments reduce their surface-to-volume area. Neighboring structures also block the wind in the heating season and unwanted sun in the summer, though this is offset somewhat by the so-called "heat island" effect, which keeps cities hotter than outlying suburbs and adds to urban cooling loads in the summer.[1] Finally, denser development also tends to include mixed land uses, so residents can walk to workplaces, stores, and entertainment.

Most cities' general plans and zoning or subdivision ordinances encourage set levels of housing and commercial building density, as well as mixed uses. In Los Angeles, for example, the general plan explicitly fosters low-density residential development and limits high-rise apartments, offices, and commercial complexes to specially designated "centers."[2] The District of Columbia limits the height of new residential and commercial buildings to twelve stories. Boulder, Colorado, controls housing density by restricting the number of building permits issued each year.[3] But while the move toward encouraging greater building density for energy conservation is catching on, the policy options open to cities will vary according to existing density, the city's development goals, the growth-management tools at hand, and the political feasibility of changing density patterns.[4]

No single pattern of "dense" development characterizes a "typical" city, so no "typical" compact development program suits all cities. In general, though, older cities in the Northeast and Midwest are relatively denser than newer cities in the South and West.[5] (See Table 3.1.) Moreover, building density varies by neighborhood, and older neighborhoods that have weathered redevelopment more than once can have an unpredictable mix of housing types.

Since building density varies widely among cities and neighborhoods, cities' "density goals" must vary too. Constructing massive high-rise buildings in the

Table 3.1. Single-Family Homes per Thousand Population. (Selected Cities)

CITY	POPULATION	NUMBER OF SINGLE FAMILY HOMES	SINGLE-FAMILY HOMES/THOUSAND POPULATION
Under 100			
Boston	636,725	33,300	52
New York	7,481,613	341,700	46
Newark	339,568	9,400	28
Chicago	3,099,391	277,800	90
100–200			
Buffalo	407,160	43,600	107
Washington, D.C.	711,518	46,300	135
Atlanta	436,057	73,300	168
Cleveland	638,793	104,800	164
St. Louis	524,964	74,600	142
200–300			
Pittsburgh	458,651	98,500	215
Philadelphia	1,815,808	415,600	229
Baltimore	851,698	172,000	202
Birmingham	276,273	62,400	226
Memphis	661,319	139,700	211
Louisville	335,954	75,400	224
Detroit	1,335,085	271,100	203
Indianapolis	782,139	173,900	222
Omaha	371,455	79,700	215
Houston	1,326,809	271,200	204
Oklahoma City	365,916	107,100	293
Denver	484,531	112,300	232
Salt Lake	169,917	36,300	214
Phoenix	664,721	175,100	263
Los Angeles	2,727,399	601,500	221
Sacramento	260,822	66,200	254
Seattle	487,091	138,600	285
Over 300			
Providence	167,724	98,500	587
Minneapolis	378,112	132,900	352

Sources: Population: City/County Data Book; Homes: Annual Housing Survey.

name of energy efficiency should not be the general rule. As a recent Urban Institute study shows, increasing building heights will reduce per capita energy use only up to a point. Beyond it, per capita consumption begins to rise with the increased need for electricity for elevators and other services.[6] Then too, high-rise structures may violate the scale of the urban landscape or otherwise impair livability. Overall, housing density will increase primarily as a function of the shift from single-family detached houses to single-family townhouses.[7]

Taking an eclectic (or neighborhood-specific) approach to zoning can allow selected small business into residential neighborhoods, reducing some work or shopping-related travel.[8] In older, central cities a complementary step would be to allow higher residential or commercial densities in redeveloped warehouse districts or low-rise commercial areas bordering higher-density residential or business areas.[9] In a city such as Los Angeles, density could be increased by using "step down" techniques—from the centers outward, building progressively lower-story multifamily or commercial structures.[10]

The administrative means of implementing compact zoning programs will also vary from city to city, depending upon the growth-management tools available. In most cases, city energy officials should work with city planning or building departments to incorporate energy elements in general plan revisions, in zoning or subdivision regulations, and in procedures for granting variances and zoning-code exceptions.[11]

COMMERICAL/INDUSTRIAL MIX

Besides knowing how their city is laid out, city officials also need to know how the local economy uses energy. Industry-based economies have different energy needs than do commercial economies. Businesses use energy mainly for space conditioning, service lighting, and appliance operation. Industries, however, use it primarily to fuel industrial processes—many of which require extremely high temperatures. As a result, industry tends to consume more energy per employee or per dollar invested than most commercial enterprises. In Bridgeport, Connecticut, energy use per industrial employee is almost 250 percent higher than per commercial employee.[12] (See Table 3.2.)

The types, as well as the amounts, of energy that businesses consume differ from what industry consumes. Commercial or service concerns use relatively large amounts of electricity for air conditioning, lighting, appliances, or refrigeration. In industry, the type of energy used depends more on the fuel sources available and the temperature level required. (See Table 3.3.)

The type and amount of energy used also vary within each sector. Food-processing industries are less energy-intensive than are primary metal produc-

Table 3.2. Employment Distribution by City, July, 1981.

| | PERCENT OF PAYROLL[1] | | PERCENT OF ENERGY USE[2] | |
CITY	GOODS	SERVICES	COMMERCIAL	INDUSTRIAL
Denver-Boulder	24	76	14.0	8.9
Dayton	32	68		49.4
Lawrence-Haverhill	42	58		37.2
Knoxville	32	68	17.7	15.3
Bridgeport	42	58	13.4	21.3
Miami	19	81		6.3
Los Angeles-Long Beach	28	72	13.3	12.4
Portland	24	76	10.0	39.0
Philadelphia	19	81		20.0

Sources: 1. U.S. DOL, Bureau of Labor Statistics, Employment and Earnings, Sept., 1981 (Table B-8).
2. Portland: Energy Conservation Choices for the City of Portland, Oregon (1977).
   Lawrence/Haverhill: The Lawrence Economic Development/Energy Program (1981).
   Other cities: Comprehensive Community Energy Management Planning (CCEMP) Audits.

ers, retail hardware stores less so than retail food stores. (See Table 3.4.) Yet, some commercial establishments are more energy-intensive than some industries.

Knowing which economic factors influence local energy consumption can help city officials develop more effective policies. One option recommended by the Portland Study Group is to persuade businesses and industries that are not energy-intensive to locate in the city.[13] Another is to encourage or require the most energy-intensive businesses and industries to improve their energy efficiency. Here, some perspective is needed because commercial users may be able to upgrade the building shell and the heating, cooling, and lighting systems without spending much money, while investments to improve the energy efficiency of production processes in industry may be much more expensive and harder to make. Accordingly, providing information and removing unnecessary barriers may be enough to encourage energy-conservation improvements in businesses, while low-interest loans, city tax breaks on energy-saving investments, or other stronger incentives may be needed to help industry.

LAND USE AND AVAILABILITY OF RENEWABLE ENERGY

In two ways, a city's physical dimensions help determine its potential for using renewable energy. First, the types and levels of energy demand depend in part on land-use considerations, so different renewable energy technologies can be used to meet different demands. Second, land-use factors influence the amount of properly oriented and unshaded space available for solar energy collection

Table 3.3. Principal Industries and Industrial Mix by City.

CITY	PRINCIPAL INDUSTRIES	TOTAL INDUSTRIAL ENERGY USE (10^9 BTU)	FUEL MIX (PERCENT INDUSTRIAL ENERGY USE)						
			ELECTRICITY	FUEL OIL	NATURAL GAS	PROPANE	COKE AND COAL	STEAM	OTHER
Boulder	Computers; paving and gravel; aircraft	1838	23		77				
Dayton	Transportation Equipment; machinery; food		24.7	8.7	45		16.9	4.7	
Greenville	Light manufacturing and assembly; food processing; agribusiness; pharmaceuticals	827	41	42	18				0.2
Wayne County, MI	Motor vehicles; steel machinery; fabricated metals		19.9	10.3	40.8		23.5		5.5
Knoxville	Apparel; textile mills; food processing; machinery		28.7	11.1	46.3	1.4	11.2		1.3

	Industries					
Bridgeport	Fabricated metals; transportation equipment; electrical equipment	11900	20.9	54.9	24.5	
South Florida	Food processing; stone clay, and glass	13682	66	.03	30	
Los Angeles	Petroleum refining; food processing; stone, clay, and glass; transportation equipment	44,700	29.5	24.6	45.6	
Portland	Paper and allied products; primary metals (aluminum); construction; lumber and wood; food processing	98,821	28.4	17.4	43.9	10.3
Philadelphia	Food processing; apparel and textile products; printing and publishing; textile mills	47,900	17.1	65.9	15.4	1.5

Sources: Portland: *Energy Conservation Choices for The City of Portland,* Oregon (1977). Other Cities: Comprehensive Community Energy Management Planning (CCEMP) Audits.

Table 3.4. Energy Intensity by Industry, 1978.

INDUSTRY	ENERGY CONSUMPTION (BTU $\times 10^{12}$)	EMPLOYEES ($\times 10^3$)	ENERGY INTENSITY (BTU $\times 10^6$/ EMPLOYEE)
Food and kindred products	980.5	1,665.6	588.67
Tobacco products	20.8	73.0	284.93
Textile mill products	326.6	898.7	363.41
Apparel and other textile mill products	69.1	1,356.0	50.95
Lumber and wood products	231.1	739.3	312.59
Furniture and fixtures	55.3	491.4	112.53
Paper and allied products	1,301.1	689.0	1,888.38
Printing and publishing	91.5	1,186.2	77.13
Chemicals and allied products	2,905.3	1,108.9	2,619.98
Petroleum and coal products	1,122.7	218.9	5,128.82
Rubber and miscellaneous plastics	261.0	770.9	338.56
Leather and leather products	21.6	255.4	84.57
Stone, clay, and glass products	1,299.8	688.7	1,887.32
Primary metal industries	2,710.9	1,213.2	2,234.50
Fabricated metal products	400.2	1,684.5	237.57
Machinery, except electrical	350.7	2,345.3	149.53
Electric and electronic equipment	254.9	2,021.5	126.09
Transportation equipment	397.7	1,991.0	199.74
Instruments and related products	79.0	644.2	122.63
Miscellaneous Manufacturing Industries	49.0	467.1	104.90

Source: Bureau of the Census, *1978 Annual Survey of Manufacturers.*

in dense areas. Specifically, three separate factors are involved: (1) the availability of adequate physical space, (2) the proper orientation of that space, and (3) the availability of solar access.

Physical Space

All renewable energy systems require certain minimum collector areas. The amount needed depends in part on system type, system scale, and climate. Thus, for example, most active domestic water-heating systems require far less collector area than combined space- and water-heating systems do. An active system designed to supply about 50 percent of the water-heating needs in a single-family home requires about 60 ft^2 of collectors in Boston. Add space heating to the load, however, and the requisite collector area could jump to over 500 ft^2. Minimum collector areas for space heating or combined systems are correspondingly reduced where loads are smaller or insolation higher. For multi-family buildings or community energy systems, less collector space per household is needed. Establishing typical collector areas for passive systems is particularly difficult since these systems are highly individualized. In general, though, the passive retrofit will not be effective unless a goodly proportion of the building's south-facing walls is unshaded and properly oriented.

Orientation

Sunlight falling on the continental United States is more intense on south-facing areas—which is why solar energy collectors need to be oriented in a general south-facing direction. (See Chapter 2.) A system's precise orientation can vary with the system's purposes, the climate, and the geography. For example, a water-heating system can be oriented farther east or west of south, depending on when most hot water is needed.

The orientation of most solar retrofits is determined largely by the building's orientation, which in turn depends upon street direction, the lot's position on the street, and the building's position on the lot—three factors controlled by zoning ordinances.[14]

A building is ideally suited for solar collection when one of its longest sides (axes) faces south. In cities, most zoning requirements specify that lot and building axes be perpendicular to streets. Accordingly, the urban solar ideal is to have streets run north to south and to have the long sides of buildings run east to west and face south. However, this arrangement has a fatal flaw in cities, since south-facing walls of most houses will be shaded by their next-door neighbors. In row-house neighborhoods, only the unit at the row's southernmost end has adequate solar access. Thus, it might be preferable to have the short side of the house facing south so that solar collectors can be placed on front or rear roofs or walls. In practice, this means, east-to-west streets are most accommodating to solar access requirements, but many other arrangements can afford adequate solar access, and other factors can sometimes offset the disadvantages of poor street placement.

First, the ranges. Some solar collectors have performed well with orientations of 45 degrees east or west of due south.[15] Thus, streets running diagonally or curvilinearly in a general east to west direction may well prove adequate. Moreover, in cities or neighborhoods where the ratio of building area to total lot area (lot coverage) is low, the chances are good that a building's south-facing roof or wall will be shaded. And where lot coverage is low, more open space is available for ground-mounting active solar collectors.

North-to-south street patterns in densely built areas may also prove acceptable where buildings have flat roofs. This proved true in a study of a low-rise rowhouse neighborhood in Baltimore.[16] The key is having enough space between collectors so that those to the south do not shade those immediately to the north. In warehouse districts, where most roofs are flat, street directions should pose no problems for collector orientation.

Where lots and buildings run parallel to streets, buildings on streets running east to west may have a long side facing south. This means that most buildings can be assumed to be "solar-accessible" if they are on streets that run generally east to west. Where streets run north to south, more information about building- and lot-orientation will be needed to evaluate solar access.

Shading

Even the most well-oriented solar collector will be useless if it is shaded. To operate properly, most collectors require exposure to sunlight from about 9:30 a.m. to 3:00 p.m. on December 21st, when the sun casts the year's longest shadows.[17] Meeting this access requirement can be difficult in densely populated, built-up areas. In lower-density residential neighborhoods, vegetation is likely to pose greater problems than buildings.[18] In one typical Denver neighborhood, about half of all shading is caused by trees, and several California cities well-endowed with shade trees have passed "solar shade control" ordinances.[19,20]

Even in denser areas, solar access for roofs may be good if building heights are uniform.[21] Shading caused by buildings is more troublesome where taller buildings to the south border lower ones to the north, or where buildings of different heights are mixed. A study of a Baltimore neighborhood with varying building heights showed that an eight-story apartment house located south of a block of row-houses shaded half of the houses during critical solar-collection hours.[22] This problem is most serious in central business districts and in areas where residential and commercial uses are mixed.[23]

POLICY CONSIDERATIONS

City policies can directly affect solar access, orientation, and even the space available for solar energy collectors. While the greatest leeway is obviously in new subdivisions, the focus must be on established areas in order to encourage substantial urban solar energy use.[24]

The space available for on-site solar collectors is fixed largely by the building's and lot's dimensions. But cities can increase the amount of usable space by enforcing setback, sideyard, rearyard, and bulk limitations in the zoning code. Seattle and some other cities now exempt solar collectors from these restrictions.[25]

In developed areas, city governments have some control over street patterns and lot- and building-orientation wherever settled land is redeveloped, vacant land developed, new subdivisions planned, or new areas incorporated. As long as traffic is not confused as a result, new subdivisions can be exempted from the requirement that all city streets follow the same directional patterns.[26] For redeveloped areas near the downtown, greater latitude can be allowed in lot positions relative to streets and building positions relative to lots.[27] Cities can also require developers to file solar access reports or give builders density bonuses or quick service on building permits if they optimize solar orientation in new tracts.[28]

In existing neighborhoods, of course, the *de facto* availability of solar access is no guarantee for the future.[29] Many lots in both residential and commercial neighborhoods are not built to their height limitations, so further development is legal. Moreover, some cities now require permits to plant shade trees.

In all, methods for developing formal guarantees or protections for solar access in newer subdivisions are well established and relatively straightforward. But in dense established areas, guaranteeing solar access is much more difficult.

Area-Wide and Lot-by-Lot Approaches

Generally, the two main approaches to protecting solar access are area-wide and lot-by-lot. Area-wide approaches are uniform, predictable, and relatively easy to administer. In contrast, lot-by-lot approaches have none of these advantages, but do have minimal impacts on normal development patterns, which makes them politically more acceptable.

The simplest area-wide strategy is to use conventional zoning techniques (height limitations, setback, sideyard, lot-coverage requirements, and so forth) to guarantee access to sunlight. Simply stating that the protection of solar access is one intended benefit from existing zoning provisions may be adequate where existing solar access is good and where most buildings are already as high as the law allows. Elsewhere, zoning for solar access may require zoning provisions. If the needed adjustments are relatively minor or the city wants to ensure solar access protection mainly where solar potential is greatest, it could impose a "solar overlay zone," allowing traditional requirements to coexist with solar zoning additions.

A more innovative approach would be to redefine height, setback, and other restrictions so as to draw theoretical "solar envelopes" around parcels of land.[30] The envelope would circumscribe the shape of a building according to the solar access needs of surrounding parcels. New construction, development, redevelopment, or modifications on "enveloped" properties would not be permitted to penetrate the envelope's boundaries. Envelopes could be designed to afford varying levels of protection—to roofs, walls, or entire lots, or to properties to the south only, or to all surrounding properties. As a form of "performance zoning," a solar envelope approach does not entail traditional uniform height, setback, or bulk requirements. Moreover, it can be applied as an overlay zone.[31]

In lot-by-lot protection techniques, restrictions on shading sources are tailored more closely to the requirements of particular collector sites than in area-wide approaches. One approach (enacted into law in New Mexico amid controversy) simply prohibits shading of solar collectors installed after a certain date.[32] A more reasonable mechanism (proposed in Massachusetts, Rhode Island, and some other states) is to require owners of built or planned solar installations to apply for formal recordations from the building department, which would also deny building permits for structures that would shade recorded collectors.

In a modified form, the solar envelope concept could also be used on a lot-by-lot basis. But this raises questions of fairness (Should a building that conforms to a solar envelope and that lies in the shadows of a larger building have its own density reduced without receiving the benefits of solar access itself?), administrative complexity (Is the city prepared to use the interactive computer programs needed to construct each envelope on the basis of site-specific data?), and predictability (What happens when the envelope changes with surrounding construction?). Of course, even the best of the lot-by-lot approach is inherently limited by time and court squabbles. Accordingly, most experts conclude that solar envelopes are best applied on an area-wide basis where residential or commercial development is moderately dense.[33]

Cities should probably not adopt any solar access protection techniques until they weigh the solar access problem. No single protection device can work in all city neighborhoods. Surveys of buildings in Boston, Philadelphia, and Baltimore indicate that substantial rooftop access is available in neighborhoods with homogenous construction.[34] In these areas, simply acknowledging a formal intent to protect solar access through existing zoning might be enough. Where new mid-rise buildings could be positioned so as not to shade each other's south-facing walls, the solar envelope technique would work. Envelope zoning could also be used where newer, taller buildings are interspersed with older, smaller ones. For protecting the solar access of community solar energy systems, a combination of approaches might work best.

Where to Find Local Information

Since cities have long regulated land use, land-use information is readily available. The key to using these sources effectively for energy planning is to define information needs clearly to streamline data-collection. For example, to encourage the use of solar water heating, a city may need only general information on solar access potential in different neighborhoods. But to offer financial incentives to low-income groups, it may need more detailed and precise data on solar access.

For help in finding and interpreting policy-related data, contact the city planning office, the community development office, or whichever city department develops and implements land-use controls.

The most comprehensive source of general land-use information is probably the city and neighborhood maps that the city or county planning departments maintain. Generally, these show street directions, the contiguity of different land uses, building densities, and commercial and industrial concentrations. Some city planning offices also maintain Sanborn Maps, which show the dimensions, types of construction, and orientations of buildings.

Detailed data on each land-use sector can also be obtained from property-tax records. Where tax records are computerized, the data on lot sizes, ages, and types of buildings permits counts and sorts. Where they aren't, one can base the calculations on data describing a representative sampling of buildings.

Few standard techniques for measuring land uses provide reliable information on vegetation shading or on solar access in mixed-use areas or border zones. To get more detailed information, some cities have commissioned aerial photographs or borrowed them from state offices.

One way to use land-use information to discover patterns of urban energy use is to rely on rules of thumb. Where single-family housing predominates, for instance, residential heating energy consumption will be relatively high, so on-site conservation and renewable energy technologies that reduce heating loads will probably reduce overall energy consumption. Alternatively, one can examine the results of community energy audits conducted in similar cities. A third option is to use state or regional data to estimate the amounts and kinds of energy consumed in different land-use sectors. A fourth strategy requires obtaining consumption data for different sectors from energy suppliers. The final alternative is to survey selected typical residential and industrial users and generalize from the results, though "typical" commercial users tend to be difficult to pinpoint.

4. City Buildings

Land-use information can help city officials discern how much energy neighborhoods and sectors use and how much they can save. But to develop effective energy-conservation policies and programs, information on the city's building stock is also needed. For example, a row-house neighborhood composed of structurally sound but poorly insulated 50-year old housing requires energy-conservation measures quite different from those suited to new and better-insulated townhouses. Substandard housing with significant structural flaws will profit most from still other measures. Policy options differ accordingly. Many state building codes set at least a minimal standard for energy efficiency in new buildings.[1] Even in the absence of codes, some builders are applying stricter conservation measures on their own. Consequently, it is much easier to find information on energy-conservation measures in newer buildings than to estimate their extent in older structures. To compensate, the emphasis of the discussion here is on existing buildings.

EFFECTS OF BUILDING STOCK QUALITY

The greatest influences on energy use in a building are the character of its shell and the design and condition of the heating and cooling system. In terms of energy use and conservation, a shell's most important features are the thermal integrity of its walls, windows, ceilings, and joints, its surface-to-volume ratio, and the materials that make up its structure.

Thermal Integrity of Building Shell

Both shell design and structural characteristics determine whether a building uses energy efficiently or wastefully. In most climates, south-facing windows add to the building's heat gains. Conversely, keeping north-side windows to a minimum reduces heat loss. The loss of heat through doorways can be cut with air-lock vestibules. In hot climates, vegetation helps cool buildings naturally, as do proper ventilation, overhangs, and shutters.[2]

Whatever its design, to use the energy efficiently a building must be well-sealed. Depending on the climate, it must also have adequate storm windows and doors. Because steep energy-price increases are relatively recent, few buildings that are not brand-new are weathertight. Indeed, before 1975 fuel savings did not seem to justify front-end expenditures for energy-saving techniques.

While builders are responding to higher prices and stricter regulations by building homes that are more energy efficient, older urban buildings have little or no insulation.[3] Proportionately fewer single-family houses in older Northeastern and Midwestern cities have attic or roof insulation than do houses in more recently urbanized areas with comparable climates. (See Table 4.1.) For example, only 39 percent of single-family housing in Baltimore has attic insulation. By contrast, 87 percent of Omaha's single-family homes have insulated attics or roofs. Even though buildings in colder climates tend to have more insulation than their counterparts in warmer places, higher heating loads and fuel prices in cold areas mean that the gap between existing and cost-effective insulation is greater where it is colder.[4]

Differences between the thermal integrity of central city versus that of suburban housing are also much greater in Eastern and Midwestern cities where

Table 4.1. Percentage of Single-Family Dwelling Units with Attic Insulation (Selected Cities with 4,000 or More Annual Heating Degree Days).

CITY	HDD	% OF SINGLE-FAMILY HOMES WITH ATTIC INSULATION
Boston	5621	59
Providence	5972	78
New York	4848	76
Buffalo	6927	54
Newark	4589	73
Pittsburgh	5930	38
Philadelphia	4865	55
Baltimore	4729	39
Washington, D.C.	4211	52
Louisville	4645	63
Cleveland	6154	53
Detroit	6228	67
Indianapolis	5577	80
Chicago	6497	73
Minneapolis/St. Paul	8159	88
St. Louis	4750	57
Omaha	6049	86
Denver	6016	78
Salt Lake City	5983	75

Source: Annual Housing Survey.

Table 4.2. Prevalence of Energy Conservation Measures in Single-Family Homes—Central City vs. Suburb (Selected Cities with 4000 or More Annual Heating Degree Days).

CITY	% HOMES WITH ATTIC INSULATION		% HOMES COMPLETELY STORM-DOORED		% HOMES COMPLETELY STORM-WINDOWED	
	CENTRAL CITY	SUB.	CENTRAL CITY	SUB.	CENTRAL CITY	SUB.
Boston	59	79	77	84	73	80
Providence	78	81	84	85	91	91
New York	76	86	78	82	75	82
Rochester	n/a	81	n/a	82	n/a	87
Buffalo	54	78	83	86	77	88
Newark	73	85	50	77	67	81
Pittsburgh	38	67	21	79	44	58
Philadelphia	55	63	65	73	66	72
Baltimore	39	67	56	61	47	56
Washington, D.C.	52	82	46	50	43	50
Louisville	63	79	72	75	61	64
Cincinnati	n/a	78	n/a	70	n/a	51
Cleveland	53	78	84	87	71	83
Detroit	67	84	87	89	82	87
Indianapolis	80	83	79	80	77	79
Chicago	73	87	77	84	78	88
Milwaukee	n/a	87	n/a	93	n/a	95
Minneapolis/St. Paul	88	93	95	92	97	98
St. Louis	57	81	72	70	67	70
Kansas City	n/a	80	n/a	71	n/a	72
Omaha	86	86	88	88	89	91
Denver	78	83	44	43	29	33
Salt Lake City	75	84	56	50	20	20
Portland	n/a	78	n/a	22	n/a	20
Seattle	72	80	20	15	10	10

Source: Annual Housing Survey.

older buildings predominate. (See Table 4.2.) According to a HUD report, roughly two thirds of all single-family inner city houses lack storm windows and storm doors, while only about half of suburban homes lack these devices.[5]

Basic energy improvements in older buildings tend to be offset by the buildings' poor structural condition. In a project aimed at insulating low-income housing, Princeton University and the Tennessee Valley Authority (TVA) found it nearly pointless to insulate some buildings until they were structurally upgraded.[6] The extent of this problem in cities can be inferred from the number of urban households reporting structural deficiencies. By the latest HUD figures, 21.5 percent of city households reported at least one structural deficiency, as compared with 17.9 percent in the suburbs.[7]

Although older commercial buildings suffer from some of the same structural problems older houses do, in general occupied commercial buildings are probably sounder structurally. Still, most lack energy-saving design features. A whole generation of "glass box" structures have large single-glazed windows on all sides and no provisions for retaining heat in winter or for minimizing overheating in summer.[8]

Surface-to-Volume Ratio

Buildings lose heat primarily by conduction or infiltration through the surface. The larger the surface-to-volume ratio, the more heat per square foot of living space lost. Put simply, smaller single-family residential and commercial buildings lose proportionally more heat than do larger multifamily and commercial structures. Shared floors/ceilings and walls in apartments and party walls in rowhomes reduce the exposure of each dwelling unit's surface to the outdoors.

Differences of surface-to-volume ratios in buildings influence the relative importance of shell-conservation improvements versus heating-system adjustments. In most single-family residences and other buildings with large surface-to-volume ratios, shell conditions greatly influence energy use. As such, they should be a primary target for energy-conservation initiatives. Where buildings have a smaller surface-to-volume ratio and more complex heating and cooling needs, the shell's condition influences energy use somewhat less than the design and condition of the heating system do.[9]

Building Materials

Neither wood nor masonry—two of the materials most commonly used for building exteriors—has high insulating properties.[10] Accordingly, the energy efficiency of both wood frame and masonry buildings depends primarily on how much insulation and weatherstripping is used. But the choice of structural materials does affect the ease and the economy with which buildings can be retrofitted with various energy-conservation measures. For example, most uninsulated wood-frame houses have behind-the-wall cavities that can accommodate blown-in insulation, while few masonry structures do. Insulating masonry structures is thus much more difficult and costly. (See Chapter 8.)

Heating and Cooling Systems

While a building's shell influences the amount of heating or cooling energy it needs, the amount of energy it uses to meet that demand is determined by its heating and cooling system. If the system's components are adequately sized and configured, well maintained, and properly controlled, a high percentage of the energy in the fuel that goes in comes out as useful heat.

Gauging the effectiveness of a building's heating system requires a thorough and systematic audit. Short of that, however, city officials can assume that heating systems of older buildings are both oversized and poorly maintained. Many older oil furnaces and gas furnaces are only 50 to 60 percent efficient, and most heating systems installed when fossil fuels were cheap are oversized—gauged to meet temperature extremes that seldom if ever occur.

In commercial and other large buildings, heating and cooling system efficiencies are affected not only by system size, but also by the system's configuration, operation, and control. Because many work areas accumulate heat cast off by people and machines, some zones actually need cooling energy year-round. Many systems, however, cool forced air to meet the needs of these "hot spots" only to reheat it for use elsewhere.[11] Also, some systems are left on too late at night or started too early in the morning. Another problem is that many control systems do not distinguish between building zones that require heating or cooling constantly and those whose demands fluctuate. In addition, the heating systems in many commercial buildings are poorly maintained: few boilers and furnaces get cleaned periodically, and much of the heat-distribution piping and ductwork is uninsulated or sagging.

Compact Development

Encouraging building owners or renters to improve energy efficiency and encouraging more energy-efficient development patterns are complementary, not competing, means of reducing energy use. Densely built housing is not necessarily structurally and mechanically energy-conserving; per capita energy use will not drop automatically with an increase in multifamily buildings. Similarly, existing multifamily buildings may need strenuous energy-conservation measures.

Opportunities to conserve energy in buildings and to make development more compact exist in most cities, but the relative emphasis placed on each will vary by city. The compact-zoning or development strategy makes most sense in cities that expect a boom in new housing. Stressing improvements to existing buildings is more appropriate elsewhere.

The Availability of Sunlight

While land-use considerations influence building orientation and sources of shading, such features as roof space, walls, and materials determine a structure's ability to use available sunlight. Since many active solar energy systems use roof-mounted solar collectors, the quality, size, and shape of a building's roof determine the system's feasibility. Also important is the system's weight: active solar-collector panels can weigh from 75 to 200 lb.[12] Most new roofs can support that additional weight, but some older roofs of inner city buildings should be reinforced before they are retrofitted. Nearly any roof will require structural bolstering before a small wind energy conversion system can be installed.

As for space, active solar water-heating systems require far less collector area than combined water- and space-heating systems do, and many city rooftops can accommodate collectors for active water-heating systems. But finding enough space for active space-heating systems can be difficult. In a typical neighborhood of single-family detached houses in Denver, for instance, an average of only 377 ft^2 of roof per house is available for solar collectors,[13] and typical row-houses have even less roof space. Flat roofs provide the most usable space since roof-mounted collectors can be oriented due south on them.

Collectors can, of course, be mounted on the ground. But few inner-city lots have room for on-site collectors for space heating. Then too, even low-rise buildings can shade ground-level structures.

For passive solar retrofits, the building's materials matter greatly. Buildings with large amounts of south-facing glass have built-in collectors already—though modifications for storing, distributing, and controlling collected heat and for preventing heat loss will be necessary.[14] If painted and glazed, buildings with south-facing brick or masonry walls can both collect and store solar heat.[15] (See Chapter 8.) In three Philadelphia neighborhoods assessed for solar potential, from 28 to 54 percent of the houses had wall materials and orientations suitable for passive solar retrofits.[16] In many urban warehouse or factory districts, flat roofs and expansive brick walls are commonplace—an auspicious combination for active solar water heating and for passive solar space heating.

Policies to encourage the use of solar designs and materials in new buildings could be combined with solar access policies. In cold climates, building codes could be altered to allow or encourage the increased use of glass on the south sides of buildings and the decreased use on north-facing walls. Where cooling loads are greater, the opposite approach could be used.[17] Another option is smoothing the way for building-code certification of innovative building materials used in passive designs.

Knowing each area's and each building type's potential for collecting and using renewable energy enables city officials to create effective incentives and regulations. To protect solar access, for instance, a city might protect south-wall access in neighborhoods with properly oriented brick or masonry housing. City governments might also think twice about allowing the removal of buildings whose design and construction make them good candidates for solar retrofits.

WHERE TO FIND LOCAL INFORMATION

To initiate policy development or to marshal data for assessing general needs, use HUD's *Annual Housing Survey,* which shows what percentage of a city's residential buildings have attic insulation, storm windows, or storm doors. This survey, however, does not indicate insulation levels, and—a more serious limitation—it provides information only on occupied, single-family homes. The DOE now collects regional data on insulation thickness (for ceilings, floors, and walls) in one-to-four family homes, but not for larger buildings.[18]

Energy-conservation potential can be estimated in various ways. Portland, St. Paul, and some other cities have used information on the age ranges of city buildings to substitute for more exact data.[19] In general, building age correlates with the energy efficiency of the building shell. According to an Urban Institute study for HUD, about 65 percent of all single-family homes without attic insulation, storm windows, and storm doors were built before 1950.[20] Houses built between 1950 and 1974 are likely to have minimal conservation features. But building age does not indicate the adequacy of a building's heating system since heating equipment can be upgraded or replaced; however, most houses built after 1950 probably still have their original furnaces.

General data on building age can be obtained from numerous sources. The best is the local building department or the state building code authority. Records of subdivision plots for newer areas or tax records for older buildings in inner cities may also help. Information on the suitability of buildings for solar applications (including that on a house's size, style, and building materials) can be obtained from Sanborn Maps or from tax records. Estimates of the structural strength of roofs should be obtained from local or state building inspectors.

Ultimately, of course, understanding how and why buildings consume energy requires auditing a representative sample of buildings. Experience indicates conclusively that on-site energy audits are essential if precise reliable information is needed. Audit forms filled out by home owners and mailed back to energy auditors all too often turn out to be inaccurate.[21] Homeowners simply cannot be expected to answer even semi-technical questions correctly.

Energy audits can be expensive and time-consuming, but often they are necessary and worthwhile. Indeed, a city would be foolish to target a revolving loan program for energy-conservation improvements on low-income neighborhoods without first gauging the soundness of the building stock. The audits will tell the city if energy improvement funds should be devoted in part to structural repairs.

Where city energy planners have to conduct their own energy audits, results from the growing number of energy audit programs should be incorporated. The planners should draw on the energy audits that HUD now encourages (or requires) in some housing-production or housing-assistance programs.[22] Other helpful sources are energy audits conducted by the city housing authority under the auspices of the Residential Conservation Service (RCS) program in conjunction with housing-modernization projects, or by urban community groups (such as the Energy Task Force in New York City and Mass Fair Share in Boston).

5. Energy Supply and Distribution

Along with land-use patterns and building characteristics, sources of energy supply and energy-distribution patterns help determine the technical and economic feasibility of using conservation or renewable energy in cities. For example, the substitution of energy-conserving or renewable-energy technologies will be more attractive to consumers currently using electricity for water heating than to those using natural gas. Electricity costs so much more than price-regulated natural gas that defraying the up-front costs of a replacement takes much less time.[1]

Other elements of the energy supply-and-distribution picture also bear upon a city's options for reducing energy use. The rate structures, capacity factors, types of fuel used by public utilities, and other supply and delivery factors all affect the economics of many conservation and renewable-energy approaches.

TYPES OF ENERGY USED

Two factors determine the types of energy that a city uses: (1) what kind of energy demand it has and (2) where it gets its energy supplies. As for demand, the most basic distinction is between thermal and electric needs. Theoretically, any end-use requiring relatively low-temperature heat represents a demand for thermal energy that can be supplied by any of a number of means. In contrast, end-uses that require very high temperatures or power translate into demands for electricity. In practice, however, thermal energy is usually used for space heating and water heating, while electricity is used for cooling, lighting, and appliances.

In most residential and commercial buildings, the relative sizes of thermal and electric demands depend largely on building type and climate. In all but the hottest climates, needs for space heating and water heating predominate in single-family homes, small apartments, and commercial buildings. Where temperatures are extremely high, demands for cooling (and thus electricity) are proportionately higher.[2] Regardless of climate, most high-rise multifamily or commercial buildings demand more electricity (for services, appliances, and lighting) than other buildings do.[3]

In industry, energy needs depend primarily upon the temperature level that the industry's processes require. A Solar Energy Research Institute survey of industries using relatively low-temperature processes indicates that natural gas or oil is most often the primary energy source.[4] But in a high-technology industrial area or in cities with large concentrations of metal-processing industries that use electrolysis, electricity demands are higher.[5]

Practice does differ considerably from theory, of course. Electric-resistance heat is commonly used to meet space- and water-heating needs that low-temperature thermal energy could meet, and because electricity generation and transmission are inefficient, some owners of all-electric homes pay utility bills that outstrip mortgage payments.[6]

To evaluate the technical feasibility of different energy-conservation and renewable-energy technologies, cities should first identify the true demand for heating versus electric energy. The key here is remembering that space-heating and water-heating loads require heat, not electricity. For example, while both Seattle and Miami rely primarily on electricity, Seattle uses most of it to heat space and water, while Miami's demands are primarily for cooling. This difference means that the two cities have different technical and policy options. Seattle can replace much of the energy it now gets from electric-resistance space heaters with building-shell conservation measures, electric heat pumps, or solar energy technologies. In Miami, the greater emphasis should be on the use of more efficient air conditioners, better planning, and the integration of energy-generating facilities. Miami might also consider renewable sources of electric energy.[7]

Another consideration is the energy supplies available for heat and electricity. These vary by region. In general, few larger, older cities use much electricity for space heating. Most Northeastern cities use oil; most Midwestern cities and a few Northeastern cities, natural gas. But electricity is being used increasingly for space heating in fast-growing cities in the South, Southwest, and Northwest.[8] (See Table 5.1.)

The price of each of the energy sources is influenced by efficiency, availability, and federal pricing regulations. Other influences are the costs of the primary fuels used for heat and electrical generation, as well as the local utility's rate structure. In general, on a delivered Btu basis, natural gas is cheaper than oil, which is cheaper than electricity. (See Table 5.2.) Energy prices in the Northwest stand as an exception to this rule since relatively cheap hydroelectricity is abundant, and natural gas costs more there than in other regions. Elsewhere, conservation and renewable-energy techniques will be more cost-effective if they displace fuel oil or electric-resistance heat instead of natural gas.

Table 5.1. Dominant Heating Energy Source—Selected Cities.

CITY	PRINCIPAL SOURCE(S) OF ENERGY FOR HEATING			CITY	PRINCIPAL SOURCE(S) OF ENERGY FOR HEATING		
	FUEL OIL	NATURAL GAS	ELECTRICITY		FUEL OIL	NATURAL GAS	ELECTRICITY
Boston	X			Chicago		X	
Providence	X			Milwaukee		X	
New York City	X			Minneapolis		X	
Rochester		X		St. Louis		X	
Buffalo	X	X		Kansas City, MO		X	
Newark	X	X		Omaha		X	
Pittsburgh		X		Dallas		X	
Philadelphia		X		Austin		X	
Baltimore	X	X		Houston		X	
Washington, D.C.		X		San Antonio		X	
Norfolk		X		Oklahoma City		X	
Charlotte		X		Denver		X	
Atlanta		X		Salt Lake City		X	
Jacksonville			X	Albuquerque		X	
Miami			X	Phoenix		X	
Birmingham		X		Los Angeles		X	
New Orleans		X		San Diego		X	
Memphis			X	San Francisco		X	
Louisville		X		Oakland		X	
Cincinnati		X		Sacramento		X	
Cleveland		X		Portland	X		
Detroit		X		Seattle			X
Fort Wayne		X		Anchorage	X	X	
Indianapolis		X					

Source: Derived from information collected by the U.S. Office of Technology Assessment.

Table 5.2. Current Energy Costs by Fuel Type
(Selected Cities).

CITY	NATURAL GAS		FUEL OIL		ELECTRICITY	
	¢/THERM	$/ MMBTU	$/GALLON	$/MMBTU	¢/kwh	$/MMBTU
Boston	.61	6.10	1.28	9.13	.083	24.44
New York City	.74	7.40	1.28	9.14	.127	37.08
Buffalo	.51	5.10	1.29	9.22	.067	19.63
Newark	.74	7.40	1.28	9.14	.127	37.08
Pittsburgh	.41	4.10	n/a	n/a	.063	18.32
Philadelphia	.60	6.00	1.24	8.82	.077	22.63
Baltimore	.53	5.30	1.24	8.82	.068	19.91
Washington, D.C.	.57	5.70	1.28	9.12	.068	19.82
Atlanta	.48	4.80	n/a	n/a	.051	14.89
Miami	.55	5.50	n/a	n/a	.075	22.10
Cincinnati/KY	.43	4.30	1.23	8.76	.054	15.76
Cleveland	.41	4.10	n/a	n/a	.082	24.15
Detroit	.50	5.00	1.27	9.05	.068	19.83
Chicago	.46	4.60	1.24	8.83	.082	24.17
Milwaukee	.55	5.50	1.19	8.48	.055	16.10
Minneapolis/St. Paul	.47	4.70	1.18	8.44	.059	17.30
St. Louis	.49	4.90	1.19	8.48	.056	16.37
Kansas City, MO	.38	3.80	n/a	n/a	.069	20.17
Dallas/Fort Worth	.46	4.60	n/a	n/a	.065	19.04
Houston	.47	4.70	n/a	n/a	.064	18.81
Denver	.45	4.50	n/a	n/a	.065	19.15
Los Angeles	.37	4.1	n/a	n/a	.071	20.76
San Diego	.41	4.1	n/a	n/a	.098	28.59
San Francisco/Oakland	.39	3.9	n/a	n/a	.053	19.43
Portland, OR	.59	5.9	1.17	8.36	.042	12.40
Seattle	.68	6.8	1.27	9.06	.024	6.94
Anchorage	.25	2.5	1.23	8.80	.054	15.72

Source: Bureau of Labor Statistics (1981).

ENERGY-USE PATTERNS

Different types of buildings use energy at different times of day, and the most economical conservation and renewable-energy options meet variable demands without oversupplying energy during slack periods. In some cases, this is a matter of choosing the right technology; in others, of adjusting energy use to coincide with the availability of renewable energy. If, for example, restaurants wash dishes mostly in the evening, solar energy collected each day could be expended then, minimizing overnight storage losses.

As for temperature levels, the key is reducing each application to the lowest point at which it is effective. This can be as simple as turning down a residential thermostat or as ambitious as installing industrial heat-recovery systems or

Table 5.3. Examples of Urban District Heating Systems.

CITY	SYSTEM CAPACITY (10^3 LBS. STEAM/HR)	NUMBER OF CUSTOMERS	FUELS USED	AVERAGE PRICE OF SYSTEM ($/$10^3$LBS.) (1978 $)
New York[1]	14,983	2,285	Residential oil	$6.76
Philadelphia[1]	3,857	670	Residential oil	5.84
Detroit[1]	2,931	843	Natural gas	5.26
Boston[1]	2,340	465	Residential oil natural gas	7.05
Baltimore[1]	990	720	Natural gas residential oil	5.47
Rochester[2]	7,986	271	Residential oil natural gas	6.00 (1980$)
Indianapolis[1]	1,722	703	Coal	4.21

Sources: 1. U.S. Office of Technology Assessment (1981).
2. Robert Botsford, *The Rochester Steam Story* (1981).

new industrial equipment that operates at low temperatures.[10] If demand for hot water peaks in early evening, the city could encourage solar energy use by letting energy consumers know how well supply matches demand peaks. Where usage patterns make it difficult to use solar technologies effectively, the city could focus instead on changing energy-use habits, much as some utilities now encourage customers to use major appliances primarily during off-peak periods.

COMMUNITY HEATING SYSTEMS

Both electricity and thermal energy can be distributed to end-uses in various ways. Which distribution system is in use affects how swiftly and economically cities can implement new conservation measures. The utility grids that supply most electrical energy serve an urban district, a whole city, or a larger region. Most thermal energy for heating space and water comes from furnaces, boilers, heat pumps, or other equipment in the building. But some center-city business districts are served instead by community heating systems that move centrally generated heat through underground piping to densely clustered groups of buildings.

New York, Boston, Philadelphia, Rochester, Minneapolis, and several other old Northwestern and Midwestern cities have such district heating systems, many of which use waste heat from nearby power plants. (See Table 5.3.) When these systems were built, they could deliver cheaper heat than individual on-site systems. But as energy costs declined, they became less competitive and fell into disrepair.[11] Now, as heating-fuel prices have risen, these systems again hold appeal. While most older district heating systems are steam-based and less economical and efficient than modern water-based systems, these older systems could be upgraded substantially, as Philadelphia, Minneapolis, Detroit, and Baltimore are now considering doing.[12,13] The keys to upgrading or building new systems are using existing lower-cost sources of waste heat, keeping piping and building connection costs low, and weatherizing the buildings being heated.[14]

Once existing heat sources have been identified, cities need to determine whether a distribution system can be economically installed and whether it will be compatible with the distribution system inside the buildings being served. Piping costs generally constitute the single most expensive part of a district heating system. Retrofitting conventional distribution technologies into city streets in dense urban areas may be prohibitively expensive.[15] But the use of low-temperature heat coupled with innovative distribution technologies could reduce distribution costs significantly, as could planning the construction of district-heating distribution systems to coincide with major sewer repairs or subway construction.[16] Whether individual buildings can be served by a district-heating system depends on the kind of fuel each building uses and the kind of distribution system it has. If the building distribution system is hydronic, hook-ups with district heating can be easy and relatively cheap. Where forced air is used, hook-ups require adding a liquid-to-air heat exchanger. Where one-way steam or electric-resistance heat is used, interconnection costs are likely to be prohibitively expensive.[17] (See Chapter 10.)

District-heating systems are economical only where heating demand is high, but oversizing the systems to accommodate buildings that use energy inefficiently is false economy indeed. Then too, installing a district-heating system could actually discourage conservation if buildings that did conserve were charged higher rates to offset the energy suppliers' revenue losses. From a policy standpoint, the city's basic challenge is to compare the economics of new or improved district-heating systems with those of upgraded building-conservation measures.

ELECTRIC UTILITIES

The economic appeal of energy conservation depends partly on how much energy is saved and partly on how much of these savings affect consumer prices. While reducing consumption of natural gas or fuel oil has predictable economic impacts, cutting electricity use often does not. The question is how utilities react to widespread conservation, which depends on their primary fuel sources, capacity pictures, daily and seasonal peaking patterns, and rate structures.

For numerous reasons, city officials must watch electric utilities' reactions to widespread conservation and renewable energy use. Many cities are annexing newly developed areas that contain high proportions of electrically heated homes. In some other cities, electric space-heating is becoming more popular as gas supplies grow less reliable or more expensive—a trend that may accelerate with gas deregulation. Equally important, opportunities abound for conserving the electric energy used for cooling, lighting, and appliance operation. (Oakland saved taxpayers over $1.5 million a year on its electric bills by installing more efficient street lights.[18])

A third factor is section 210 of the Public Utility Regulatory Policies Act (PURPA), which compels utilities to purchase electricity produced by cities and individuals.[19] Lowell, Massachusetts; Paterson, New Jersey; and Idaho Falls, Idaho—these and other cities are now refurbishing old hydroelectric facilities to create revenue. Moreover, as photovoltaic cells and other decentralized renewable sources of electricity become more readily available, this trend should accelerate.[20]

Finally, utilities themselves are beginning to invest in end-use conservation improvements to curtail demand and in renewable-electricity sources to improve the economy and flexibility of supply. (See Chapter 11.) But even though public utilities can make or break urban energy-conservation efforts, most cities have little influence over utility policy. A few utilities are municipally owned, but investor-owned utilities regulated by state Public Utility Commissions (PUCs) or Public Service Commissions (PSCs) supply most cities with power. Given this ownership pattern, the best hope for cities is to work with local utilities to develop mutually beneficial conservation strategies. But where relationships are less cordial, cities can try to initiate state-level PUC reviews of utility policies that work against municipal or individual energy-conservation efforts. City energy officials can also advocate state regulatory policies that compel public utilities to pursue energy-conservation options more vigorously.

To understand a utility's situation, examine the utility's fuel sources, capacity factor, peaking patterns, and rate structures. First, fuel sources. Typically, a utility that relies on expensive or unreliable fuel sources will be more positive about energy conservation and renewable energy use than one whose fuel sources are secure and cheap. (See Table 5.4.) In practice, few utilities rely on a single primary fuel source. Some cities rely on four primary fuels, and some utilities cover their baseloads with one fuel and use others to meet intermediate or peak loads. (See Table 5.5.) Then too, with the price and availability of some fuels uncertain, many utilities now expect to switch from one primary fuel to another. New York currently uses oil to generate about two thirds of its electricity, but its plans are to burn relatively more coal. In the Northwest, some utilities are shifting from hydropower to nuclear-pumped storage for peak power.[20]

How a utility's fuel sources affect the economics of energy conservation depends on two key variables: (1) fuel price and (2) fuel availability. Hydroelectricity costs far less than electricity produced from oil, so conservation or renewable energy options that save or displace electricity hold more appeal in the Northeast than in the Northwest. Consider too that most oil-based utilities are increasingly worried about foreign supply cutoffs, which makes domestic coal and nuclear power more attractive.[22]

An electric utility's capacity—the maximum amount of power it can generate at any one time—is another factor influencing a utility's financial condition. In principal, utility capacity should cover the maximum expected customer demand and a "reserve margin" designed to accommodate unpredicted demand surges or to offset power outages. More precisely, the utility's base capacity should meet such regular year-round demands (as those for lighting and appliance operations); its intermediate capacity should meet such regular

Table 5.4. Primary Fuels for Electricity Generation (Selected Cities).

CITY	PRINCIPAL PRIMARY FUEL(S) FOR ELECTRICITY GENERATION				
	OIL	GAS	COAL	NUCLEAR	HYDROPOWER
Boston	X			X	
New York	X			X	
Rochester			X	X	
Buffalo	X				
Newark	X		X	X	
Pittsburgh			X		
Philadelphia		X	X	X	
Baltimore			X	X	
Washington, D.C.			X	X	
Norfolk	X		X	X	
Charlotte			X	X	
Atlanta			X		
Jacksonville	X			X	
Miami	X			X	
Birmingham				X	X
New Orleans	X				
Memphis			X		
Louisville			X		
Cleveland			X		
Detroit			X		
Fort Wayne			X		
Chicago			X	X	
Milwaukee			X	X	
Minneapolis/St. Paul			X	X	
Kansas City			X		
Dallas/Fort Worth		X	X		
Houston		X	X		
San Antonio		X	X		
Oklahoma City		X	X		
Denver			X		
Salt Lake City			X		
Albuquerque			X		
Phoenix			X		
Los Angeles	X	X			
San Francisco	X	X			
Sacramento	X	X			
Portland				X	X
Seattle					X
Anchorage	X	X			

Source: Derived from information collected by the U.S. Office of Technology Assessment.

Table 5.5. Comparison of Fossil-Fuel* Inputs to Steam-Electric
vs. Peaking Facilities by Region.

REGION	TOTAL FOSSIL-FUEL BTU INPUT TO STEAM-ELECTRIC UNITS (BTU)	% COAL	% OIL	% GAS	TOTAL FOSSIL-FUEL BTU INPUT TO STEAM-ELECTRIC PEAKING FACILITIES (BTU)	% OIL	% GAS
New England	513,062.8	6.4	92.4	1.2	1160.0	88.1	11.9
Mid-Atlantic	1,892,813.9	58.8	31.4	9.7	54803.4	47.3	92.7
East North Central	3,209,263.1	95.0	4.1	10.9	28376.6	25.9	73.0
West North Central	1,450,751.1	92.1	0.5	7.1	7276.4	49.4	50.6
South Atlantic	3,116,342.7	74.3	20.4	5.3	48802.9	49.8	50.2
East South Central	1,527,726.7	91.8	2.3	5.9	5342.9	27.3	72.7
West South Central	3,324,294.1	30.4	2.2	67.4	65510.2	2.1	97.9
Mountain	1,379,495.8	88.1	1.2	10.8	14903.9	1.5	98.5
Pacific	1,041,031.5	8.5	31.4	60.4	20086.0	54.3	45.7

*Excludes nuclear and hydropower.
Source: Energy Information Administration, *Cost and Quality of Fuels for Electric Utility Plants,* June, 1981.

predictable increases over baseload as normal air-conditioning demands on average summer days; and its peak capacity should meet the highest predictable demand, such as air conditioning on the hottest summer days. Just as a utility's current capacity was determined by demand projections and investment decisions made years ago, demand projections and investment decisions made now will determine a utility's future capacity.

When demands outstrip predictions, a utility must operate continuously at or near peak capacity, which is the most expensive form of operation.[23] Most utilities forced to operate at peak capacity for very long either build new baseload plants or purchase additional power from another grid. But when regular demand falls below predicted levels, the utility must maintain unneeded facilities, some of them brand new and all of them expensive. More generally, setting and maintaining capacity is complicated. One problem is infrequent peaking. If demand peaks only once or twice a year, utility customers end up paying for "peak insurance" since prices must cover the costs of maintaining rarely used peak facilities. Then too, older plants in the utility's grid must at intervals be retired or refurbished, while others may need to be converted so that they can make use of different primary fuel. As the utility's mix of generating facilities changes in terms of age, size, and efficiency, the utility may have to replace outmoded or worn-out facilities even if demand or growth in

Table 5.6. Utility Capacity Situations (Regional Power Pools).

REGIONAL RELIABILITY COUNCIL	RESERVE MARGIN (%) JULY 1980	CAPACITY OVER 20% R.M.	RESERVE MARGIN (%) FEB. 1981	CAPACITY OVER 20% R.M.
Northeast Power	43	23	37	14
Mid-Atlantic Area	28	8	48	28
East Central Area	27	7	35	15
Southeastern	27	7	28	8
Mid-America (MO, WI, IL)	21	1	34	14
Southwest	26	6	65	45
Mid-Continent Area (ND, SD, MN, IA, NE)	21	1	51	31
Texas	36	16	50	30
Western Systems	21	1	27	7

Source: Derived from U.S. Office of Technology Assessment data (1981).

demand shrinks. Luckily, however, a utility may be able to buy extra capacity from or sell it to a larger regional grid, reducing the need to construct new power plants or to charge customers for idle excess capacity.

In terms of capacity utilization, the utilities serving large U.S. cities vary tremendously. In many older Northeastern and Midwestern cities, excess capacity is developing as populations decline and higher prices force demand down. In the Southwest and West, demand increases stemming from regional growth are outstripping capacity. (See Table 5.6.)

Capacity factors can encourage or impede energy-conservation efforts across the country, irrespective of the availability or cost of primary fuels. In Philadelphia, for example, recent efforts to use more energy-efficient heat pump systems in a redeveloped downtown area were stymied when the local utility, which has substantial excess capacity, offered the complex a very low rate for electric-resistance heating.[24]

Closely related to capacity is another factor that affects local opportunities to implement conservation and renewable energy programs: usage peaks in the utility's service area. Electricity use peaks both daily and seasonally, and law requires the utility to meet those demand peaks. Usually, the highest daily peaks occur during peak seasons—typically, say, the hottest hour on the hottest day of the year. But in some areas, the year's highest daily peak occurs in an off-peak season. For example, in Southern Florida, the highest daily peak is in winter because the highest hourly demand for electricity occurs there on unusually cold winter mornings.[25]

Keeping peak electricity consumption down benefits both the utility and the energy consumer. Conservation and new energy supplies make the most impact if peak energy is saved. Replacing peak electricity reduces the need for new peak power plants, while curbing off-peak power use does not. Similarly, if conservation or renewable-energy measures merely reduce the load factor (electricity used divided by peak capacity), the utility's fuel costs will decline while the operation and maintenance costs of existing peaking facilities or the capital costs for new peaking facilities remain the same. When this happens, utilities have an incentive to make up their revenue losses by charging customers who reduce off-peak energy use a higher rate or by increasing all rates.

In general, poor seasonal matches between an energy-conservation or renewable-energy technology and the peaking patterns of a utility do not occur unless the technology replaces substantial amounts of electricity during the utility's off-peak season. For example, solar heating would mismatch with a utility's seasonal peak only if the utility was summer-peaking and if it served large numbers of electrically heated homes. Conversely, a photovoltaic system, which provides its peak output in summer, could be "out of step" with winter-peaking utilities.

Many utilities try to cut operating costs by helping customers to better manage energy use and shave peaks, primarily by improving load management in order to offset the rising costs of construction, operation, and primary fuels. One load-management strategy is to use a computer to match supplies to demand more efficiently. Another is to offer "interruptible" rates to certain classes of customers. A third is to encourage the use of off-peak energy storage.

Where widespread on-site conservation measures would adversely affect the local utility's capacity and load factor, setting urban energy policy will be difficult at best. If the utility is municipally owned, the city can prevent it from encouraging inefficient uses of electricity, but the utility can still ask to raise the unit prices of delivered electricity to compensate for idled capacity or reduced load factors resulting from conservation. If price increases are unavoidable, the city could help the needy offset utilities' raised rates. Alternatively, the city could help defray the maintenance costs of idled capacity until an older plant is retired and overall capacity falls. In the case of privately owned utilities with excess capacity, the city will probably have to work with the Public Utility Commission or the utility's other customers to encourage energy conservation efforts. Here, experimentation is the key since both local problems and local resources vary so much. As for cities whose predicted demands exceed current utility capacity, they can cooperate with utilities in long-range planning, or become qualified power producers under the provisions of PURPA 210, or follow the Los Angeles municipal utility's lead in providing consumers with no-interest loans for the purchase of solar water heaters.[26]

Utility rate structures also greatly influence how much money consumers can save by taking conservation or renewable energy measures. Yet, city officials must understand three related factors to understand rate structures.

The first is that the issue of rate structuring is related but not identical to the larger issue of energy pricing—whether utilities should charge average or marginal prices for energy. Average prices combine the higher costs of more expensive fuels (say, foreign oil or nuclear power) and new generating equipment with the lower costs of less expensive fuels (say, hydropower) and existing equipment.[27] In contrast, marginal pricing reflects the unit costs of producing additional energy from the utility's most expensive sources. While most utilities theoretically charge "averaged" prices, in fact some are closer to average and others to marginal pricing.

Second, few utilities employ a single rate structure. Instead, most apply different structures to different classes of customers or to the same class of customers under different seasonal conditions or usage-patterns. Third, the relation between energy conservation and rate structures can change since the use of energy-conservation or renewable-energy techniques can alter the demand peaks that a utility must meet.

Some of the more common types of utility structures include the following:

- Flat Rates—The same unit price is charged for every unit of energy consumed, regardless of time of day or peak periods.
- Declining Block Rates—Per unit costs fall as energy use increases. This structure was developed as a marketing device when utilities wanted to encourage increased use.
- Inverted Rates—Per unit costs increase as consumption increases.
- Lifeline Rates—Relatively low unit prices are charged until a given level of consumption is reached; then, rates increase at steeper levels as demand rises. This structure provides basic utility services at prices the poor can afford.
- Time of Day Rates—Higher unit prices are charged for daily peak energy.
- Interruptible Rates—The utility has the option of cutting back service during extraordinarily heavy peak periods while the user (usually a business or industry) is charged lower unit prices overall in return for its willingness to accept cuts in the heaviest period.
- Demand Rates—Higher unit prices are charged to customers whose demands are unpredictable. A utility's justification for this rate structure is that a customer with unpredictable demand may decrease the utility's load factor without eliminating its need to maintain capacity.

Plainly, some of these rate structures support energy conservation while others hinder it. For example, the declining block rate penalizes energy conservers.

In contrast, the inverted rate rewards those who conserve energy. More specifically, however, how rate structures affect conservers depends on how particular conservation measures "fit" with particular rate structures. For example, if such a measure reduced peak demand, time-of-day pricing could reward attempts to conserve. But if it saved only off-peak energy, the consumer's cost savings would be less substantial. In this second case, either the rate structure or the use of the conservation technology would have to be adjusted to maximize cost savings.

Lifeline rates require less careful coordination between energy conservation technology and rate structure. However, this rate structure, like the inverted rate, does nothing to encourage the use of off-peak power because unit prices depend upon how much, not when, energy is consumed.[28] How lifeline rates affect energy conservation also depends on which type of rate structure is imposed on the amounts of energy used in excess of lifeline consumption levels.

Where energy-conservation measures save the most expensive energy, the city can confine its role to helping finance off-peak storage devices. But where utility rate structures discourage energy conservation, the city can ask utilities or (if the utility refuses) PUCs to rework the offensive rate structures, citing Title One of the Public Utility Regulatory Policies Act of 1978, which requires PUCs to determine whether current rate structures foster or hinder energy-saving activities.

WHERE TO FIND LOCAL INFORMATION

Extensive information on how the city uses energy in its municipal operations is usually available from either local centralized sources (such as an office of municipal, general, or emergency services) or from city departments (such as those for schools or policy). However, obtaining any but the most general information—total gallons of oil or gasoline purchased, for example—can prove impossible.

General information on the amounts and types of energy consumed by different classes of residential, commercial, or industrial customers can be obtained most directly from electric and gas utilities, fuel oil dealers, or gasoline suppliers, though suppliers' data is sometimes hard to get, difficult to interpret, and too general for some uses. In some cases, city or state agencies charged with monitoring air pollution can provide some information on the types of fuel large residential, commercial, or industrial facilities use.

Of course, land-use information on housing types and densities, commercial/ industrial mixes, and sectoral relationships can be used to estimate general energy-use patterns. For more precise and better organized information, state Public Utility Commissions or Energy Offices may be more valuable sources.

To get more detailed information on energy consumption (including times of use and temperature levels), cities will probably have to survey or audit a cross-section of residential, commercial, and industrial energy consumers. Most utilities will not release consumption figures without first obtaining permission from individual customers, and few utilities supply more than general figures on times of use.

Information on the utility's fuel sources, capacity, peak patterns, rate structures, and load-management programs can be complex. The best approach is to synthesize information from several sources: the utility, the state PUC, and state or local utility activist groups.

6. Socio-Economic Factors

Energy problems are not just technical problems. For at least three reasons, the causes and effects of urban energy problems, and their solutions, cannot be isolated from the city's social structure or its economy. (1) Understanding the city's vulnerability to energy shortages and price increases can enable the city to anticipate and counter particularly severe blows. (2) Socio-economic factors affect the speed with which individuals and businesses adopt energy-conservation technologies. For each income level, different strategies are needed. (3) Energy programs should address such larger urban needs as job-training and business development.

While each city faces different social and economic problems, some generalizations do hold true. One is that little empirical evidence supports recent impressionistic accounts of urban revival. Many cities designated as distressed 10 years ago have since lost more inhabitants, jobs, and businesses.[1] Recent HUD assessments of housing quality, demographics, and job- and business-development all show cities' plight worsening over time in relation to surrounding suburbs. A second generalization is that inner cities are socially and economically worse off than cities as a whole. Older Northeastern and Midwestern inner cities have lost people, jobs, and businesses while in Midwestern cities, economic growth is confined primarily to newly annexed areas.

Encouraging household energy conservation will be difficult if residents do not have the money or incentive to adopt the measures recommended. Yet, energy improvements may be needed most in poor areas, and most rental housing is less energy efficient than owner-occupied dwellings.[2] Some 51 percent of all urban dwellings units are rented, as compared with about 33 percent in the suburbs and 22 percent in exurban or rural areas.[3] (See Table 6.1.) In many large central cities, the percentage of rentals is higher still. In Boston, for instance, 73 percent of all housing is rented; in New York, 71 percent. These urban rental units differ from rental housing in the suburbs or in new developments. For one thing, they are probably older.[4] (See Table 6.2.) In 1976, more than 7 of every 10 buildings with more than 50 dwelling units were located in central cities.[5]

In rental units, energy inefficiency is difficult to combat. Of the many arrangements for splitting energy payments between the owners and residents, none provides a clear financial incentive to conserve either by improving the

Table 6.1. Tenure of Occupied Housing by Location

TENURE	U.S. TOTAL	CENTRAL CITY	SUBURB	OUTSIDE SUBURB
Owned (%)	65	51	67	78
Rented (%)	35	49	33	22

Source: Boynton, ed., *Occasional Paper in Housing and Community Affairs* (1979).

Table 6.2. Percentage of Occupied Dwellings by Location and Tenure Status (1976).

TENURE	MEDIAN AGE INSIDE CENTRAL CITIES (YEARS)	MEDIAN AGE IN SUBURBS (YEARS)
Owned	26	18
Rented	35	17

Source: Boynton (1979).

building or by making voluntary lifestyle changes. If tenants pay for all energy costs, the landlord has no incentive to upgrade or weatherproof the building. Tenants have an incentive to reduce their energy consumption, but they may need the landlord's permission to make energy-efficiency improvements, and they will not profit from the increased property value of major conservation retrofits. Consequently, few can do more than set back their thermostats, turn off lights, and make other minor behavioral changes.

If landlords pay all energy bills, they should want to improve the building's energy efficiency. Yet, many building owners simply raise rents to cover increased energy costs. Even where rent controls prevent building owners from passing on fuel costs, the same restrictions can also discourage banks from lending them capital for conservation improvements.[6] In either case, tenants whose rent payments cover energy costs have no incentive to curb their energy use.

Whether they own or rent housing, few people on low incomes can afford to make energy conservation purchases. As a recent HUD paper concluded, "the larger a city is, the more likely it is to have a large percentage of its residents in poverty."[7] Because the poor spend proportionately more on food, fuel, housing, and basics, they have little left over for making longer-term energy-saving purchases. Even low-cost or no-cost conservation measures address only a sliver of the housing problems that many low-income urbanites who live in structurally unsound buildings must face.

As an Urban Institute study conducted for HUD indicates, about 60 percent of energy-inefficient single-family homes are occupied by people who earn less than $8,000 per year, and a larger percentage of this housing is substandard compared with more energy-efficient homes.[8] (To help this group, cities can make energy-conservation improvements part of public housing modernization, a move now specifically encouraged by HUD.)

Effective policies addressing the "renter's dilemma" must influence both building owners and tenants. After all, the building owner has a longer-term interest in energy savings and more freedom to make structural or mechanical improvements. Conversely, tenants can make important lifestyle adjustments that can multiply the benefits of building-shell or heating-system retrofits.

Numerous policy strategies have been proposed for improving the energy efficiency of rental buildings.[9] Some states (Oregon and Minnesota, for example) now make energy-conservation measures eligible for funding under low-interest loan programs for financing housing rehabilitation. Other states or cities are using federal or local money to develop targeted energy-conservation financing programs aimed at rental buildings. Portland is using a $3 million Urban Development Action Grant (UDAG) from HUD to subsidize 8-percent bank loans for energy improvements in homes, apartment buildings, and businesses. Rochester, Springfield (MA), and St. Paul plan to use their UDAG money for similar purposes.[10]

Local tax breaks can also serve as conservation incentives. Since 1955, New York City has offered a flexible and effective tax abatement program—"J-51." Under this program, owners of rental property, condominiums, and cooperative apartments can write 8.33 percent of the cost of building improvements off of their city taxes each year for up to 20 years. If they take this option, they are also exempted from increased property tax assessments during the period of the tax abatement. (The advantage to the city is that all improved rental properties stay under the city's rent-stabilization program while taxes are abated.) Like Pittsburgh's direct grant program, New York's tax abatement helps landlords obtain bank financing for rehabilitation. Banks are more willing to make property improvement loans, knowing that the city will require property owners to offer proof of construction expenditures.[1]

Much as they provide tax incentives, cities can use their rent-regulating authority to encourage energy conservation. The Cambridge (MA) rent-control board is using a two-pronged approach: (1) landlords may increase rents at the rate of the finance charge on loans for energy-conservation improvements, which speeds up amortization, and (2) not all landlords can pass through the full costs of fuel to tenants.[12] In 1978, Minnesota began requiring all rental buildings constructed before 1976 to comply with one set of conservation standards by 1980 and another, more stringent, set by 1983.

Recognizing that energy retrofitting of rental buildings should save tenants money, some cities now systematically encourage tenant involvement in energy-saving practices. Some conduct or sponsor tenant-education programs or do-it-yourself audit programs for tenants. In New York, ever more buildings acquired by the city under tax foreclosure are being leased back to tenant-managers, who are provided with energy conservation information.[13] Other cities or states have programs for helping ensure that tenants benefit from structural improvements to buildings. Through its "Rent Break" program, Pittsburgh will underwrite improvements in rental buildings if landlords agree to hold down rents for a specified period.

The obvious benefits of approaching both landlords and tenants have led some states, localities, and community groups to launch special programs. Sometimes these involve negotiations between landlords and tenants, while sometimes the thrust is toward providing energy-conservation funding and information assistance to landlords who sign rent-stabilization agreements. For example, the Urban Coalition of Minneapolis (UCM) brought rental housing under the DOE low-income weatherization housing program that it administers: it made energy-improvement subsidies contingent upon temporary rent freezes.[14]

Besides helping renters, central cities could realize economic development advantages through careful urban energy planning. Gasoline prices, urban mass transit systems, and the concentration of business could help stay the flight of consumers to the suburbs. In addition, the city's residents and businesses constitute a substantial market for energy-conservation improvements, a market that could be served by new businesses and new jobs located in the city.

Since not all businesses and industries are equally vulnerable to energy price increases or shortages, the city needs to target energy programs toward the businesses and industries that spend and stand to save most on energy.[15] It should also encourage energy savings through zoning for denser development or mass transit improvements. Linking downtown shops to outlying areas through the mass transit system will not, for instance, have the same economic-development consequences as will fostering new commercial centers.[16] Cities also need to know how many and what kinds of local jobs the use of each conservation or renewable-energy technology will create or destroy.

Two rules of thumb govern the links between social policy and energy policy. First, policies for protecting potentially vulnerable businesses and jobs should be shaped with two variables in mind (1) the type of business and (2) the kinds of conservation and renewable measures that it can use. Second, policies for developing energy-related jobs and businesses must take account of potential markets for different conservation or renewable-energy measures and the local skills mix.

WHERE TO FIND LOCAL INFORMATION

Begin any search for information on the city's socio-economic characteristics by contacting housing, community development, planning, economic development, and human resources offices. In particular, records of welfare or fuel-assistance payments can be used to help target conservation programs to low-income persons. Some state employment offices predict the employment impacts of various large-scale or widespread energy development. Information on the unemployment of skilled workers can be obtained from local state unemployment records, CETA prime sponsors, or local labor organizations.

HUD's Annual Housing Survey provides general information on tenure by city. HUD also collects data on different cities' per capita income levels. Consult the Department of Labor's regional or area offices for job-related data.

II. TECHNOLOGIES

Introduction

Energy technologies are evolving too rapidly to catalog inclusively. But there are now enough well-documented, careful, and realistic studies to form a foundation for urban energy planning. In the analysis that follows, studies based on performance data are accorded more weight than theoretical studies, and studies without "hard data" are used only if they reflect reasonable assumptions about future energy prices, system costs, financing charges, and system performance. Overall, the stress here is on how to decide whether a particular energy technology is technically, socially, and financially suited for use in a particular neighborhood or city. This question is really many questions, of course. Accordingly, five categories of issues are raised in Part II:

1. Technical Approaches

- How does the technology work? Is it commercially ready? Can its performance be predicted reliably? What technical problems can be anticipated? Can a city do anything to solve or reduce them?
- Are the current-model technologies likely to become obsolete? Are the second or third generation of existing technologies worth waiting for? How will the conservation and renewable-energy technologies now available fit in with those under development?
- Is a city's climate more suited to the use of some technologies than others?

2. Economics

- How can the technology compete economically with the conventional energy sources that now do the same job? Is the technology more competitive with some conventional energy sources than with others?
- Is the technology's economic appeal expected to increase?

3. City Applications

- How can the technology be used in different parts of a city? Which technologies are most appropriate for which areas?
- Can the technology be used in existing buildings? Or in new buildings only? Which structural changes are needed to make buildings more compatible with particular energy-saving technologies?

4. Institutional Constraints

- What are the land-use impacts of the technology? How much urban space will its widespread use require?
- How will different technologies fit with existing energy-supply and distribution networks? Can the objectives of public utilities be anticipated and answered?

5. Socio-Economic Considerations

- Can city residents afford the technology without financial assistance?
- What is the technology's potential for creating (or displacing) local jobs? For aiding (or upsetting) local economic development?

Most of these issues or questions are fairly straightforward, but do note that discussions of economics in this book center on the needs of local homeowners, building owners, or industries, since consumer appeal will determine whether the city's role will be limited to education and promotion or will also include financing and other incentives. In general, one criterion mentioned frequently is payback—the time it takes before the savings in conventional energy costs attributed to a conservation or renewable-energy technology equal that technology's purchase price. But in many cases, payback can be much less important than other measures. A building owner may, in many instances, be more concerned about impacts in monthly or yearly cash flows. In this book, economic appraisals of indisputably cost-effective technologies are confined mainly to present costs and simple measures of cost effectiveness, while more expensive or less well-known technologies are examined as fully as available data allow. As for the non-monetary costs of continued reliance on conventional energy sources—the costs of reliance on imported oil or nuclear power, including impacts on the national balance of payments, environmental quality, inflation, worker health, unemployment, and national security—these are both vitally important and virtually impossible to quantify. Indeed, calculations based on marginal energy resource costs alone surely underestimate the true long-term economic appeal of conservation and renewable-energy resources.

7. Supplying Domestic Hot Water

Domestic water heating accounts for a large portion of U.S. energy consumption. The residential sector uses about 3 percent of the nation's total annual energy budget, and water heating in commercial buildings takes roughly 2 percent more. In much of the country, water heating soaks up more energy in the home than lights and appliances or cooking—or even space cooling.

Residential hot water use per dwelling unit has little to do with building type. (Far more important is the number of people, each of whom uses 15 to 20 gallons of hot water daily.) But the *percentage* of total energy consumption used for water heating does differ by building type. In general, multifamily buildings use less energy than single-family dwellings do to heat a square foot of space, so they use *relatively* more energy for heating water. (See Table 7.1.)

Generalizing about hot water use in commercial buildings is more difficult since building and business types in the commercial sector are so diverse. It is true, however, that commercial and institutional facilities use relatively small percentages of energy for heating water. In a study of commercial strips, warehouses, and central business districts in Baltimore and Denver, researchers

Table 7.1. Percent of Residential Energy Consumption Used for Domestic Hot Water in Different Building Types (Selected Cities).

CITY	SINGLE-FAMILY DETACHED	SINGLE-FAMILY ATTACHED	MULTI-FAMILY LOW-RISE	MULTI-FAMILY HIGH-RISE
Baltimore	10.2%	16.0%	14.5%	18.9%
Dearborn	9.7	14	14.5	15.4
Dayton	17.8	19	25.7	15.5
Greenville	13.3	15.6	16.5	(no data)
Boulder	17.6	17.7	17.5	18.5
Seattle	13.6	(no data)	20.6	23.1

SOURCES: Baltimore: TWISS (1980).
Dearborn: CCEMP Energy Audit–Wayne County (1980).
Dayton: CCEMP Energy Audit (1980).
Greenville: The Greenville Energy Program (1980).
Boulder: Community Energy Management Plan (1980).
Los Angeles: The Energy/LA Action Plan (1980).
Seattle: Energy Ltd. (1980).

found that at most 3 percent of the energy used in these areas was for heating water. Yet, small percentages can translate into larger amounts. In the three commercial areas surveyed, annual hot water demands ranged between 300 and 16,000 MMBtu. In addition, laundries, restaurants, hotels, and some other businesses use fairly large percentages of their energy budgets on water heating.

Another influence on water-heating demand is the difference between delivery and inlet temperatures—on average, 55° and 140°F, respectively. Since 140°F is too high for the normal run of household applications, reducing the design delivery temperature can be an important conservation strategy.

The heating source also helps determine demand. Most water-heating systems in the U.S. draw on natural gas or electricity. A diminishing number of oil-fired water heaters are also in use.[1] In all of these systems, incoming city water is heated upon contact with a heat exchanger inside the water tank. In gas systems, the heat exchanger is a flue heated by combustion; in electric water heaters, electric resistance coils immersed in water supply the heat. Hot water is then distributed through the plumbing system to points of use.

Since different energy sources contain different amounts of energy per unit, and because not all heating systems are equally efficient, different amounts of each energy source are needed to heat the same amount of water under the same design and climatic conditions. (See Table 7.2) Thus, to provide the 25 MMBtu that a typical four-person family uses for water heating, about 50 MMBtu of fuel oil (357 gallons) would be required, but only about 41 million MMBtu of natural gas (417 therms). As for electricity, it is very efficient to use, but not to produce and distribute: about two thirds of the oil or coal used to produce electric power is lost in generation and transmission. Thus, supplying 25 MMBtu of electricity for water heating requires about 75 MMBtu of primary energy. If fuel oil were the primary generating fuel, about

Table 7.2. Btu Contents and System Efficiencies for Conventional Energy Sources.

ENERGY SOURCE	BTU PER ENERGY UNIT	TYPICAL BURNER EFFICIENCY
Natural Gas	100,000 Btu per *Therm* (100 ft³ of Natural Gas)	50–60% Existing 65–75% New
Fuel Oil #2	142,000 Btu per gallon	50% or Below—Existing 55–65% New
Electric-Resistance Heat	3,413 Btu per kwh	85–100% On-Site 25–33% Primary Fuel

535 gallons would be needed to provide 25 MMBtu of electrically heated water to each family each year.

The amount of energy needed for heating water can be reduced in various ways. Most are available and proven, and many are inexpensive. Several of these techniques and approaches can be combined. Most important, they are startlingly effective: even relatively simple energy-saving techniques could reduce water-heating energy needs by at least one quarter and perhaps by one half.[2]

SIMPLE CONSERVATION APPROACHES

Numerous simple approaches or practices can help reduce energy consumption for water heating. Most are either adjustments at the point of use to cut energy demand or efficiency improvements in the hot-water delivery system.

Technical Approaches

Demand reductions can come from such simple usage changes as running dish- and clothes-washers only when they are full. They can also come from such simple technical "fixes" as reducing water pressure, reducing the delivery temperature of water from 140° to 130° or 120°F, or installing flow restrictors in showerheads and faucets. (Together, the last three measures could reduce on-site residential energy consumption by nearly 50 percent.[3])

Temperature reductions can also improve combustion efficiency. The gas or oil burner can be adjusted to operate less intensely for longer periods. Furnaces are more efficient when operating at "steady states" than when cycling on and off. (See Chapter 8.) The need to maintain continuously burning pilots can be eliminated by replacing them with electric-ignition devices, which spark the burner on when storage tank temperatures drop too low.

One of the more extensive improvements in combustion efficiency now under study is the flue damper, which keeps heat from flowing out of the building once combustion has stopped. Flue dampers are being used in gas space-heating units with increasing frequency, though they may be less useful in gas water heaters since flue dampers must remain open long enough to allow all residual gases to be exhausted. (See Chapter 8.) Then too, other more cost-effective combustion improvements may reduce some of the impacts of flue dampers. For example, operating the burner longer at lower temperatures improves steady-state efficiencies but also requires an open flue for longer periods.[4]

More experimental techniques for improving combustion include the following:

- Reducing fuel inputs to gas and oil systems.
- Replacing flues to provide more heat-exchanger surface.
- Adjusting the amount of air used in the combustion process.

Water-heating system efficiencies could also be improved by reducing heat losses from tanks and piping. Most hot-water tanks and pipes are under-insulated, but blanket-type insulation can easily be cut to fit both. Preformed tank jackets and pipe coverings can also be purchased at hardware stores. Taking these measures to reduce heat losses saves energy both directly and indirectly. Since temperatures have been set high in the past to compensate for stand-by and distribution losses, lowering settings makes additional savings possible.

Economics

As indicated in Table 7.3 many of these measures are quite inexpensive. Hot-water insulation kits sell for around $20. Flow restrictors for showers cost between $5 and $15. While materials and equipment for larger buildings obviously cost more, savings are commensurately larger. Insulating techniques will not become obsolete, and most insulation lasts as long as the water tank. A final advantage to these measures is that they are necessary first steps toward adopting more extensive and effective energy conservation strategies.

Questions still exist about the technical and economic effectiveness of many combustion improvements, especially those to water heaters already in use. Flue replacements and, in some cases, electric-ignition devices for gas furnaces are probably most economical when factory-incorporated into new water heaters.

City Applications

All of these techniques work in every urban setting and in every type of city building. In substandard housing, their use will immediately reduce energy bills because their effectiveness does not depend on the building's structural condition. In multifamily buildings, where water-heating constitutes a larger percentage of energy use than in a single-family building, their use is especially appealing. Indeed, in both large multifamily or office buildings, system size and the difficulty of getting large numbers of tenants to change their water-usage habits make system improvements particularly important.

Table 7.3. Estimates of Costs, Energy Savings, and Payback of Basic Water-Heating Energy Conservation Measures.

MEASURE	ESTIMATED COSTS (DOLLARS)	ESTIMATED END-USE ENERGY SAVINGS (MMBTU)	ESTIMATED SIMPLE PAYBACK
Temperature Reduction	None[2,3]	6–7[2,3,4]	Immediate[3]
Cold Water Washing in Laundry	None[3]	4–8[3,4]	Immediate[3]
Low-Flow Showerheads	5–15[2,3]	8.2[4]	Less Than One Year[3]
Water Tank Insulation	15–30[1,3]	1.5–3.0[1,2,3,4]	1–2 Years (Electric) 2–4 Years (Gas)
Spark-Ignition Stack Damper	150[1]	2[1,2]	
Stack-Heat Recovery	No data	10% Energy Savings[4]	
Reduce Fuel Input Rate	No data	1[2]	
Reduce Excess Combustion Air	No data	3[2]	
Improve Stack-Heat Transfer	No data	1[2]	
Increase Gas Furnace Efficiency (from 72 to 78%)	30[1]		
Increase Gas Furnace Efficiency (from 72 to 82%)	55[1]		

Sources: 1. *A New Prosperity: Building a Sustainable Energy Future*, 1981
2. OTA, *Residential Energy Conservation*, Volume II (1979).
3. Mitre Corporation, *Energy Technologies for New England* (1980).
Sources: 4. Joe Carter and Robert G. Flower, "The Micro Load," *Solar Age*, September, 1983

Institutional Constraints and Socio-Economic Considerations

Utility policies will be crucial in determining the level of financial savings these measures provide. (See Chapter 5.) Combined utility load-management techniques and consumer education with the widespread adoption of these simple approaches could help flatten peak loads. Some utilities are, in fact, urging customers to consider changing their usage patterns and to insulate tanks and pipes. Utilities are likely to respond negatively to techniques only if most of their clients have electric water-heating systems and if the utilities themselves have excess capacity.

Since all these approaches are relatively low cost, all but the poorest of city residents can afford them. Moreover, payback is so rapid that their appeal extends to renters who pay their own utility bills. Of course, such low-cost and easy-to-install measures are unlikely to be the source of many new jobs, except as components of extensive energy-conservation rehabilitation or retrofitting projects.

CONSERVATION TECHNOLOGIES

After making the cost-effective conservation adjustments discussed above, cities should consider new technologies designed to use nonrenewable energy more efficiently or to use renewable energy to heat water.

Technical Approaches

At least five major appliance firms now sell heat pumps that can be used solely for water heating.[5] The units operate much like conventional heat pumps, extracting heat from ambient air and then delivering it to the point of use—in this case, the domestic hot-water tank. The heat pump itself operates on electricity, but it can provide two or three times the number of Btu supplied by conventional electric-resistance heating. One manufacturer claims that the standard coefficient of performance (heat energy output divided by heat energy input) of his hot-water heat pump exceeds 2.70 and promises substantial Btu energy savings over all competing fuel types.[6]

Integrated appliances, which use excess heat produced in one operation to help raise the temperature of domestic water, are another option. For example, air conditioners extract heat from the inside air only to expel it from the building. Clothes dryers also vent heat to the outdoors or the immediate area. Refrigerators too cast off heat. In contrast, new integrated appliances can be built to do two jobs instead of one.[7] Alternatively, heat-recovery units (heat exchangers plus a piping system to move heat to the water tank) can be retrofitted onto existing appliances. Although many of these appliances or techniques are still being developed, some show promise of halving energy consumption for water heating.[8]

A related strategy is to recapture heat from used hot water ("grey water") from showers, dishwashers, and clothes washers before it is drained out of the building and to transfer that heat to the hot-water tank. In general, the approach is to install a holding tank to collect discharged grey water, a heat exchanger to remove heat, and a piping system that transfers that heat to the domestic water supply.

The devices described above supply additional heat to a water-heating system already in use. An alternative option is to downsize the entire system by adding or substituting one or more smaller water heaters (of 10 to 15 gallon capacity, as opposed to the standard 60 to 80 gallon capacity). Tank temperatures could be varied: lower for shower and tapwater, higher for dishwashers.

A similar approach would be to use "tankless" (or demand) water heaters—heat exchangers with small gas burners or electric elements that raise the temperature of incoming city water to meet specific needs. Some tankless systems can raise 2.5 gallons per minute from 70° to 100°F and thus meet residential applications on demand. Some of these units can be pre-set to raise temperatures by a given number of degrees. Others are thermostatically controlled to heat water to a given delivery temperature.[10] Proponents of demand water heaters contend that gas-fired units can save 30 to 50 percent of the energy needed to maintain conventional gas water heaters to supply the same delivery temperatures. Compared to electric water heaters, these units save even more.

Not all these technologies are equally advanced or commercially available. Hot-water pumps will soon be widely available, while the only integrated appliance currently marketed is a combined air conditioner/water heater.[11] Demand water heaters enjoy wide use in Japan and Europe, though they are available only on a limited basis in the U.S.[12] In general, the impact of improved efficiency will differ from one integrated appliance to another. For example, an air conditioner able to extract hot air more efficiently will improve the cost-effectiveness of an integrated air conditioner/water heater, whereas a more efficient refrigerator makes a poor candidate for part of an integrated appliance because it gives off less heat.

Economics

The economics of water-heating conservation technologies varies by climate, the nature of the existing water-heating system, and the type of building in which they are used. Since none of these technologies is inexpensive, these performance-affecting factors should be considered carefully. (See Table 7.4.) For example, the Carrier Corporation calculates that its integrated Hot Shot® unit can save up to 51 percent on hot-water energy bills in Houston, but only 13 percent on Boston.[13] Likewise, since the hot-water heat pump processes indoor ambient air, it will be most cost-effective where indoor air is naturally warm most of the year.[14]

Obviously, each of these approaches will be more cost-effective if it displaces expensive fuel. Currently, this means that in most areas these devices will be more economical when they save electric-resistance energy and less so when they reduce consumption of natural gas. However, with natural gas price deregulation, energy-saving investments will become more attractive compared to all conventional fuel sources. For example, if an integrated air conditioner/water heater cut natural gas use for water heating by 51 percent in Houston,

Table 7.4. Estimated Installed Costs of Energy-Saving Water-Heating Technologies (Residential).

TECHNOLOGY	ESTIMATED INSTALLED COSTS (DOLLARS)
Hot-Water Heat Pumps	600–700[1]
Integrated Water Heaters	200–500[2]
Heat-Recovery Units	100–200[2]
Demand Water Heaters (w/thermostat)	550–600[3]
(w/o thermostat)	530–550[3]

Sources: 1. Evan Powell, New Energy Saving Heat Pump Water Heaters, *Popular Science;* April, 1980.
2. Office of Technology Assessment, *Residential Energy Conservation,* vol. 1, (1979).
3. Kit Mann, Demand Water Heaters, *Home Remedies,* (1981).

first-year savings would be about $45 at current gas prices. But if deregulated natural gas prices approximated crude oil prices on a dollar-per-million-Btu ($/MMBtu) basis, first-year savings could more than double.[15] For the same reasons, replacing or supplementing natural gas water-heating systems with technologies that require electric-resistance energy entails problems. (See Table 7.5.) Overall, the technologies that recapture and use waste heat hold the most promise for reducing natural gas consumption.

City Applications

Most of the hot-water heat pumps on the market today are sized for residential applications (80 gallons a day or less), but at least one firm is marketing units

Table 7.5. First-Year Cost Savings and Simple Paybacks of Hot-Water Heat Pumps vs. Electric-Resistance and Gas-Fired Water Heaters (Residential Application).

SYSTEM TYPE	FIRST-YEAR ENERGY COSTS[1] (DOLLARS)	FIRST-YEAR ENERGY COSTS COMPARED WITH COSTS OF HOT-WATER HEAT PUMP (DOLLARS)	SIMPLE PAYBACK: HOT-WATER HEAT PUMP[2]
Hot-Water Heat Pump	125.88	—	—
Electric-Resistance Water Heater	343.86	+217.98	3 years
Gas-Fired Water Heater	132.75	+ 6.87	Not in System Life-time

[1]Assuming electricity at 6¢/kwh and natural gas at 45¢/therm. Performance estimates from Table 3.1.
[2]Assuming installed costs of $650 and annual fuel escalation rates of 12 percent for electricity and 15 percent for natural gas.

large enough to process the 2,000 gallons of water or more a day used in multifamily or commercial buildings.[16] Locating either size heat pump in the living space presents no problem in warm climates where these units can help cut cooling energy needs by transferring warm air from the indoors to the hot water tank. But in colder climates, heat pumps should be installed in basements or utility areas to minimize the amount of heat extracted from the living space.

While waste-water recovery systems and integrated appliances can be used in single-family residences, these technologies may ultimately prove best suited for use in large buildings that have both high demands for air-conditioning and large quantities of discharged waste water. Although the integrated appliances now available are sized for use in single-family homes, waste-heat recovery systems now on the market can be used in multifamily buildings and in commercial facilities.

Buildings that discharge large amounts of grey water can take advantage of economies of scale and the high thermal efficiencies of large drainage tanks. In many facilities, the tank area need be no larger than the building's conventional hot-water tank. Then too, tanks installed under vacant land will not interfere with future above-ground construction.

Other waste-heat recovery systems could use hot air vented from apartment-house laundry rooms or extracted by office cooling systems. Both of these heat sources will be available most of the year in most climates. As it now stands, many offices use their cooling systems even in winter to offset heat gains from lighting, machines, and appliances.

To use waste-heat recovery systems, some buildings may have to have their heating and cooling systems altered. In standing buildings, the efficiency of the whole energy system would have to be upgraded if these systems are to be used, and engineers could integrate waste-heat recovery systems within a larger energy-conserving design for new construction. Reducing lighting and using more energy-efficient appliances could cut cooling loads and thus reduce the amount of waste heat discharged by the cooling system. Alternatively, waste heat could be recovered for space heating, especially if hot-water loads were small or partly met by solar heating.

Using downsized water-heating systems or demand water heaters makes sense in both single-family residences and businesses. Already, demand water heaters are being used in California in car washes.

Institutional Constraints and Socio-Economic Considerations

The land-use impacts of these new appliances will be minimal, though electric utility concerns are likely to be similar to those discussed in conjunction with conservation adjustments. If many larger buildings cut their demands for electric energy, the problem could be greater, but so is the potential for integrating the use of these technologies with load-management approaches to help shave usage peaks.

These technologies are not outrageously expensive, but they do cost more than simple conservation adjustments, so the poor will need financial assistance to afford them. Then too, many current model water-heating technologies are suited primarily for use in single-family residences. Even when larger models are developed, whether they are adopted in large buildings will depend on landlords, not tenants.

These drawbacks aside, the widespread use of these technologies would increase business and job opportunities for appliance dealers, installers, and perhaps even local manufacturers. Most such appliances would be added on to hot-water delivery systems, replacing energy rather than other appliances.

SOLAR HOT WATER

Solar water heaters may seem novel, but their prototypes were widely used over 50 years ago in the United States. Many thousands of homes in California and Florida had solar water heaters until very low rates for electricity and natural gas prompted large-scale consumer shifts. Now, higher utility costs are beginning to make solar water heaters attractive again—and not just in hot, sunny regions.

In some ways, solar water heating represents a natural extension of conservation adjustments and technologies. True, some conservation technologies such as hot-water heat pumps will compete with solar water-heating systems in some applications. But many others, demand water heaters and waste-heat recovery systems among them, can be highly compatible with solar energy use. Together, energy-conservation technologies and solar water heating could reduce current water-heating energy needs by up to 75 percent in some applications.[17]

Technical Approaches

As the solar water-heating industry becomes more competitive and diverse, the range of solar products continues to widen. Most solar-heated water is now provided by thermal energy systems. In these systems, the sun either heats the water or heats a transfer fluid that does.

Typically, a solar water-heating system consists of a collector that absorbs solar heat, a transfer medium that carries the heat to a storage tank, and a distribution system that delivers solar-heated water to points of use. The principal differences among the many system types available are in the kinds of

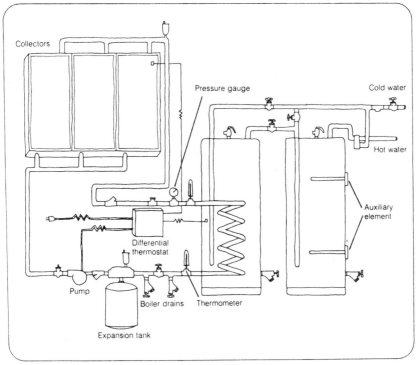

Typical domestic solar water heater

collectors used, the transfer media employed, and the methods used to move solar heat from collectors to storage to points of use. Most collectors are metal, but some are made of rubber, plastic, or other nonmetallic materials. Some systems even use hot-water tanks as solar collectors. For transfer media, systems may use city water, distilled water, antifreeze solutions, or oils or other synthetic compounds. To move heat, active systems use pumps or other mechanical means, while passive systems rely on the natural movement of heat supplemented occasionally by low-horsepower mechanical devices.

Active Systems: Flat-Plate Collectors with Metal Surfaces. Most solar water heaters sold today are some type of active system using flat-plate collectors with metal surfaces. The solar collectors are modular panels about 20 ft^2 each. The number of panels included varies with building type, hot water needs, and climate. Usually, single-family residential systems employ two to four panels totaling 30 to 90 ft^2. Collectors can be installed on rooftops, atop such structures as carports, or at ground level as long as they have adequate southern exposure. (See Chapter 3.)

Collector panels consist of two major components: (1) an absorber plate and (2) a cover plate. Absorbers are usually painted flat black so that they can absorb heat efficiently, while the surfaces of some models are also treated with special coatings to increase absorptivity. Across the absorber surface are rows of metallic tubing (also painted) that provide flow channels for the transfer medium. In most solar water-heating systems, this transfer medium is a liquid.

The collector surface is covered with one or more sheets of glazing material—usually glass, fiberglass, or plastic compounds—that allow sunlight to pass through but keep much of the absorbed heat from being lost to the air. More than one sheet of glazing is needed in colder climates since more heat will be lost from the collector surface as the difference between its temperature and the air temperature increases. To prevent additional heat loss, the collector is insulated on the back and the sides.

As sunlight strikes the collector surface, heat is absorbed by the transfer fluid, which begins flowing when the temperature of the absorber surface reaches a predetermined level: it ceases flowing when temperatures fall below it.[18] How the piping system containing the transfer fluid is configured depends on which kind of transfer fluid is used. When ordinary city water is employed, the piping system continually circulates it between collectors and a conventional hot-water tank, drawing cooler water from the bottom of the tank, heating it as it passes across the collector surface, and then returning it to the storage tank. At night or whenever it could freeze, the water is drained from the collectors either into a holding tank or out of the building. If a system uses distilled water or water containing a corrosion inhibitor, the piping will circulate it between the collectors and a separate storage tank, where heat transfer to domestic water occurs as city water is pumped through a heat exchanger immersed in the tank. To prevent freezing, the transfer fluid will drain back into the storage tank when temperatures fall.

Since these systems must drain collectors at freezing temperatures, many used in northern climates employ an antifreeze solution as the transfer medium.When the system is not operating (at night or on cloudy days), the transfer fluid is simply halted. To keep the antifreeze solution out of the domestic hot water tank, these winterized sytems employ "closed-loop" piping: the antifreeze solution circulates only between the collector and a heat exchanger, which transfers collected heat to the domestic water. Heat exchangers in systems that circulate a toxic antifreeze solution are required by most building codes to have double walls.[19]

Instead of an antifreeze solution, some systems use oil compounds as transfer fluids. Ostensibly, absorber tubing and piping will not corrode if oil is circulated through them, but oils absorb heat less efficiently than water solutions: some laboratory tests indicate that collector performance can fall by 10 to 15 percent when oils are used as heat-transfer liquids.[20]

In some systems, solar-heated water can be stored in the building's normal domestic hot-water tank. In these tanks, the gas burner or electric heating element will raise the temperature of the solar-heated water to delivery levels if it is not already hot enough. Other solar water-heating systems employ two domestic hot water tanks for storage. In the first, city water is heated when it comes into contact with the system's heat exchanger. This tank serves as a continuous source of water for a second tank, which pumps water to the end-use on demand.

Active Systems: Nonmetallic Surfaces. Collectors are the most expensive components of the solar water-heating system, so most efforts to reduce system costs focus on reducing collector costs. One way to cut costs is to substitute rubber or plastic for the metal absorber used in conventional flat-plate systems.

On one type of system, a lightweight synthetic rubber material, EPDM (ethylene-propylene-diene-monomer), is molded into a black mat containing small water tubes to accommodate transfer fluid. For temperature applications in hot regions or for swimming pools, the collector mat alone may suffice. Otherwise, glazing and insulation are needed. Some manufacturers sell the collector in preglazed, preframed panels. But collectors are also available in rolls that can be sized, cut, glazed, and framed at the building site. The rubber system can use ordinary city water as a transfer medium in all climates. Freezing is not a problem because the transfer fluid is drained from the system during cold periods, and freezing does not damage molded rubber tubing anyway. However, if the collectors are not drained, the rubber tubing and the copper piping headers will have to be connected indoors to the building's plumbing system because the headers and piping cannot withstand freezing. Rubber systems do not require heat exchangers and other components needed in systems that use antifreeze, and their storage and distribution components are identical to those in flat-plate metallic systems.

Another low-cost collector currently being marketed is made of clear molded plastic and uses a black transfer fluid to absorb solar heat.[21] This collector is both simpler and cheaper than a metallic flat-plate collector. The absorber, fluid channels, cover glazing, and manifolds are all made of a single polymer material, so they weigh five to six times less than metallic collectors. In this system, the black liquid transfer fluid moves from the collectors through conventional metallic piping to a heat exchanger in a storage tank, where the heat is transferred to the household's water. But this system also needs a holding tank, so the economic advantage of using lower-cost collectors is partially offset.

Passive Systems: Thermosyphon.   Thermosyphoning water heaters make use of natural heat movement to heat and to store heated water. In such systems, solar storage tanks are elevated above the collectors, rather than placed in the basement. As the water in the collector warms, it flows naturally upward to the storage tank, where the warmest stored water also rises to the top. A pipe from the top level of storage carries water to the house either directly or (if the thermosyphon is used for preheating) through a main hot-water tank. As warmer water flows from the collector to storage, colder water from the bottom of the storage tank flows to the collector for heating.

Thermosyphon systems come in two basic arrangements. The collector panel and the storage tank can be combined, with the storage tank positioned horizontally above the collector panel. Alternatively, the collector panel and storage tank can be separate units connected by pipe runs. The advantage of the first arrangement is that little heat is lost in moving heated water to storage. The advantage of the second is that lower-cost components can be used. Thermosyphon systems can be built on-site using relatively cheap conventional building materials, and some new thermosyphon system designs make use of lower-cost collectors.[22] In warm climates, the collector can be mounted outdoors on site-built frames; in colder areas, inside greenhouses or other glass-covered areas. In either configuration, storage tanks can be filled from a central storage tank inside the building or directly from the city water supply.

With thermosyphon systems, heat loss and freezing are serious concerns. Consequently, systems without special heating elements can probably be used only in relatively warm climates or warm seasons unless the storage tanks are in a greenhouse or some other conditioned space and the collector features a draining mechanism.

Passive Systems: Integral or Batch Water Heaters.   Integral water heaters use the same component to collect and store solar heat, so no mechanism for transferring heat between collector and storage is needed. The simplest integral systems can be nothing more than black polyethylene bags or black plastic pipes that are manually filled in the morning, heated during the day, and then emptied as needed.

A more effective type of integral water heater is a so-called "breadbox" system, named for its oblong shape. Its three main components are a water tank, an insulated container that surrounds the tank, and cover glazing for reducing heat loss. Typically, a standard 30-or 40-gallon hot-water tank painted flat black or coated with a selective surface is used as the collection/storage device for each breadbox unit.

To maximize energy collection, some breadbox systems direct larger amounts of sunlight onto the tank's surfaces via reflective devices. A curved surface of foil or plastic mirrors can be mounted inside the box or behind the tank. Alternatively, the tank itself can be mounted on a support frame atop reflector surfaces. The tank's absorptivity can also be improved by coating it with a selective surface paint or by glazing it.[23]

Breadbox systems can be integrated with conventional water-heating systems in several ways. For example, in the sunniest locations, breadboxes could work alone in buildings with relatively small hot water needs. They could also serve as preheating tanks, drawing incoming city water directly, raising its temperature, and then piping it to conventional water tanks for "boosting" as necessary. As yet another alternative, a "stand-alone" system employing an electric-resistance coil as back-up could be used. However, because freestanding systems are entirely out of doors, losses of purchased electric heat from the hot-water tank could offset solar savings. Overall, using breadboxes as preheating tanks is probably the most efficient way to use them under most circumstances.

In a breadbox system that is not pressurized, the tank will need to be refilled when the hot water has been used up. In pressurized systems, which are more common, cold water is pulled into the tank as hot water is drawn so that some hot water gets mixed with cold. Such mixing can be minimized, however, by plumbing two or three tanks together, drawing hot water off from one and cold water into another.

Unlike most of the other systems discussed here, which can be either roof- or ground-mounted, most breadbox systems are too heavy and bulky to be roof-mounted on any but new houses.[24] Filled, breadbox systems weigh approximately 40 lb per square foot.

Breadbox systems have not been used much in colder climates because the storage tanks are kept outside and because the systems also seem to require fairly large amounts of direct sunshine to be effective. (As a rule of thumb, in any month with more than 20 cloudy days breadbox systems will lose heat.[25])

If performance advantages are the measure, no single system works best under all circumstances. Most operating data indicate that flat-plate metal collectors can deliver more Btu per square foot than any other other collector type. But, in some cases, the difference is slight. Moreover, flat-plate metallic collectors cost the most per square foot of the collectors discussed here, and many of the other systems compare favorably with them on a cost-per-delivered-Btu basis.

Today, a well-made and properly installed active flat-plate system of conventional dimensions (30–90 ft^2) can be expected to provide about half of the annual hot water needed by a family of four in a temperate or cold climate, assuming modest conservation measures are taken. In warmer, sunny areas, the same system will meet relatively more of the demand for hot water. Thermosyphon and breadbox systems can meet upward of 50 percent of a typical family's annual hot water needs in a warm climate.[26] In colder climates, contributions from these systems may be quite high in the temperate and warm months.[27] But system shut-downs in freezing months are common—definitely an offsetting factor.

The performance of all systems can be vastly improved if simple conservation adjustments are made. Reducing temperature settings and minimizing heat losses will improve the overall performance of *any* water-heating system. For example, both New England Electric and HUD demonstrations of solar water-heating systems found that the poor performance of many systems installed in the programs' early stages was due in part to the simple failure to insulate tanks and pipes.[28]

One concern yet to be resolved is how well the system wears over 10 to 20 years. For example, experiences in HUD's Residential Solar Demonstration Program indicated that some cast-iron piping and storage components in space-heating systems using liquid storage corroded after only two or three years.[29] Although this problem should be less critical in water-heating systems (because they use mostly copper piping), it does warrant continued attention. Another open question is how well nonmetallic systems can withstand lengthy exposure to high temperatures and other environmental forces.

Economics

Single-Famly Applications. Solar domestic water-heating systems are economical today under certain conditions. Consumers whose water is heated by electric-resistance heating or fuel oil will see their systems "pay off" much sooner than those who use gas-fired water heaters, at least until gas prices are deregulated. Under current terms for home-improvement financing, only lower-cost systems come close to being economically attractive. All systems, however, are hurt by the high cost of money: where solar water heaters are purchased with savings, they are economically attractive; but if the solar energy system is paid for with a standard home improvement loan, it is marginally economical in most applications. Yet, special financing packages can markedly improve the economics of both lower-cost and "standard cost" systems.

As Table 7.6 indicates, the economics of solar water-heater retrofits (the largest potential market in cities) varies from city to city. In four representative cities, no system achieves positive cash flow before the loan is paid off, and payback is shortest in Boston homes with electric-resistance heat. All other paybacks are 10 years or longer.

Table 7.6. Economics of Solar Domestic Hot-Water Systems—Single-Family Retrofit Applications Using Flat-Plate Metal Collectors.

CITY	COLLECTOR AREA	AUXILIARY ENERGY	YEARS TO POSITIVE CASH FLOW	YEARS TO RECOVER DOWN PAYMENT	PAYBACK
Boston	62 ft^2	Electric-Resistance	6	8	8
		Fuel Oil #2	6	11	11
		Natural Gas	6	12	12
Wash., DC	52 ft^2	Electric-Resistance	6	11	11
		Fuel Oil #2	6	11	11
		Natural Gas	6	15	15
Grand Junction, CO					
	30 ft^2	Electric-Resistance	6	11	11
		Fuel Oil #2	6	11	11
		Natural Gas	6	17	17
Los Angeles	32 ft^2	Electric-Resistance	6	10	10
		Fuel Oil #2	N/A	N/A	N/A
		Natural Gas	6	19	19

Assumptions:
1. Fuel prices: See Table 3.3.
2. Collector installed costs: Fixed at $3,000.
3. Fuel price escalation rates: 12% for electricity, 13% for fuel oil, 15% for natural gas.
4. Retrofit financed with 16%, 5-year home improvement loan.
5. Federal solar tax credit of 40% received in year 2.
6. No additional property tax owing to solar retrofit.

To get an idea of system adjustments that could improve the economics of solar water heating, consider three alterations on the base case described above. (See Table 7.7.) In the first one, assume a substantially reduced water-heating load with consequent reductions in solar energy system costs. In the second, the "standard" hot-water demands assumed in the base case are met with currently available lower-cost solar collectors. In the final adjustment, the "mini-load" and the lower-cost collectors are combined. The most significant results are the relatively more inviting paybacks for both the "low-cost" and "low-cost plus mini-load" cases. However, even the use of lower-cost systems does not itself offset the high cost of money. Some additional points from the comparisons also deserve notice. First, extensive prior conservation alone does not make solar energy purchases less appealing, as long as the costs of active solar systems can be reduced in proportion to load reductions.[30] Second, load reductions affect the economics of low-cost systems somewhat more adversely, perhaps because the tax credit is lower if the front-end costs are.

Table 7.7. Economics of Solar Domestic Hot-Water Systems—Single-Family Retrofit Applications—Variations.

WASH., DC	COLLECTOR AREA	YEARS TO POSITIVE CASH FLOW	YEARS TO RECOVER DOWN PAYMENT	PAYBACK
Electricity				
Base Case	52	6	11	11
Mini-Load	29	6	11	11
Low-Cost	56	6	6	6
Mini-Load and Low-Cost	30	6	6	6
Fuel Oil #2				
Base Case	52	6	11	11
Mini-Load	29	6	12	12
Low-Cost	56	6	6	6
Mini-Load and Low-Cost	30	6	7	7
Natural Gas				
Base Case	52	6	15	15
Mini-Load	29	6	16	16
Low-Cost	56	6	10	10
Mini-Load and Low-Cost	30	6	11	11

Assumptions:
1. Current fuel prices and fuel escalation rates, loan terms, and tax incentives identical to Table 7.6.
2. Mini-load reduces hot-water demand by lowering delivery temperature from 140° to 120°F and reducing consumption from 70 to 55 gallons per day.
3. Low-cost system is $20 per square foot of collector, installed.

Since interest rates have become so high, comparing alternative financing schemes is important. As Table 7.8 shows in a comparison of three financing packages, reducing financing charges can improve system economics markedly. The base case features a 5-year, 16-percent loan. The second resembles a program in Palo Alto that provides 20-year loans at 6 percent for conservation and renewable-energy improvements in existing homes. The third uses a TVA program (20-year loans at 10.5 percent) as a model.

Not surprisingly, for the systems with electric-resistance and fuel auxiliary, package two is the first to result in positive cash flow. But because interest accumulates over a much longer period under this option, payback is actually longer than in the base case for the systems with oil and gas back-up and remains unchanged for the system with electric-resistance auxiliary. Nonethe-

Table 7.8. Impacts of Special Financing on Economics of Solar Water Heaters—Single Family Retrofits (Wash., DC).

	POSITIVE YEARS TO CASH FLOW	YEARS TO RECOVER DOWN PAYMENT	PAYBACK
Electricity			
Package #1[1] (Base Case)	6	11	11
Package #2[2]	2	2	11
Package #3[3]	5	2	13
Fuel Oil #2			
Package #1	6	11	11
Package #2	4	2	12
Package #3	6	2	14
Natural Gas			
Package #1	6	15	15
Package #2	10	10	16
Package #3	11	16	18

1. 5-year home improvement loan at 16% interest.
2. 20-year loans at 6% interest.
3. 20-year loans at 10.5% interest.

less, since most consumers make purchases primarily on the basis of cash flow, package two probably holds the most appeal.

This relative appeal aside, even special financing programs may not be enough to offset the high initial costs of all standard flat-plate systems. Package three reduces the number of years to positive cash flow only slightly in the electricity and fuel-oil cases. Thus, only lower-cost systems will enjoy wide appeal for single-family retrofits unless extremely low interest rates are available.

As Table 7.9 shows, systems competing with electric-resistance heat and fuel oil reach positive cash flow in the first year, regardless of how the system is financed. Even where natural gas is available, system buyers achieve positive cash flow in two years under the more generous of the two special financing packages.

Multifamily Applications.   In general, owners of multifamily or commercial buildings will find solar water-heating systems very attractive as long as the solar energy system increases in value with the physical plant as a whole. As Table 7.10 indicates, this factor is critical in determining the economic appeal

Table 7.9. Combined Effects of Special Financing Plus Use of
Low-Cost System[1] on Economics of Solar Water Heaters—
Single-Family Retrofit (Wash. DC).

	YEARS TO + CASH FLOW	YEARS TO RECOVER DOWN PAYMENT	PAYBACK
Electricity			
Package #1 (Base case)[2]	6	6	6
Package #2[3]	1	2	5
Package #3[4]	1	2	6
Fuel Oil #2			
Package #1 (Base case)	6	6	6
Package #2	1	2	6
Package #3	1	2	7
Natural Gas			
Package #1 (Base case)	6	10	10
Package #2	2	2	10
Package #3	4	2	11

1. Low-cost system at $20/ft², installed.
2. 5-year home improvement loan at 16% interest.
3. 20-year loan at 6%.
4. 20-year loan at 10.5%.

of active solar water-heating systems for multifamily retrofits. With a 5-year loan at 16 percent, no system achieves positive after-tax cash flow until after the fifth year. But if the solar energy system appreciates at 10 percent per year along with the rest of the building, the net present values of all systems in buildings with either electric-resistance or fuel-oil auxiliary heating become positive in the first or second year. This happens because the net present value includes both the monetary value of fuel savings and that portion of proceeds from the property's sale attributable to the appreciated solar energy system.

If the solar energy system does accrue value, internal rates of return are appealing—at least 20 percent for buildings with electric-resistance or fuel-oil auxiliary in all locations and at least 30 percent for buildings with electric back-up, everywhere except Washington. Because solar water-heating systems in multifamily buildings tend to be much larger and far less uniform than single-family installations, system costs will vary with location. Systems installed in sunny regions can deliver 50 percent of the building's domestic hot water with much less collector area than can systems in less sunny areas.

As Table 7.11 shows, conventional financing adjustments that cannot make fifth-year cash flows positive can nevertheless reduce negative cash flows significantly. In particular, reducing loads by making prior conservation improve-

Table 7.10. Economics of Solar Domestic Hot-Water Systems—Multifamily Retrofit Applications—Flat-Plate Metallic Collectors.

CITY	COLLECTOR AREA	AUXILIARY ENERGY TYPE	YEAR IN WHICH NET PRESENT VALUE IS POSITIVE	IRR AFTER 5 YEARS (%)	AFTER-TAX CASH FLOW (5TH YEAR)
Boston	1500 ft^2	Electirc-Resistance	1	31.8%	−2349
		Fuel Oil #2	2	21.9	−4501
		Natural Gas	3	18.6	−5263
Wash., DC	1300	Electric-Resistance	1	24.9%	−3384
		Fuel Oil #2	2	22.8	−3769
		Natural Gas	3	16.7	−5069
Grand Junction, CO	750	Electric-Resistance	1	31.2%	−1344
		Fuel Oil #2	1	28.5	−1609
		Natural Gas	3	17.2	−3074
Los Angeles	800	Electric-Resistance	1	33.3%	−1170
		Fuel Oil #2	N/A	N/A	N/A
		Natural Gas	3	15.3	−3553

Assumptions:
1. System sized to deliver 50% of hot water needs to multifamily building of 32,000 ft^2 (100 residents).
2. Financed through 5-year loan at 16% interest.
3. Current energy prices and price escalation rates are those used in Tables 7.6—7.8.
4. System costs are $4,000 plus $25 per square foot of collector.
5. Federal Business Energy Tax Credit of 15% is applied.

Table 7.11. Economics of Solar Domestic Hot-Water Systems—
Multifamily Retrofit Applications—Variations (Wash., D.C.).

	YEAR IN WHICH NPV IS POSITIVE	IRR (5 YEARS) (%)	AFTER-TAX CASH FLOW (AFTER 5 YEARS)
Electricity			
Base Case	1	24.9%	−3384
Mini-Load	2	23.3%	−2315
Low-Cost	1	29.6%	−1888
Mini-Load + Low-Cost	1	38.7%	−842
Fuel Oil #2			
Base Case	2	22.8%	−3769
Mini-Load	2	21.2%	−2551
Low-Cost	1	26.6%	−2278
Mini-Load/Low-Cost	1	34.3%	−1072
Natural Gas			
Base Case	3	16.7%	−5069
Mini-Load	3	15.4%	−3348
Low-Cost	3	18.3%	−3595
Mini-Load/Low-Cost	2	22.9%	−1851

Assumptions:
1. Current fuel prices, fuel escalating rates, loan terms, and tax incentives identical to Table 7.10.
2. Mini-load reduces delivery temperature from 140° to 120°F and cuts usage from 70 to 55 gallons per day.
3. Low-cost system is $20/ft^2 of collector installed.

ments does make the overall economics of both the base case and the "low-cost" scenario better. Reducing the hot-water load for the base case lowers the internal rate of return slightly, but by no more than 1.6 percent in any case. Only in buildings with electric auxiliary heat does reducing the load defer achievement of positive net present value. But, both those slight drawbacks are offset by the improvement in fifth-year cash flow. Similarly, while using less expensive collectors does not make cash flows positive in the fifth year for standard and mini-loads, it does improve both internal rates of return and cash flows significantly.

The much greater impacts of financing terms are dramatically illustrated in Table 7.12. Package one is the 5-year, 16-percent base case. Package two incorporates provisions of a Palo Alto program that provides 10-year, 8-percent loans for energy improvements to multifamily buildings. Package three is modeled on the provisions of HUD's 241 loans (10 years at 7 percent). With either package two or package three, cash flow is dramatically better than in the base case, while package three provides about $1,000 more in 5-year cash flow benefits than does package two.

Table 7.12. Impacts of Special Financing on Economics of Solar Water Heaters—Multifamily Retrofits.

	YEAR IN WHICH NET PRESENT VALUE IS POSITIVE	INTERNAL RATE OF RETURN AFTER 5 YEARS (1%)	AFTER-TAX CASH FLOW (5TH YEAR)
Electricity			
Package #1[1] (Base Case)	1	24.9%	−3384
Package #2[2]	1	39.5%	+ 714
Package #3[3]	1	39.9%	+1731
Fuel Oil #2			
Package #1	2	22.8%	−3769
Package #2	1	36.2%	+ 329
Package #3	2	35.8%	+1346
Natural Gas			
Package #1	3	16.7%	−5069
Package #2	2	26.8%	− 972
Package #3	3	24.5%	+ 45

1. 5-year loan at 16%.
2. 10-year loan at 8%.
3. 30-year loan at 15.5%.

Commercial Buildings. In general, it is much harder to predict the economics of solar water heaters for commercial facilities than for residential applications. Some rules of thumb do hold up, however. In smaller commercial buildings, savings roughly equal those for single- or multifamily buildings with comparable alternative fuels. In laundries, restaurants, hotels, motels, and other heavy users of water, savings on hot-water bills can be substantial, which suggests that solar water heating can be economical when the costs of alternative fuels are high. For example, in DOE's Non-Residential Solar Demonstration Program, such heavy users achieved savings of up to 65 percent.

Public Buildings and Uses. In Table 7.13, the long-term cash flows associated with the purchase of a new multifamily building in a typical city-financed housing program are examined. By the tenth year, cash flows for all systems in all cities are positive. The largest surpluses are in buildings using electric-resistance or fuel-oil auxiliary heating. In 9 of 11 cases, positive cash flows in the 15th year are $10,000 or greater.

One way to measure the public economic benefits of using alternative energy

Table 7.13. Long-Term Cash Flows for a Solar Water-Heating System—New Multifamily Construction[1] ($ 1981).

CITY		ADDED OPER NET INCOME YR-10[2] (DOLLARS)	DEBT SERVICE + MIP[3] YR-10 (DOLLARS)	BEFORE-TAX CASH FLOW YR-10[2] (DOLLARS)	ADDED NET OPER INCOME YR-15 (DOLLARS)	DEBT SERVICE + MIP YR-15 (DOLLARS)	CASH FLOW YR-15 (DOLLARS)
Boston	electric-resistance	16,864	3,801	13,063	29,942	3,801	26,141
	fuel oil #2	9,765	3,801	5,964	18,333	3,801	14,532
	natural gas	7,733	3,801	3,932	16,142	3,801	12,341
Washington, D.C.	electric-resistance	10,202	3,343	6,859	18,203	3,343	14,860
	fuel oil #2	9,314	3,343	5,971	17,501	3,343	14,158
	natural gas	5,074	3,343	1,731	10,794	3,343	7,451
Grand Junction, CO	electric-resistance	9,108	2,084	7,024	16,274	2,084	14,191
	fuel oil #2	8,613	2,084	6,530	16,210	2,084	14,126
	natural gas	3,645	2,084	1,561	7,920	2,084	5,836
Los Angeles	electric-resistance	10,474	2,198	8,276	18,682	2,198	16,484
	fuel oil #2	n/a	n/a	n/a	n/a	n/a	n/a
	natural gas	2,578	2,198	379	5,773	2,198	3,575

1. The sum of added rental income plus energy savings resulting from the solar installation in a given year.
2. MIP = Sum of added maintenance, real estate, insurance, and miscellaneous expenses due to the solar installation.

Table 7.14. Costs of Solar Domestic Hot-Water/Space Heating Systems Versus Estimated Marginal Costs of Competing Fuels ($1980/MMBtu).

ENERGY SOURCE	1980 $/MMBTU	
	Raw Fuel	End-Use[1]
Regulated Gas		
Existing Hook	2.10	3.50
New Hook	4.60	7.67
Deregulated Gas	6.80	11.33
Oil ("Social Cost")	10.90	21.80
Electricity		
Existing Lines	19.90	
New Lines	25.70	
Active Solar Space Heating and DHW		
Hartford	18.17–27.02	
Washington	17.46–25.13	
Los Angeles	12.63–21.24	
Active Solar Space Heating and DHW		
(With 40% federal tax credit)		
Hartford	11.90–17.70	
Washington	11.40–16.40	
Los Angeles	8.30–13.90	

1. End-use costs include effects of 60% furnace efficiency in gas-fired systems and 50% furnace efficiency on fuel-oil systems.
Source: A New Prosperity: Building a Sustainable Energy Future, (1981).

technologies is to compare the costs of these technologies with the marginal costs of producing new energy from conventional sources (that is, the cost per kwh of electricity produced by new power plants or costs per therm of natural gas from newly drilled wells). In general, higher marginal prices are closer than lower average costs to the social costs of using depletable or dangerous fuels. In Table 7.14, costs per million Btu of energy produced by solar water heaters are compared with estimates of current marginal costs per million Btu of electricity, imported oil, and deregulated natural gas. Even without the 40-percent federal tax credit, solar costs can fall below the marginal costs of electricity delivered through existing lines. With the tax credit, even the highest solar costs are less than the marginal costs of delivered electricity, and the lowest solar costs are comparable to those of imported oil.

Solar energy's long-term competitiveness with conventional energy sources depends on how fast solar-energy equipment costs increase relative to the marginal costs of conventional fuels. For example, if the costs of solar-energy equipment increase only at the rate of inflation while the marginal costs of electricity, oil, and gas increase at higher rates, the marginal costs of conventional fuels will eventually exceed the costs of delivered solar energy. While solar energy collectors will not become clearly preferable to other sources of energy until they become cheaper, the Solar Energy Research Institute predicts that costs per MMBtu of delivered solar heat will decrease markedly by the year 2000 in all three cities compared in Table 7.13.

City Applications

Single-family residential and small commercial buildings in various climates can employ many of the solar water-heating technologies described here. However, most larger residential and commercial structures with solar water-heating systems currently use active flat-plate systems with metallic collectors, which have relatively high thermal outputs. (The ultimate effectiveness of nonmetallic collectors in large installations cannot be evaluated without more experience.)

Since few solar water-heating systems in single-family residences need more than 100 ft^2 of collector area, most detached units and row-houses will have enough roof or ground space for collectors. For example, analysis of row-house neighborhoods in Baltimore indicate that average dwelling units there had approximately 210 ft^2 of south-facing roof space, more than enough to accommodate solar collectors for a water-heating system.[31]

Like the conservation technologies described earlier, solar water-heating systems can be used effectively even where buildings' shells are energy inefficient. However, solar water heating should be added when energy-efficiency improvements are made to the building. The efficacy of this dual approach was borne out early in 1976. A burned-out tenement on East 11th Street in New York City was renovated with both energy-conservation measures and solar water heating. Heating costs were reduced by 50 percent by adding energy-conservation features, and needs for conventional water-heating were reduced by 70 to 80 percent with the use of an active solar water-heating system.[32]

In many city buildings, roof-mounted, flat-plate metallic collectors or thermosyphon or breadbox systems cannot be installed until the roof support structure is checked to make sure that it can handle the added weight. While no roofs have collapsed under collector weight so far, the roofs of some older inner-city buildings have had to be reinforced before collectors could be installed.

Multifamily buildings are particularly well suited for solar water-heating systems. Less collector area per dwelling unit is needed in large buildings than in single-family homes, and larger storage tanks can operate more efficiently because their smaller surface-to-volume ratio means they lose less heat. As a result of these combined factors, as much or more hot water can be delivered to multifamily units at less cost than in single-family applications. And most multifamily buildings are as high or higher than surrounding buildings, so unwanted shading is not a problem.

Table 7.15. Patterns of Domestic Hot-Water Use in Commercial Building Types.

Group One (over 50% of DHW use after 2:00 p.m.)
Clinics
Hospitals
Hotels/Motels
Nursing Homes
Group Two (30–50% of DHW use after 2:00 p.m.)
Gymnasiums
Offices
Secondary Schools
Shopping Centers/Stores
Warehouses
Group Three (Less than 30% of DHW use after 2:00 p.m.)
Auditoriums/Theaters
Community Centers
Elementary Schools

Source: U.S. Department of Energy, *Standard Building Operating Conditions,* Technical Support Document for Notice of Proposed Rulemaking on Energy Performance Standards for New Buildings, Nov., 1979.

The suitability of solar water heaters for commercial buildings depends primarily on those buildings' hot-water requirements. Most businesses and institutions use 40 gallons or more a day—enough to make a solar water-heating system worthwhile. But whether solar water heating is effective in commercial installations also depends on when the hot water is used. If needs are greatest in late afternoon or evening, solar collection and storage are well suited to meet them. But if most of the demand occurs earlier, overnight storage is required.

A general idea of usage patterns in commercial buildings emerges from the hot-water demand profiles for 12 commercial and institutional building types that DOE examined when developing the Building Energy Performance Standards. (See Table 7.15.) The first group of buildings uses 50 percent of its daily hot water after 2:00 p.m. The second group uses between 30 and 50 percent at that time; the third, less than 30 percent. In general, building types in the first group are particularly well suited to using solar water-heating systems, those in the second group have an acceptable usage pattern, and those in the third group (only three of the twelve building types) may find solar systems of questionable value. Clearly, solar water-heating and hot-water usage patterns in commercial buildings are generally compatible.

Facilities that use large amounts of hot water can also profitably combine solar energy systems with heat recovery from grey water. In the Aratex Laundry project in Fresno, for example, a waste-heat recovery unit preheats incoming city water before pumping it to a storage tank where its temperature is boosted by solar heat. On demand, water from the storage tank is pumped through an auxiliary steam heater to points of use. Sequentially, both the waste-heat recovery and solar-energy systems preheat laundry-service water. Together, the two systems have lopped 48 percent off annual energy demand for water heating.[33]

Institutional Constraints and Socio-Economic Considerations

Unlike the use of conservation techniques and technologies, the use of solar water-heating systems requires significant attention to land-use issues. Solar collectors must more or less face south, and they must be substantially free of shade. Guaranteeing both forms of solar access on a neighborhood-, area-, or city-wide basis may require adjusting land-use controls.

Proper orientation is the less serious concern, partly because experience now indicates that collector orientation can deviate substantially from true south without seriously diminishing system effectiveness.[34] In addition, many types of collectors can be oriented properly on buildings with almost any street orientation. If the building is improperly oriented, metallic collectors can be mounted on support structures to compensate. Indeed, many water-heating packages include adjustable supports, as do rubber collectors.

With large multifamily and commercial installations, compensating for poor orientation is harder than it is with single-family residences. Yet, many of these facilities have flat roofs or open ground-level spaces that allow some flexibility in orientating collector arrays.

As noted in Chapter 3, the collector's surface area should be shade-free between nine in the morning and three in the afternoon on December 21st. As for sources of shading, low-density residential neighborhoods are more likely to have problems with trees. In high-density or mixed-use areas, tall buildings pose the greater problem. Except where widespread constant shading occurs, smaller water-heating systems can be manipulated to minimize shading problems. Support structures can be used to orient collector panels away from natural or built obstructions.[35]

Given this much, two conclusions about land use seem clear. Few land-use controls may be necessary to provide de facto solar access protection in many built-up city neighborhoods. On the other hand, to guarantee access in the future, cities may need to enact some of the zoning or land-use controls discussed in Chapter 3.

How a utility will respond—another key concern—depends upon where the utility gets its fuel and whether it has excess capacity, as well as on the number of solar-energy systems installed in the utility service area. Where the use of large numbers of solar water heaters reduces utility load factors without reducing peak demand, or where utilities operate significantly below capacity, solar users may see their unit costs of energy rise. (See Chapter 5.)

Some utilities have tried to cut peak demand and reduce the need for new power plants by encouraging their customers to use solar water-heating systems. The Tennessee Valley Authority (TVA) has initiated a large-scale program to combine solar water heating with load management through the use of off-peak storage. In late afternoons and evenings, TVA's customers use solar energy collected during the day. In another program, Los Angeles' municipal utility is offering no-interest loans for solar water-heating systems that cost up to $4,000 in one- to four-family buildings and up to $10,000 in larger buildings.[36]

Even relatively low-cost solar water-heating systems represent major purchases for middle-income people. Given only conventional financing provisions, few poor families could afford even the least expensive solar water heaters. For example, a professionally installed lower-cost system for a single-family home in a warm climate costs at least $400, even with federal and state tax credits. In colder climates, costs may exceed $2,000 (after tax credits) for professionally installed single-family systems.

A second impediment to widespread solar use in many cities is the high proportion of rental housing in many urban areas. Since solar water heaters are costly, private owners of rental buildings probably will not rush to install them on their properties without incentive or regulatory inducement, excellent performance records notwithstanding.

Where cities do motivate residents to make solar energy investments in low-income and rental housing, local economies will benefit from new job and business opportunities. According to the National Association of Solar Contractors, installing one solar water heater in an existing single-family home requires two person-weeks of labor. Thus, twenty-six retrofit installations would generate one job per person per year. Depending on the level of investment and on the nature of the local economy, either existing businesses will expand or new energy businesses will emerge. The manufacture of domestic water-heating systems can also be local.

Jobs in both the manufacture and installation of solar water-heating systems can be targeted to unemployed persons. Most manufacturing jobs are likely to be for relatively low-skilled workers. In contrast, installation jobs will probably pay better and provide more opportunities for advancement. But because installation jobs are more complicated—requiring a combination of carpentry, plumbing, and electrical skills—more training and supervision are needed. Where neither can be provided, insulation and weatherization installation and greenhouse construction probably represent better job opportunities than solar water-heater installation for unskilled workers.

8. Heating Buildings

About 20 percent of the nation's energy goes to heat buildings. In most of the United States, over half the energy consumed in residential buildings is for heating. Even in relatively warm Greenville, North Carolina, 62 percent of all residential energy goes for heating. In Los Angeles, homes use just under half of their total energy for space heat.

Recently, much has been written about the inherent energy efficiencies of multifamily buildings.[1] (See Chapter 3.) Moreover, analyses show that most single-family detached homes have the greatest opportunities for cutting residential heating energy use. Yet, in older central cities, many multifamily buildings are not energy-efficient either. As Table 8.1, indicates, space heating in Baltimore and Dayton requires about the same percentage of total energy in single-family detached as in single-family attached buildings. In Baltimore, the percentage of energy used for space heating is actually greater in low-rise

Table 8.1. Percentage of Total Energy Load for Space Heating in Different Housing Types (Selected Cities).

CITY	HOUSING TYPES[1]			
	SFD	SFA	MFLR	MFHR
Baltimore	48.4%	47.5%	50.5%	41.7%
Dearborn	74.3	62.6	53.5	51.5
Dayton	66.9	65.8	56	50
Greenville	67.8	63.9	35.3	n/a
Boulder	70.9	56.7	49.2	46.8
Portland	60	n/a	47	n/a

1. *Codes:* SFD—Single-family Detached
SFA—Single-family Attached
MFLR—Multi-family low-rise
MFHR—Multi-family high-rise
Sources: Baltimore: Twiss (1980).
Dearborn: CCEMP Energy Audit.
Dayton: CCEMP Energy Audit.
Greenville: The Greenville Energy Program (1980).
Boulder: Community Energy Management Plan (1980).
Portland: Energy Conservation Choices for the City of Portland, Oregon, 1977.

Table 8.2. Percentage of Total Load Devoted to
Space Heating—Commercial Building Types
(Boulder).

COMMERCIAL BUILDING TYPE	% OF LOAD DEVOTED TO SPACE-HEATING
Food Stores	29.8
Eating and Drinking	17.8
General Merchandise	40.4
Other Retail and Services	53.0
Wholesale	70.4
Office	62.7
Hotel/Motel	19.8
Auditorium/Arenas	58.9
Movie Theatres	81.3
Hospitals	53.8
Terminals	64.4
Religious/Social	80.2
Cultural/Museums	60.0

Source: City of Boulder, *Community Energy Management Plan,* (1980).

apartment buildings than in single-family homes. Space-heating energy use per square foot of living space is also comparable: 41.8 MMBtu/ft^2/year in single-family detached homes and 40.3 MMBtu/ft^2/year in apartments.[2]

As for space heating in commercial buildings, energy consumption patterns are harder to assess. Most commercial structures have heavier demands than homes for lighting and refrigeration. But many use large amounts of space-heating energy too—just how much depends on the building and the business. One indication of the range is an analysis of commercial buildings in Boulder: some used as little as 14 percent of their energy for space heating, while others used over 80 percent. (See Table 8.2.)

Even in similar buildings in similar climates, the energy used for heating varies with the energy source employed, since different types of energy sources have different Btu contents, and different types of heating systems have different efficiencies. Moreover, the principal energy sources for space heating vary by region. (See Chapter 5.) Oil is the primary heat source in much of the Northeast and in parts of the Midwest and Northwest; natural gas in the Mid-Atlantic, Southern, and Western regions; and electricity in many parts of the West and South. (See Table 8.3.)

Heating-system efficiencies and Btu contents for different fuels used in space heating are virtually identical to those for domestic hot-water applications. Accordingly, a house needing 50 MMBtu of heat per year would consume

Table 8.3. Dominant Heating Energy Source—Selected Cities.

CITY	PRINCIPAL SOURCE(S) OF ENERGY FOR HEATING		
	FUEL OIL	NATURAL GAS	ELECTRICITY
Boston	X		
Providence	X		
New York City	X		
Rochester		X	
Buffalo	X		
Newark	X		
Pittsburgh		X	
Philadelphia		X	
Baltimore	X		
Washington, D.C.		X	
Norfolk		X	
Charlotte		X	
Atlanta		X	
Jacksonville			X
Miami			X
Birmingham		X	
New Orleans		X	X
Memphis			X
Louisville		X	
Cincinnati		X	
Cleveland		X	
Detroit		X	
Fort Wayne		X	
Indianapolis		X	

CITY	PRINCIPAL SOURCE(S) OF ENERGY FOR HEATING		
	FUEL OIL	NATURAL GAS	ELECTRICITY
Chicago		X	
Milwaukee		X	
Minneapolis		X	
St. Louis		X	
Kansas City, MO		X	
Omaha		X	
Dallas		X	
Austin		X	
Houston		X	
San Antonio		X	
Oklahoma City		X	
Denver		X	
Salt Lake City		X	
Albuquerque		X	
Phoenix		X	
Los Angeles		X	
San Diego		X	
San Francisco		X	
Oakland		X	
Sacramento		X	
Portland	X		
Seattle			X
Anchorage	X	X	

Source: Derived from data collected by the Office of Technology Assessment.

Table 8.4. Estimated End-Use and Primary Energy Consumption in Existing Residences (by Region).

REGION	END-USE ENERGY/SPACE HEATING (MMBTU/YR)	PRINCIPAL ENERGY SOURCES	PRIMARY ENERGY (MMBTU/YR)
New England	113.63	Fuel Oil	227.26
Mid-Atlantic	100.43	Natural Gas	167.38
East North Central	140.18	Natural Gas	233.63
West North Central	127.10	Natural Gas	211.83
South Atlantic	51.03	Electricity	170.10
East South Central	71.22	Natural Gas	118.70
		Electricity	
West South Central	80.67	Natural Gas	134.45
Mountain	90.63	Natural Gas	151.05
Pacific	53.69	Electricity	178.97
		Fuel Oil	107.38

Source: Derived from data reported by General Accounting Office (1981).

about 83 MMBtu of natural gas, 100 MMBtu of fuel oil, and 150 MMBtu of the primary fuel used to produce electricity. (See Table 8.4.)

ENERGY SAVING TECHNIQUES

Which approach to saving space-heating energy works best varies with each buildings's size, age, use occupancy profile, and condition. The three complementary approaches to building energy conservation discussed in this chapter—(1) conservation improvements to the building shell, (2) conservation improvements to the heating system, and (3) the use of renewable-energy technologies—work in both new and existing buildings, though retrofit applications are emphasized since most of the housing stock that will exist in the year 2000 will already have been built.[3]

Energy Conservation in Building Shells

Technical Approaches and Issues.   The building shell affects how much hot or cold outside air gets inside the building and how much of the building's heating or cooling energy stays inside. Of the many structural characteristics that affect the building shell's thermal integrity, the three most important are overall tightness, the amount and quality of insulation, and the window type and quality.

Each shell characteristic operates in conjunction with the others, and effective building retrofits must address all together. Implementing one does not make the others less important. On the contrary, the energy-saving effects of the combined measures reinforce each other. For example, with the addition of storm windows and attic, wall, and floor insulation to basic sealing and caulking, a single-family home in the Northeast will consume up to 85 percent less energy for heating than a similar home without these features.[4]

Reducing Air Infiltration and Heat Leakage.   Inadequate insulation is not always the only or even the primary cause of energy waste. Air infiltration through cracks or gaps is a prime offender. Such gaps can be found at building joints between wall surfaces and between the interior walls and fire walls.[5] Multifamily buildings with many windows could also be highly vulnerable, particularly since such large buildings lose heat through service or delivery areas, or through outside vents that have no dampers.[6]

Infiltration is worse in older buildings where walls, roofs, or foundations are likely to be cracked or damaged. Even in new, well-insulated and well-sealed buildings, heat escapes through gaps between insulation batts and blankets, through such uninsulated "patches" in insulated areas as trap doors in attics, or through cracks where pipes or electrical outlets penetrate walls or ceilings. Indeed, "by-pass" leaks can reduce the energy savings expected to result from added ceiling insulation by 30 to 70 percent.[7]

In new houses, infiltration levels can be greatly reduced. Better sealing and framing practices and the addition of extra insulation account for the improvement. Greatly reducing air infiltration, however, raises concerns about the build-up of such indoor air pollutants as carbon dioxide, formaldehyde, and radon.[8] If infiltration rates are brought to below 0.6 air changes per hour, stale air must be removed with an air-to-air heat exchanger or by some other means.[9]

Methods for reducing infiltration in existing buildings are also technically well-established. Indeed, leaks can be plugged in many different ways. Reducing infiltration losses in multifamily and commercial buildings is relatively more complicated and expensive. Besides caulking, weatherstripping, and sealing, reducing infiltration in such buildings may require adding air-lock vestibules at mail entrances or in service or delivery areas.

Although sealing leaks is relatively easy, identifying them can be difficult. Cracked windows, large gaps between door frames and door panels, and many other leaks are hard to ignore. But trained energy conservation specialists ("house doctors") and even specialized diagnostic equipment are needed to find some heat leaks. In one test for leaks, a high-powered blower pulls air out of the house and an infrared scanner reveals the more obscure heat-loss paths.

Insulation Improvements.   Plugging leaks does not, of course, eliminate conductive heat loses from walls and roofs. For this, adequate insulation is needed. Few buildings are insulated to cost-effective levels, and about one quarter of the nation's one-to-four family houses have no attic insulation.[10] Yet, precisely assessing the insulation levels and the effective levels of thermal protection of buildings is difficult.

Insulation effectiveness is usually measured by an R-value, which indicates how well a material resists conductive heat losses. The higher the R-value, the better insulated the structure. Typically, new single-family homes now have R-19 insulation in the ceilings and R-11 insulation in the walls.[11] In contrast, houses built between 1973 and 1976 commonly used R-13 in ceilings and R-11 in walls.

Effective insulation levels in older buildings are even harder to set. One variable is the building's age, another the uncertain effects of time on insulation quality. One rule of thumb the Lawrence Berkeley Laboratories uses is that partially insulated single-family buildings built before late 1973 have R-11 insulation in ceilings and R-8 insulation in walls.[12] By comparison today, insulation to at least R-38 in ceilings and R-30 in walls can be cost-effective in cold climates.

Upgrading insulation can significantly reduce a home's heating needs. Figures from the National Association of Home Builders (NAHB) Research Foundation indicate that in a cold climate increasing ceiling insulation from R-11 to R-38 could save nearly 15 MMBtu of energy per year.[13] For the record, a small number of new "superinsulated" houses (which have insulation even in the eaves, window frames, and electrical boxes) can actually do without a central heating system.[14]

Insulation can be added to buildings in various ways. In residential and commercial buildings with attics, insulation batts or blankets can simply be placed over dropped ceilings. This strategy is effective in single-family buildings, and in multifamily and commercial structures as well. An analysis of conservation options in Baltimore predicted that adding R-20 insulation to a previously uninsulated roof in a typical high-rise apartment building could reduce the heating load by 20 percent.[15]

Where cavities exist between interior and exterior walls, particle insulation can be blown in or foam insulation can be pumped in. (This approach is generally suitable only for uninsulated wood-frame buildings, though, since masonry walls have few or no air cavities.) Blown-in insulation can also be used in attics. In contrast, foam insulation is a poor choice for attic use because high temperatures can cause foam to deteriorate.[16]

In rehabilitating masonry buildings or attic-less structures, insulation can be added by installing walls or ceilings inside existing ones and placing insulation in the newly formed cavities. While this new approach is expensive and reduces living space, it can save substantial amounts of energy. For example, a rehabilitation project in New York City cut its heating demand by an estimated 37 percent when it took conservation measures that included adding insulation behind new walls and new dropped ceilings.[17]

Another approach is to re-side the exterior, placing insulation outside the existing wall and adding a new layer of siding. This approach is probably economical only in a wood-frame building scheduled for re-siding.[18]

Improved Windows.   Large amounts of heat can be lost through window surfaces by conduction, convection, and radiation. According to the U.S. Congressional Office of Technology Assessment (OTA), typical single-glazed windows have R-values ranging from 0.9 to 2.0—quite low.[19]

In new buildings, heat losses through windows can be minimized by reducing window areas on the coldest sides of the building, by installing two or three layers of glass, or by using new heat-retentive window materials or designs. In existing buildings, the key to cutting losses is adding more glazing and thermal shutters or shades. A more ambitious approach is to retrofit the building with replacement glass that has improved thermal qualities.

The simplest retrofit is adding storm windows, which 40 percent of existing one-to-four family buildings lack.[20] However, storm windows are not always the most cost-effective window retrofit. In some single-family homes in milder climates, adding insulating shades may be more economical.[21] In larger buildings, using reflective shades and making the heating system more efficient may be more economical choices.

With or without storm windows, the use of blinds, shutters, or shades can also reduce conduction (and infiltration) losses by interposing another layer of material between heated air and the window's cooler surface in the winter and between cooler air and warm window surfaces in summer. Both engineering and economic analyses have indicated that shutters or blinds must be tight-fitting to be effective.[22] Where air gaps exist, heat can be let out (in winter) or let in (in summer) by convection.

Window technology for use in older buildings and new construction is advancing rapidly.[23] Most improvements are in designs, glazing techniques, or shade and shutter materials. Glazing materials now being developed allow greater penetration of solar energy into the living space, and double layers of this glass can reduce conduction heat losses without sacrificing heat gain in winter. Some blinds allow one side of the blind to absorb heat in winter and another to resist it in summer. In addition, some shading devices are so thermally efficient that they can be considered insulation, usable either on the inside or the outside of a building.

Since both outside air temperatures and climate control needs change daily and seasonally, effective window designs allow heat gain and retention when it is cold and prevent heat gain in hot weather. This means that they integrate heat collection, insulation, and shading functions.

Economics. Which conservation improvements to building shells are most cost-effective depends upon the type and location of the building, as well as upon the number of energy-efficiency features already installed. Although few conservation retrofits are cheap, a carefully chosen package of improvements can be quite cost-effective. (See Table 8.5.) In many cases, the simplest retrofits are the most immediately cost-effective. For example, the Urban Institute calculates that caulking, weatherstripping, and adding clock thermostats at a cost of only about $200 could save up to 25 percent of a home's annual heating energy.[24] Paybacks for even an extensive retrofit "package" (storm windows plus roof, wall, and floor insulation) average 7.0 years nationally in owner-occupied housing. In colder regions with high utility costs, paybacks can be much shorter (4.0 years in the Northeast and 5.9 years in the Midwest), although in warmer areas they can take much longer (8.2 years in the South and 12.6 years in the West). Paybacks for adding ceiling insulation and storm windows are under 3 years everywhere except the West. (See Table 8.6.)

As noted earlier, using payback as a criterion for cost-effectiveness assumes that retrofits are paid for with cash. If borrowed money is used, the more appropriate indicator is usually year-to-positive cash flow. Estimates of the time it takes to reach a positive cash flow are not as numerous as payback estimates. According to the Urban Institute, most single-family households

Table 8.5. Estimated Costs of Selected Shell-Conservation Measures (1980 $)

MEASURE	AREA (FT^2)	ESTIMATED INSTALLED COSTS (DOLLARS)
Uninsulated ceiling retrofit to R-19	1350	470.00
—to R-27	1350	580.00
Uninsulated walls retrofit to R-11	1232	860.00
Single-pane storm windows	200	900.00
Double-pane storm windows	200	1,400.00
Infiltration reduction		240.00

Source: A New Prosperity: Building a Sustainable Energy Future, (1981). p. 52.

Table 8.6. Average Retrofit Costs and Estimated Paybacks for Basic Conservation Packages Installed in Single-Family Owner-Occupied Homes (Nationally and by Region).

	RETROFIT[4] COSTS (1980 $)	FIRST YEAR[4] SAVINGS (1980 $)	ESTIMATED[4] PAYBACK
Package One[1]			
National	200.00	78.00	2.6
Northeast	200.00	157.00	1.3
North Central	200.00	82.00	2.4
Southern	200.00	70.00	2.9
West	200.00	29.00	6.9
Package Two[2]			
National	390.00	119.00	3.3
Northeast	310.00	209.00	1.5
North Central	280.00	143.00	2.0
Southern	245.00	90.00	2.7
West	810.00	66.00	12.2
Package Three[3]			
National	1,590.00	225.00	7.0
Northeast	1,420.00	353.00	4.0
North Central	1,385.00	235.00	5.9
Southern	1,710.00	209.00	8.2
West	1,845.00	146.00	12.6

1. Clock thermostat; caulking and weatherstripping; filling holes and air leaks; wrapping heating ducts and hot-water pipes; insulating hot-water heater.
2. Package one plus: attic insulation (R-19 in Northeast and North Central regions; R-11 in South and West); storm windows.
3. Package one plus: wall insulation (R-13); floor insulation (R-19 in Northeast and North Central; R-11 in South and West); attic insulation (R-30 in Northeast and North Central regions; R-11 in South and West); storm windows.
4. Calculated taking account of items already in place.
Source: Urban Institute (1980), pp. 107, 111, 116.

installing stringent conservation packages (storm windows plus insulation on roof, walls, and floors) could achieve positive cash flow in a year or two if they claimed the full federal energy tax credit—perhaps sooner with available state tax credits.[25]

While these payback or cash-flow periods are economical for homeowners, they probably hold less appeal for occupants of single-family rental units. Even if tenants pay their own utility bills, they probably do not rent one place long enough to recoup conservation investments. They are also unlikely to borrow

Table 8.7. Average Retrofit Costs and Estimated
Paybacks for Basic Conservation Packages
Installed in Single-Family Rental Homes
(Nationally and by Region).

	RETROFIT[4] COSTS (1980 $)	FIRST YEAR SAVINGS (1980 $)	ESTIMATED PAYBACK (YEARS)
Package One[1]			
National	200.00	59.00	3.4
Northeast	200.00	132.00	1.5
North Central	200.00	69.00	2.9
Southern	200.00	53.00	3.8
West	200.00	22.00	9.1
Package Two[2]			
National	525.00	113.00	4.6
Northeast	510.00	210.00	2.4
North Central	425.00	138.00	3.1
Southern	390.00	98.00	4.3
West	920.00	66.00	13.9
Package Three[3]			
National	1,880.00	192.00	9.8
Northeast	1,720.00	331.00	5.2
North Central	1,640.00	225.00	7.3
Southern	1,980.00	174.00	11.4
West	2,025.00	121.00	16.7

1, 2, 3, and 4: See Table 8.6.
Source: Urban Institute, pp. 111, 113, 116.

money to pay for improvements in homes they do not own. A final consideration is that single-family rental units may require more energy-conservation retrofit measures than owner-occupied homes, so costs and paybacks are commensurately greater.[26] (See Table 8.7.)

Much less information is available on the cost-effectiveness of shell improvements in large multifamily and commercial buildings, and available energy-conservation studies tend to focus on adjustments to the commercial or apartment building's heating systems, as distinct from shell improvements. Still, costs for reducing infiltration will be relatively low except where a vestibule or some other air-lock entryway must be added. The costs of improving glazing and insulation will depend on the size of the building, though the costs per square foot of materials will probably not be appreciably more or less than for the same material used in single-family homes. Installation costs, however, may increase with the size of the building.

Conservation Improvements in Heating Systems

While the building's shell affects the amount of heat needed, its heating system determines how much energy it takes to produce that heat. Improvements to both the building shell and the heating system are necessary for an effective energy-conservation retrofit.

Most heating systems used in single-family buildings are relatively simple. In an oil- or gas-fired unit, the fuel is burned in a combustion chamber to produce heat. That heat is passed over a heat exchanger that warms an air or water-transfer fluid that is then circulated by ducts or radiators to the house.

Space-heating furnaces are roughly as efficient as gas- or oil-fired water heaters—about 60 percent in older gas models, 50 percent in older oil furnaces, and up to 80 percent in the newer models of either type.

In a centralized electric heating system, an electric heating element is immersed in a boiler or located in a duct in the forced air system. In zoned electric-resistance systems, electric elements are dispersed throughout the house in baseboards. Most of the heat that these systems produce goes to the house so that electric-resistance heating has a very high (95 percent or more) efficiency inside the building. But if generation and transmission losses are counted, the overall efficiency of electric heating falls below one third. For this reason, an increasing number of electrically heated buildings now have electric heat pumps that extract heat from the air and transfer it to the building. Essentially air conditioners working in reverse, heat pumps can usually provide about twice the heat contained in the electricity used to run them. Therefore, they can reclaim about one half of energy lost in generation and transmission.

In centralized heating systems in larger buildings, heat sources are larger, and control and distribution systems more complicated.[27] Sometimes, the heating and cooling systems work together. In most cases, the heat source is a fossil-fueled boiler that heats water or air, or produces steam that is then circulated throughout the building. Control over operations is through a central thermostat or through thermostats in individual zones or rooms. When decentralized thermostats are used, the distribution system must provide a wide range of temperatures to meet varying demands. Some systems (terminal-reheat systems, for example) raise the temperatures of centrally distributed forced air by using individual heating units in each zone. Others (dual-duct systems and multizone systems, among them) produce heated- and cooled-air flows at the central plant and then mix them to produce the desired temperature. Still others (including induction systems and fan-coil units) mix centrally heated or cooled air with recirculated, conditioned, or fresh air at individual units in each room or zone.

Technical Approaches.   Heating systems can be made more efficient in a surprising number of ways. Moreover, most techniques for improving the efficiencies of heating systems are available now.

Maintenance.   A simple, but often overlooked, strategy is maintenance. Furnaces or boilers should be cleaned and serviced periodically according to manufacturers' specifications. Periodic maintenance is particularly important in larger buildings, where boilers are large or more numerous. According to a Greater Washington (D.C.) Board of Trade estimate, simply removing built-up soot from a boiler's fireside could save up to 15 percent on annual heating costs.[28] The National Electrical Contractors Association (NECA) and the National Electrical Manufacturers Association (NEMA) recommend 14 maintenance steps in all kinds of boilers and from 3 to 6 additional steps in specific types of heating systems, all of which can enhance combustion efficiencies. (See Table 8.8.) In particular, keeping a daily log of pressure and temperature levels can indicate if tubes or nozzles need cleaning or if pressure adjustments are necessary, and checking smoke levels in stack gasses can indicate whether the burner is working efficiently.[29]

Upgrading Efficiencies.   More complicated and effective technologies and approaches—most of them commercially available—increase overall system efficiency by improving the combustion, distribution, or control subsystems. The extent to which each is feasible in a given heating system depends on the system's overall condition and the operating environment.

Furnace combustion can be improved in two ways: (1) reducing fuel use but not heat output or (2) increasing output without increasing fuel use. In many building heating sytems, fuel can be saved by downsizing oil burners (by installing a smaller nozzle so that the burner receives less fuel) or gas furnaces (by reducing the number of burner racks used). Downsizing is feasible in most systems because most can produce more heat than is needed. Oversized systems burn fuel more quickly than smaller units, so heat losses from combustion and distribution are commensurately greater. Also, because larger furnaces heat up quickly, they tend to operate for shorter periods and to lose fuel as the furnace or boiler starts and stops. Downsizing both lowers heat losses by reducing the amount of heat generated at any one time and reduces fuel losses by increasing the time of operation.[30]

Other efficiency improvements are in combustion techniques. In most new oil furnaces, a retention-head burner is used to enhance the flue-air mixture by recirculating the combustion air, thus allowing more complete combustion and increasing furnace efficiency by 20 to 30 percent. Yet another approach is to

Table 8.8. Suggested Maintenance Procedures for Furnaces and Boilers Used in Larger Buildings.

BOILER/FURNACE TYPE	PROCEDURE
general	Inspect for scale deposits on waterside surfaces.
	Inspect fireside of furnaces and tubes for soot deposits.
	Check door gaskets for tight seal. Replace if necessary.
	Keep daily pressure and temperature log to determine need for tube and nozzle cleaning, pressure, or linkage adjustments, etc. Keep log entries of firing rate, realizing that stack temperatures may vary 100°F during load change.
	Inspect stacks. If haze is present, a burner adjustment may be necessary.
	Inspect linkages for tightness.
	Observe fire when shut down. If it does not cut off immediately, check for faulty solenoid valve.
	Clean and inspect nozzles and cup of oil-fired furnaces regularly.
	Check stack temprature. If more than 150°F above steam or water temperature, clean tubes and adjust fuel burner.
	Inspect boiler insulation, casing, refractory, and brickwork for hot spots. Seal as necessary.
	Replace all obsolete or little-used pressure vessels.
gas	Clean mineral or corrosion build-up on gas burners.
fuel oil	Check and repair oil leaks at pump glands, valves, or relief valves.
	Replace oil-line strainers when dirty.
	Check oil heaters to see that manufacturers' recommended oil temperatures are being maintained.
coal-fired	Inspect stokers, grates, and controls for efficient operation. If ashes contain considerable unburned coal, adjustment probably necessary.
central furnaces, make-up air heaters, and unit heaters	Clean all heat-exchanger surfaces. Check and adjust air-to-fuel ratio as necessary.
	Inspect burner couplings and linkages.
	Inspect casing for air leaks. Seal as necessary.
	Replace or repair faulty insulation.
	Follow suggested guidelines for fan and motor maintenance.
radiators, convectors, baseboard and finned-tube units	Inspect for obstructions in front of unit and remove whenever possible.
	Be sure air collected in high points of hydronic units is vented.
	Keep all transfer surfaces clean.

Table 8.8. Suggested Maintenance Procedures for Furnaces and Boilers Used in Larger Buidlings.

BOILER/FURNACE TYPE	PROCEDURE
electric heating	Inspect electrical contacts and working parts to ensure they are in good working order.
	Replace heater elements when dirty.
	Check and adjust controls as necessary.
	Keep heat transfer surfaces clean and unobstructed.
	Keep air movement unobstructed.
	Periodically inspect heating elements, controls, and fans to ensure proper functioning.
	Check reflectors on infrared heaters for cleanliness and proper direction.
	Determine whether electric heating equipment is operating at proper voltage.
	Check controls for proper operation.

reduce fuel use in gas furnaces by replacing the gas pilot light with an electric ignition device like those most oil burners already have. Combustion efficiencies can also be improved by reducing the amounts of heat lost through the furnace walls, or through the flue.

Since most furnaces use indoor heated air for combustion, some heated air will always be lost through the flue unless the furnace uses outside air in combustion. The sealed combustion furnace does just that. Connected to the outside via an intake duct, this furnace operates without indoor air. Even though the outoodr air is colder, and energy must be spent raising it to combustion temperatures, this drawback is minimal if outdoor air is circulated through a heat exchanger in the flue.

Just as correctly sizing a single boiler is essential in a small building, properly sizing and controlling a multiple-boiler system is necessary in larger buildings. Heating systems in larger buildings should use as few heat sources as possible—not several at partial capacity.

Using waste heat can also improve combustion efficiencies in large systems. For example, a heat exchanger known as a boiler-stack economizer can recapture heat from the boiler stack and preheat water being fed into the combustion chamber. Other technologies now on the market can use different sources of waste heat in the boiler room to preheat either the oil or the combustion air.

Adjusting furnaces can improve residential heating efficiencies by as much as 25 percent. (See Table 8.9.) But while how much energy each improvement saves is known, how much a combination of improvements saves is not. For instance, while a downsized gas burner will lose fewer Btu in exhaust gasses,

Table 8.9. Estimates of Potential Energy Savings from
Furnace-Efficiency Improvements (Residential Systems).

IMPROVEMENT	ESTIMATED ENERGY SAVINGS
Furnace downsizing	10–20 MMBtu/yr[1,2]
Burner efficiency adjustments	5 MMBtu/yr[2]
Convective stack dampers	20 MMBtu/yr[2]
Variable-firing-rate burner	30 MMBtu/yr[2]
Hydronic reset	+ 10–15% seasonal efficiency[1]
Sealed combustion furnace	up to +5% seasonal efficiency[1]
Temperature reduction (to 65–68°F)	2–4% fuel use/degree reduced[1]
Temperature setback (55°F 8 hr/day)	Up to 15% fuel use[1]

Sources:
1. MITRE, *Energy Technologies for New England.*
2. OTA, *Residential Energy Conservation.*

how this reduction affects the amount of energy and money saved by using automatic flue dampers has not been extensively analyzed.

A distribution system is efficient when the building receives most of the heat that the furnace generates. Most distribution inefficiencies stem from poor flow rates in the transfer medium and excessive heat losses from piping or ductwork. Sometimes, water or air flows are not what they should be because pumps or blowers do not work properly. In other cases, the distribution system is poorly designed—full of bends and turns. Most distribution heat losses occur because piping or ductwork is poorly insulated. In a typical Washington, D.C., apartment house, annual losses in 100 ft of uninsulated pipe could waste as much as $2,400 (1980 dollars) worth of oil heat.[31]

Distribution systems can be improved mightily by checking and maintaining pumps and blowers periodically, by making small adjustments in the configuration of pipes or ducts, and by insulating piping and ductwork. In addition, managers of large buildings should make sure that water and air flows are no higher than necessary. It may well pay to convert a steady-flow system to one that reduces or increases the flow rate as heating needs change.

In a two-way steam-distribution system, efficiency can be improved by making sure that condensate return-lines (pipes returning water that forms as steam temperatures and pressures drop) work properly. However, retrofitting one-way steam systems to allow condensate return may be prohibitively expensive except as part of complete building or heating-system rehabilitation.

Control adjustments are relatively straightforward in smaller buildings, where a single central thermostat most likely controls the heating system. Using a clock thermostat that automatically reduces the temperature setting at night and resets it as morning approaches can maintain comfort levels with less energy. While larger buildings are likely to have more complicated controls, improvements to these controls are also likely to be commensurately more effective. Clock thermostats make sense in larger buildings—as do more expensive devices and techniques. One strategy is to reduce heating in empty offices or unoccupied areas of apartment buildings at night. Similarly, laundry rooms, rooms adjacent to boilers, and other areas where heat gains are unusually large need much less heat than other areas. In some cases, heat use can be balanced and overheating avoided if individual thermostats are installed in each dwelling unit or office.

Control adjustments can be particularly effective in systems where heated and cooled air are mixed. Sometimes, lowering the delivery temperature of the heated air and raising that of the cooled air will mean that no mixing is necessary. Even when some is, efficiency improves as mixing decreases.

Heat Pump Advances.   In many parts of the country, air-to-air heat pumps are effective and economical alternatives to electric-resistance heating. Heat pumps, which can also be used for cooling, take heat from the outside air, raise its temperature, and transfer it to the house. Most deliver about twice the amount of heat as contained in the electricity they use. This means that most heat pumps have seasonal coefficients of performance (COPs) of 2.0 for heating. For cooling, the seasonal COP is significantly lower than those of the newer central and room air conditioners available.[32] (See Chapter 9.)

Although heat pumps have been commercially available for years, they have not been widely used. Questions about their reliability linger and electric rates have until recently seemed low.[33] However, heat pumps are now beginning to appear in more new homes and commercial buildings, and they can be integrated with either hydronic or forced-air distribution systems on existing buildings.

Since air-to-air heat pumps draw heat from the ambient air, they are more useful where outdoor temperatures are relatively high. For this reason, it often pays to increase the temperature of the heat source—by, for example, ground-coupling heat pumps to take advantage of temperatures that tend to be warmer than ambient air in the winter months.[34] In industrial applications, heat pumps could use low-temperature waste heat as a heat source. Combining heat pumps with solar energy collectors may, in some cases, be both efficient and economical. Where heat-source temperatures are too low, heat pumps need to be backed up by another heat source.

Most advances in heat-pump technology are aimed at delivering more heat to the load or reducing the energy input or both. One approach is to make heat-pump components more efficient. For instance, heat pumps move heat to the indoors or outdoors through heat-transfer coils containing fluids that absorb or reject heat as temperatures and pressures change.[35] One way to improve the pump's efficiency is to make the heat-transfer mechanism more efficient either by increasing coil area or speeding up the rate at which heat is transferred to or from working fluids.[36] Where, for example, heat pumps are used for space heating only, coils and other components can be designed and sized for an optimal heating COP.

A second general approach is to change the design fundamentally. For example, one gas-fired heat pump works on the principle of absorption rather than compression. In a conventional heat pump, the refrigerant rejects heat as pressure increases. In this new system, the refrigerant's latent heat is chemically absorbed by another fluid, so no compressor is needed. Natural gas could also be used to run heat engines that would in turn operate heat-pump compressors.

Still a different technique is to alter the design of the electric heat-pump system. One alteration has replaced the fluorocarbon refrigerant fluid (used in vapor or Rankine thermodynamic cycles) with air (Brayton cycle). The advantage of this approach is that the heat transfer to and from the working fluid is not limited by the vaporization and boiling points of refrigerants.[37]

At least some versions of these advanced heat-pump designs are or will soon be on the market. However, most analysts agree that improvements will be evolutionary, not revolutionary, and even improved heat pumps will probably not be able to carry the entire heating load in extremely cold climates without some back-up. Then too, the improvements themselves may have technical or economic drawbacks, and encouraging reliance on fossil fuels instead of electricity may be unwise where electricity can be obtained from hydropower and other renewable resources.

Heat-Recovery Systems. Buildings that generate large amounts of waste heat—such as larger multifamily or commercial structures—make good candidates for using waste-heat recovery systems. Most such systems use heat exchangers to transfer heat from an exhaust air stream or some other heat source to the forced-air or hydronic heating system, usually before the air or water gets to the furnace or boiler. In effect, waste heat is used to preheat the distribution medium so that less energy is needed to raise the medium to delivery temperatures.

Which type of system works best in a building depends on which types of heating system and heat source are used. Where a forced-air heating system is used and supply and exhaust ducts are located close together, an air-to-air heat exchanger is the simplest option. A similar device, the rotary heat exchanger, uses a rotating drum or wheel installed between the two ducts to transfer both sensible heat and water vapor, so the device can serve as both a preheater/pre-cooler and a humidifier/dehumidifier.[38] While these systems are difficult to use if the exhaust air stream contains toxic or hazardous fumes, the problem can be minimized by including purge components in the drum or by adding special filters to the intake valve.[39] Where the hazards of exhaust air or the distance between exhaust and intake ducts make using air-to-air or rotary heat exchangers impossible, the so-called "run-around" system, which circulates a working fluid between heat exhaust and intake air ducts, can be used to enhance cooling in summer by transferring heat from intake to exhaust.

Heat can also be recovered from such liquid or vapor heat sources as grey water, hot condensate from steam distribution, or heated refrigerant gas. The basic process here is to circulate working fluid between two heat exchangers, connecting the water source and the heat distribution system.

Combined Heat and Power—On-Site Cogeneration. Large-scale heat engines used in industrial facilities can "cogenerate" heat and power simultaneously. Smaller cogenerating units for use in smaller buildings are also beginning to appear commercially. In the heating mode, these engines could work much as conventional boilers do, heating a liquid or air transfer medium that then cir-culates through the building. The critical difference is that part of the fuel input also runs an electric generator that provides on-site power.

Probably the first available residential-sized cogenerating unit is Flat's Total Energy Module. TOTEM can cogenerate 15 kw of electric power and 133,000 Btu of heat at a 135°F distribution temperature. The system's total fuel effi-ciency is 95 percent, compared with 60 to 75 percent for new gas furnaces and under 30 percent for most large power plants. The unit, packaged as a single module that can be installed easily in a basement, can be retrofit to either forced-air or hydronic distribution systems.[40]

Another promising invention by Sunpower, Inc., uses a gas-fired Stirling engine both to drive a heat pump and to produce electricity. Essentially, this device simply improves upon the fossil-fueled heat pump by making on-site power production possible. The Sunpower Total Home Energy System should be available by 1984.[41] (See Chapter 11.)

Energy Management and Control (EMC) Systems. In many large buildings, different areas need different amounts of heat at different times and for differ-ent durations. Then too, many large facilities have their heating and cooling systems configured together. Accordingly, some simple control adjustments can greatly improve building energy efficiency, and the development of more elab-orate control systems—some employing microprocessors and computers—holds out even greater promise.

Different types of control systems are suitable for different types of buildings. The determining factors include building size and the nature of the heating, ventilation, and air conditioning (HVAC) system. The simplest control systems can control the regular start-ups and shut-downs of equipment according to predetermined schedules. They can also help to optimize equipment operations based on climatic conditions. More elaborate controls will manage several pieces of equipment or even several systems simultaneously.

A computerized EMC system can control the amount of heat particular building areas receive and time the delivery. More sophisticated EMC systems use time- or environmentally activated controls located throughout a building. Sophisticated control systems can also monitor and adjust the operation of combustion and distribution subsystems. For example, an EMC system can oversee the operations of boilers, furnaces, or heat pumps in a central plant, indicating when, and often why, particular units need repairs. Control systems can also manage the preheating (or precooling) of intake air by a heat-recovery technology and alter air or hydronic flow rates to track heating loads more closely.

The more elaborate HVAC control systems can even manage over-all electricity demand so that buildings do not use expensive peak electricity unless absolutely necessary. To "shed loads," the control system's computer traces current electricity consumption. When consumption surpasses a pre-set limit, the system shuts off unnecessary services or shifts consumption to uses that cannot be curtailed. (See Chapter 11.)

Economics of Heating System Improvements. In small buildings, some improvements in the combustion, distribution, and control subsystems are cost-effective for building owners no matter what kind of system is employed. How economical each option is depends on the equipment's nature and condition and the building's location and size. The costs of various system adjustments vary widely, from less than $25 (for furnace maintenance or simple oil-burner downsizing) to over $500 (for adding variable-firing-rate burners). The cost-effectiveness of furnace adjustments will also vary with the furnace type and the combinations of improvements made. (See Table 8.10.)

Little has been written about the cost-effectiveness of combinations of heating-system improvements. But common sense can sometimes fill the void. Downsizing is virtually always cost-effective, and sometimes achieves payback in less than one year. For finding out whether more elaborate adjustments are needed, other criteria come into play. If, for example, a boiler or furnace is due for replacement, installing an energy-efficient unit two or three years before necessary makes more sense than making a major improvement with a five-year payback.

Table 8.10. Estimates of Costs and Cost-Effectiveness of Selected Heating System Improvements—Small Buildings.

CONSERVATION MEASURE		ESTIMATED COSTS (DOLLARS)		ESTIMATED COST/ EFFECTIVENESS
Cleaning/Tuning		50		Less than one heating season[1]
Clock thermostat		60		1–2 heating seasons[1]
Downsizing		20[1,2]		1–2 heating seasons[1]
Sealed-combustion furnace		100–500[1,2]		2–5 years[2]
Electric-convective stack dampers		200[2]		
Hydronic-temperature reset		500[1]		3–7 years[1]
Water-fuel emulsion gun		150–200[2]		
Condensing flue—gas		*New*	*Retrofit[2]*	
furnace	oil	1,100	400	
	gas	800	300	
Near-condensing				
furnace	oil	650	250	
	gas	500	200	
Variable-firing-rate furnace		100–200[2] higher than conventional		
Thermally purgable furnace		400 (new)[2]		

Sources:
1. MITRE, *Energy Technologies for New England* (1980).
2. OTA, *Residential Energy Conservation* (1979).

The economic attractiveness of particular furnace or combustion improvements also depends on the heating load's size. In colder climates with longer heating seasons, more Btu are saved per year so paybacks are shorter. For example, a flame-retention burner that reduced fuel consumption about 16 percent would pay for itself in less than 18 months in a climate with a heavy heating load, but not for nearly four years in a milder climate.

Improvements that require converting from one conventional fuel source to another should be evaluated carefully in view of the alternative—upgrading the efficiency of the existing system. The recent rash of conversions from oil burners to gas furnaces, for example, may not be cost-effective if natural gas prices are deregulated. By the same token, the economics of converting from electric-resistance heat to an electric heat pump will depend on the costs of heat-pump equipment, the extent of any alterations needed in the building's distribution system, and the climate. Where it is very cold, building-shell improvements may be the better buy.

If retrofits can be timed to include shell conversion and furnace improvements, substantial energy savings and attractive paybacks or cash flows should be possible. If, for example, heavy insulation is added at the same time as solar features, oil- or gas-furnaces could be downsized significantly. Ostensibly, shell improvements could in some cases entirely eliminate the need for a central furnace, making it possible to obtain all of the heat from wood stoves, modular room-heating units, or water heaters large enough to heat space as well as water.[42]

In small buildings, any distribution or control improvement short of completely reworking the distribution system is relatively inexpensive. Energy savings from such improvements will result in swift paybacks. In large buildings, the cost of improving heating systems is greater, but resulting energy savings are commensurately greater too. Buying and installing an EMC system can cost upwards of $200,000 for very large facilities. Nonetheless, the National Electrical Manufacturers Association (NEMA) and the National Electrical Contractors Association (NECA) claim that paybacks for such systems can be as short as two to three years.[43] A study by the Solar Energy Research Institute found that of 25 retrofits in larger commercial buildings, 20 achieved payback in less than four years and all did so within seven years.[44] For larger improvements, an engineering team should be called upon to make energy and economic analyses. However, NECA and NEMA's guidebook for owners and managers of large buildings identifies 178 "minimal-cost" heating-system conservation measures at a glance.[45]

Cities should begin their economic analysis of these energy-conservation options (which can lop 10 to 80 percent off energy use in commercial buildings) by comparing the costs-per-Btu of typical conservation measures with the current or marginal costs of conventional fuel sources. In a sampling of eight building retrofits, the Solar Energy Research Institute pegged the costs-per-million-Btu saved at less than $2.00 (1980) in six and no higher than $6.00 in any of them.[46] In contrast, average costs today for one million Btu of natural gas are $4.60. For oil, the figure is $8.90 and for electricity, $19.04.

Institutional Constraints and Socio-economic Considerations. Materials and installation techniques used in building-shell retrofits must meet normal building and fire-code requirements. Of course, compliance with such codes contributes to system quality and consumer protection.

Electric utilities' reactions to large-scale shell-upgrading efforts will depend on the nature of the utility's fuel source and capacity, as well as on how many households make the changes. In a computer simulation, OTA found that residential installation of insulation would not measurably affect utilities unless at least one third of the customers in the service area used electricity to heat space.[47] Heating system improvements in oil- or gas-heated homes are not likely to arouse opposition either. Widespread conversions from electric-resistance heating to electric heat pumps are a different matter, though some electric utilities (such as Seattle City Light) may support such conversions as part of load-management programs.

Since most low-income buildings are energy inefficient, great opportunities exist for reducing their energy consumption through fairly basic conservation retrofits. For example, fully retrofitting a five-story tenement in lower Manhattan—including both shell conservation and heating-system improvements—will cut energy consumption by roughly one half.[48] A Philadelphia project to retrofit oil burners in low-income neighborhoods reaped savings of over 20 percent in some houses.[49] Insulating, caulking, and weatherstripping low-income housing in Fresno, California, resulted in fuel savings of about 27 percent per dwelling unit.[50]

Still, the poor can afford few such improvements. The Urban Coalition of Minneapolis calculates that the average costs for a retrofit in its weatherization program was $1,450: $450 for materials and $1,000 for labor.[51] In general, basic shell retrofits (caulking, weatherstripping, clock thermostats, storm windows, and attic insulation) in single-family houses cost about one percent of the average annual income earned by families in the U.S. For families earning $8,000 a year, however, the percentage of annual income needed to pay for these exact same improvements ranged from 25 percent in the East to 14 percent in the West. Moreover, in many low-income buildings, it is pointless to retrofit the shell unless the structure and heating system are rehabilitated—a considerable additional cost.[52]

SOLAR ENERGY AND ENERGY CONSERVATION

Solar heating might appear fascinating but unnecessarily sophisticated in view of the startling effectiveness of energy conservation techniques and approaches, the high first costs of some solar technologies, and the need for a back-up system in most solar homes. Yet even with the vast savings possible through conventional energy conservation, solar heating constitutes an important addition—even an economical alternative in some cases—to some of the more extensive energy-conservation measures. While many buildings could be heated most cheaply if strict conservation measures were applied without adding solar energy techniques and devices, combining the two is in many cases the ideal.

Six factors can make a conservation and solar match worthwhile. (1) Using renewable-energy techniques allows a building to be a "safe" energy producer, meeting much of the "post-conservation" energy demand without causing pollution or depleting scarce resources. (2) In some cases, a package of solar equipment or designs and moderate conservation measures may be more economical than extensive conservation measures alone.[53] (3) Under certain utility rate structures, solar systems may be better able to replace expensive peak electricity than strict conservation measures alone. For example, comparing energy consumption in solar homes with that in well-insulated homes using electric-resistance heat or electric heat pumps, the Public Service Company of New Mexico found that the passive homes consumed less energy than either of the other two during peak periods. If a time-of-day rate was developed, the utility predicted, an average passive home's seasonal heating costs would have been $50 less than an average electric-resistance home and $90 less than for a home using an electric heat pump.[54] (4) The line between energy conservation and solar energy techniques is sometimes thin. For example, many new window technologies can be used in passive solar energy systems, and heat pumps also combine conservation and renewable-energy approaches since they use naturally heated ambient air as a heat source.[55] (5) Further developments in certain energy-conservation technologies may improve their economic appeal. (6) Some buildings are so well suited for solar energy technologies that solar deserves special consideration. Buildings in sunny but cold climates, buildings that easily lend themselves to a passive solar retrofit, buildings that need heat primarily after peak solar-energy collection periods, and buildings with such expensive back-up energy systems as electric-resistance heat are all prime candidates.

So various are the possibilities, of course, that no single assessment of the merits of combined solar energy and energy-conservation approaches is possible. But the technical overview and evaluation criteria set forth here should provide a solid basis for decisionmaking.

Passive solar heating at the Milford Reservation Center, Milford PA (*Courtesy of Kelbaugh & Lee Architects, Princeton NJ*).

Technical Approaches

Despite the basic differences between active and passive solar-heating systems, both types perform essentially the same functions. Both collect solar heat and either distribute it to the building immediately or store it for use as needed.

Passive systems distribute heat to a building by nonmechanical methods—heat conduction, convection, and radiation—while active systems use mechanical devices such as pumps and fans. Thus, most passive solar-energy system "components" serve some structural purpose besides collecting or moving heat. In contrast, active solar-energy system components are pieces of hardware: solar collectors, storage tanks, heat exchangers, and the like.

Passive Solar Heating. Thousands of U.S. buildings now use passive solar-heating systems, and tens of thousands of additional "sun-tempered" buildings collect heat through large expanses of south-facing glass. The first such buildings were built by "solar pioneers." Then, forward-looking architects and builders picked up the basic ideas. Now, however, tracts of houses and clusters of offices with passive heat are appearing in all parts of the country. Moreover, the U.S. Department of Housing and Urban Development (HUD) supported the construction of 91 house designs that tract developers could use, and the

Passive solar office building for the *St. Petersburg Times,* Clearwater FL.

U.S. Department of Energy (DOE) is supporting the construction or retrofit of passive heating systems in 30 typical commercial buildings of varying dimensions.[56,57]

Direct Gain. Perhaps the simplest passive solar approach is to keep a living space warm by admitting solar energy directly through south-facing glass and storing some of the heat for use later in dark-colored mass: walls or floors or tile, brick, or masonry. In most passive systems, heat is distributed to other parts of the building by means of carefully placed walls, doorways, and staircases that facilitate natural heat flow. Occasionally, a separate air plenum specifically designed to move collected solar heat from one end of the building to the other is added.

Two potential problems endemic to direct-gain systems are (1) heat loss at night or heat loss on cold, cloudy days through the glass surface and (2) overheating of the living space, especially in the summer. Yet, heat losses can be cut by using double- or (in cold climates) triple-glazing.

Movable insulation (including thermal shutters) or reflective shades also prevents excessive heat gain, though at the cost of some sunlight and an obstructed view. Alternatively, overhangs specified in the building design can shade the glazing from excessive heat gain in the summer, yet allow full exposure to sunlight in winter. Vents in the glazing can also shunt out excess heat.

Indirect Gain. An indirect-gain system interposes an absorption and storage component between the glazing and the building's interior. Typically, this storage component is a Trombe wall made of grouted cement blocks or some other type of masonry, or a water-wall composed of single- or multiple-containers of water.[58] Usually, both the mass wall and the water containers are painted flat black to increase absorptivity. In new buildings, the storage wall can be inside the building shell with the glazing surface and framing serving as the building's exterior. In some designs, the building's exterior wall absorbs and stores energy—an appealing feature in retrofits, where glazing covers the existing brick or concrete exterior.[59]

Indirect-gain systems run less risk of overheating than most direct-gain designs. They are also somewhat more flexible in terms of heat distribution. If necessary, collected heat can be distributed immediately to the building through vents at the top and bottom of the wall. Air heated in the space between the wall and the glazing rises and exits through the vent to the living space. Circulation stops only when the temperature of the air between the wall and the glazing roughly equals that in the living space. When heat is not needed immediately, the vents can be closed.

Whether or not collected solar heat is used immediately, it can be distributed

California State Offices, Sacramento CA (*Courtesy of State Architect's Office*).

much as the heat collected in a direct-gain system is. Additionally, plenums aided by small fans can also be used to connect the space between the glazing and the storage wall with areas farther from the glazing.

Indirect-gain systems must also feature mechanisms for preventing heat loss at night and overheating in the summer. Generally, the best means are those used to prevent these problems in direct-gain systems: insulation, movable shutters and shades, and overhangs and vents.

The general effectiveness of indirect-gain systems and the ease with which they can be installed are both being improved. Under development now are various modular wall units that combine glazing, absorption, and storage components in single packages. Most are sized so that they can be combined into fairly large (300–400 ft^2) configurations or used in smaller designs.[60]

Some manufacturers are also developing mass wall units that make use of a phase-change storage medium—salt compounds or other materials that melt as they gain heat and release heat as they recrystallize. Such materials tend to absorb much more heat per unit of volume than water or masonry materials, so less space is needed for the storage component. These phase-change materials can be incorporated in packaged collector/storage wall systems in numerous ways, including conventional-looking ceiling panels or floor tiles that can serve an aesthetic function as well as storing heat.[61]

Although indirect-gain components can be prepackaged, they need not be. All can be assembled at the building site using conventional construction materials. Mass storage walls, for instance, can be built of conventional concrete blocks and some water-wall systems use drum barrels painted flat black.

Isolated Gain—Sunspaces.   Both direct- and indirect-gain systems collect and store solar radiation in a portion of the heated space. But, in isolated-gain systems, heat is collected in an area that is not part of the building's mechanically conditioned space.

The most common type of isolated-gain system is the sunspace—a greenhouse or unheated atrium. Additionally, a very different type of passive system that operates on isolated-gain principles is the thermosyphoning air panel.

Sunspaces collect solar heat through south-facing glass, then store it in the floors or in the wall separating the sunspace from the living area. Stored heat can be distributed via radiation or convection from a vented mass wall into the house proper, via conduction through a storage mass floor extended from the sunspace under the adjacent conditioned area, or via convection through air plenums.

A more elaborate version of this last approach is the so-called "double-envelope" design. Essentially, this solar approach to new housing is based on the construction of two building shells, the outer one heavily insulated. The interstice is used to move heat collected by an integral greenhouse located on the building's south side. Air heated in the greenhouse rises to the top of the space, where it enters a plenum extending upward through the roof, down the north wall, under the flooring, and back to the greenhouse—bathing the interior walls with greenhouse-heated air.

As in all solar gain systems, designs and materials used in a sunspace system must prevent nocturnal heat loss and summer overheating. Thus, in all but the mildest climates, sunspaces or greenhouses using window glass, acrylic, or fiberglass must be at least double-glazed.

It is usually difficult to fit an attached greenhouse with an overhang suitable for preventing heat gain. But the greenhouse structure itself can be used to control heat penetration. Unfortunately, most attached greenhouses have sloping roofs, so they are vulnerable to large amounts of summer sun. Sometimes, a portion of the roof is permanently covered and insulated. In other designs, movable shutters or shades that can be partially drawn in the summer or fully closed at night or on cloudy days are used. Reflecting shutters and blinds can also block out excess heat in the summer in sunspace systems.

Since so many sunspace designs use greenhouses, a relatively large industry now manufactures components for greenhouse construction—glazing materials, molded movable insulation, blinds and reflectors, and storage components—as well as packaged greenhouse kits. Increased availability aside, even greenhouse retrofits using preassembled parts can seldom be justified economically on the basis of energy savings alone.[62] For one thing, much of the heat gained by the greenhouse will be used to heat the added space. Yet, the addition of attractive living space and the opportunities for cultivating plants and foods provide incentives too.[63] Where costs are a major factor, less expensive greenhouses can be assembled on-site using conventional, inexpensive building materials.[64]

Isolated Gain—Thermosyphoning Air Heaters.   In some ways, this system represents a radical departure from other passive solar designs. Instead of using glazing in the building shell to collect solar energy, it employs modular flat-plate solar collectors mounted vertically on wall surfaces or at angles sloping from the walls to the ground. As the collector panels heat air, it flows upward by natural convection and enters the living space through vents or ducts. At the same time, cooler air flows back to collectors through a vent that connects the bottom of the collector to the heated space. Easier to retrofit than most other passive designs, the collectors can also be configured in large arrays to heat bigger areas or in smaller arrays of one or two panels to heat single rooms.

Factory-built panels are available for use in these systems. Alternatively, panels can be site-assembled using fairly common and inexpensive building materials. Vertical-wall collectors represent one of the most common urban retrofits in use. Indeed, the Urban Solar Energy Association in Boston plans to sell a kit for assembling a thermosyphoning air panel for around $440.[65]

Earth Sheltering.   Strictly speaking, earth sheltering is not a passive solar heating system, but rather a design that works well in combination with passive solar design to enhance energy conservation in nonsolar buildings.[66] In earth-sheltered design, earth serves as a barrier against wind and violent weather even as it moderates temperature fluctuations in the ambient air. Thus, infiltration heat losses are reduced, conduction heat losses through ceilings and/or walls are minimized, and the heat-storage capacity of the building is improved.

In some earth-sheltered designs, relatively small mounds of earth are compacted against walls ("earth-berming"). In others, walls and roofs are recessed several feet underground. The chief technical concerns in earth-sheltered construction are ensuring adequate structural strength and preventing moisture accumulation and water penetration. Techniques for doing both are well-established, although they require careful attention to construction details. The structural integrity of earth-sheltered housing can be guaranteed by employing

Cuando wall—passive solar retrofit to large apartment house, New York City (*Courtesy, New York Energy Task Force*).

materials that have greater mass than those used in normal building construction. Earth-sheltered housing can be waterproofed with vapor barriers and other sealing techniques.

Passive System Performance.   Since many of the earliest passive solar buildings were built for private clients (often the architects themselves), systematic monitoring of passive solar homes has only recently begun. Then too, most solar applications combine more than one system type. These considerations make it particularly difficult to compare performances among system types, let alone in different building types or climates. It is inarguable, however, that good passive solar-heating systems can be strikingly effective, especially in new buildings. Indeed, performance monitoring of a few passive houses built with HUD grants in the late 1970s indicates that passive solar energy is meeting from 75 to 80 percent of the annual space-heating loads.[67] (See Tables 8.11, 8.12, and 8.13)

Assessing passive solar potential in existing buildings is harder. The interior and exterior design, the material characteristics of the building, and the building's pattern of heating demands must all be taken into account. For example,

Table 8.11. Measured Performance of Some Passive Solar Homes (Single-Family—New).

INSTALLATION LOCATION AND DATE COMPLETED	PERCENTAGE OF BUILDING HEATING-ENERGY USE PROVIDED BY PASSIVE SOLAR
Sommersworth, NH (1979)	66
Ames, IA (1979)	32
Prescott, AZ (1978)	81
Royal Oak, MD (1976)	79
Atascadero, CA (1976)	100
Santa Fe, NM (1976)	96
Bozeman, MT (1977)	· 85
Yakima, WA (1977)	63
East Pepperell, MA* (1979)	35

*Super-insulated house with very low heating load.
Source: Solar Energy Research Institute, A New Prosperity: Building a Sustainable Energy Future, 1981, p. 85.

a mass-wall retrofit of a Washington, D.C., commercial building (which uses more heat in the later afternoon and early evening than in the late evening) meets about 60 percent of the building's yearly heating needs.[68] Other buildings less suitable structurally for large-scale retrofits and more in need of heat after dark will not be able to obtain such impressive results. The solar contribution in a renovated low-income apartment building in St. Johnsbury, Vermont, using fish-tank storage behind south-facing windows is only about 12 percent annually, though this approach is so cheap that the heat is a good buy.[69]

Just how different the contributions from passive solar retrofits can be from one building to another is indicated by projected savings from HUD grants for passive retrofits in multifamily buildings and DOE grants for passive additions. (See Tables 8.14 and 8.15.)

Although no serious performance or reliability concerns have emerged, one potential problem is overheating, especially in direct-gain and sunspace systems. For example, some overhangs positioned to provide shade at the summer solstice are not at the correct angles to work when the sun angles are lower, and some passive systems that do not have movable insulation often overheat during the day and then lose excessive amounts of heat at night.[70] Another observed problem has been the inadequacy of heat-distribution techniques in some passive designs. Sometimes this problem stems from poor interior structural layout or from attempts to move heat over long distances without the aid of fans. This criticism has been levelled particularly often at double-envelope designs.

Table 8.12. Predicted Performance of Passive Solar Designs for New Tract Houses (HUD Residential Solar Demonstrations).

LOCATION	HOUSE TYPE	HOUSE SIZE (NET HTD. AREA-FT^2)	HOUSE PRICE (DOLLARS)	SYSTEM DESCRIPTION	HEATING LOAD (10^6 BTU/YR)	PREDICTED SOLAR FRACTION (%)
Mat-su Borough, AK	2-story, earth-bermed	1466	85,000	2-story greenhouse, trombe wall	57.9	62
Gaithersburg, MD	2-story contemporary	2737	190,000	2-story atrium, skylights	91.2	27
Myersville, MD	2-level earth-sheltered	2036	120,000	greenhouse, clerestory windows	46.1	63
Shelton, CT	2-story Cape Cod	1900	85,000	2-story greenhouse, active air collectors	61.9	49
Putnam Valley, NY	ranch style, earth-bermed	1473	60,000 + land	thermopane south wall, clerestory windows	42.0	90
New York City, NY	2-story contemporary	2550	125,000	2-greenhouses with trombe wall backing; clerestory windows	146	38
East Wilton, ME	salt box, earth-bermed	1204	55–60,000	trombe wall, triple-glass	65.8	71

Location	Type		Cost	Solar features		
Turner, ME	2-story Cape Cod	1320	85,000		68.7	45
Kent, OH	2-story contemporary	2158	130,000	2-greenhouses, skylights	76.2	46
Ypsilanti, MI	2-story contemporary	2460	123,000	greenhouse, south apertures	134	51
Walden, CO	3-bedroom contemporary	1660	75,000	2-story greenhouse, backed by trombe wall	168.3	36
Truckee, CA	octagonal, earth-sheltered	1650	140,000	greenhouse	63.2	86
Springfield, OR	contemporary ranch house	1719	100,000	greenhouse, floor-to-ceiling windows	64.8	60
Pocotello, ID	2-story contemporary, earth-bermed	2662	92,000	greenhouse, trombe wall, skylight	106.6	77
Santa Fe, NM	contemporary adobe	1583	85,000	greenhouse, trombe wall	90.0	83
Tucson, AZ	3-bedroom rambler	2632	167,000	double-glazed panels, active air system	71.4	70
Santa Rosa, CA	2-story; double-envelope	1815	169,000	sliding glass doors, skylights greenhouse	30.1	80

Source: HUD Project Summaries (1980).

Table 8.13. Predicted Performance of Passive Designs for New Commercial Buildings.

LOCATION	BUILDING DESCRIPTION	HEATED SPACE (FT²)	SYSTEM DESCRIPTION	HEATING LOAD MMBTU/YR	SOLAR FRAC. %	COOLED SPACE (FT²)	COOLING LOAD MMBTU/YR.	SOLAR FRAC. %	LIGHTING LOAD MMBTU/YR.	SOLAR FRAC. %	DHW LOAD MMBTU/YR.	SOLAR FRAC. %
Troy, NY	Single-story contemporary visitor center.	4,316	Large sunspace with thermal storage in its walls and floor; upper section serves as warm air plenum. Skylights on north side provide natural daylighting.	166.8	64	4,316	12.7	80	36.0	61	NA	NA
Syracuse, NY	Three-story educational building.	47,000	Super-insulated (roof R-80, walls R-30) and earth-bermed where appropriate. Three different reflector-aided day-lighting systems providing some heat as well. Air-to-air heat exchanger ventilation.	88.0	51	47,000	39.0	NA	432.8	21	17.3	29
Princeton, NJ	Single-story contemporary office building.	64,000	Three sets of large atriums provide heat and day-lighting. Atrium heat drawn to basement rock bed for storage. Unvented water-wall serves as a thermal mass and absorber.	3084.8	72	64,000	2393.6	88	2272.0	76	742.4	50
Philadelphia, PA	Three-story program center.	5,100	1170 ft² trombe wall supplies heat to the main gathering space. 500 ft² of glazing along south wall for direct gain.	200.0	50	-0-	-0-	NA	10	—	15.6	50

Location	Type	Description										
Alexandria, VA	Single-story auditorium/ gymnasium.	Two nonconvecting trombe walls tapered in thickness from 8 in. at bottom to 4 in. at top. Two linear clerestories admit light absorbed by walls and floor. Thermosyphon DHW tied to trombe wall.	8,900	127.0	52	8,900	87.1	100	68.6	65	87.9	26
Mt. Airy, NC	Single-story contemporary public library.	Shaded clerestories equipped with light shelves provide heat and daylighting. South glazing comprises 17% of gross floor area (1 in. insulating glass). Concrete roof, floors, and masonry walls provide storage.	13,570	83.0	50	13,570	161.0	51	118.0	6	.002	55
Memphis, TN	Single-story double-shed type greenhouse; board and batter siding.	Design allows for clerestory effect. Vaulted skylights on sloped roof. South side lined with sliding glass doors. Backup heat from a bin of decaying organic matter, heat ducted through plenum.	6,000	250.0	90	6,000	250.0	60	-0-	99.9	-0-	NA
Bessemer, AL	Single-story contemporary public school.	Rooftop clerestories provide heat and daylighting to each of 18 classrooms. Water-filled cement pipes beneath the clerestories absorb radiation and prevent glare.	27,452	326.5	45	27,452	240.3	17	235.9	25	7.6	NA

Table 8.13. Predicted Performance of Passive Designs for New Commercial Buildings (DOE, 1982) (Continued)

LOCATION	BUILDING DESCRIPTION	HEATED SPACE (FT²)	SYSTEM DESCRIPTION	HEATING LOAD MMBTU/YR	SOLAR FRAC. %	COOLED SPACE (FT²)	COOLING LOAD MMBTU/YR.	SOLAR FRAC. %	LIGHTING LOAD MMBTU/YR.	SOLAR FRAC. %	DHW LOAD MMBTU/YR.	SOLAR FRAC. %
Columbia, MD	One-story church educational building.	5,493	Each of the building's 3 sections is equipped with a large clerestory with light shelf attached and insulating shades. Insulated masonry walls serve as thermal mass.	62.1	45	5,493	14.5	11	21.4	11	NA	NA
Wells, MN	One-story modified "V"-shaped bank building.	5,538	2240 ft² of double glazing on south and southeast walls. Clerestories above the offices and lobby. More daylighting provided by lower windows along teller and bookkeeping wings. Gypsum ceilings, walls, concrete floors.	334.0	67	5,538	146.0	20	255.0	65	1.5	NA
Grand Junction, CO	Two-story airport terminal.	66,700	3380 ft² of south clerestory contributes light and heat. South windows below admit radiation absorbed by mass floor. A vented trombe wall heats car rental area. South wall acts as hybrid collector.	1,239.0	24	66,700	688.0	100 with evap. cooling	1,084.0	50	52.0	—

Location	Building type	Area	Description									
Littleton, CO	Three-story contemporary office building.	17,950	Glare-free daylighting provided by a daylight space above a translucent ceiling. Considerable south glazing with insulating shades. When shades down, air channelled to ceiling plenum and convective loop.	301.1	40	17,950	164.4	50	235.0	NA	—	NA
Glenwood Springs, CO	Three-story college multi-use facility.	26,948	Built into hillside. Large atrium acts as a collector from which all zones draw heat. Black DHW preheat tanks suspended in atrium. Clerestory windows and trombe walls in several zones. Greenhouse with water-column storage in one zone.	578.4	56	31,037	49.3	81	478.0	44	101.1	40

Source: Department of Energy, 1981.

Table 8.14. Predicted Performance of Passive Solar Retrofits in Multifamily Buildings (HUD Residential Solar Demonstration Program).

LOCATION	UNITS	SYSTEM DESCRIPTION	HEAT LOAD (MMBTU/YR)	PREDICTED SOLAR FRACTION (%)
Boston	12	Glazing retrofitted for direct gain and sunspace. Exterior brick retrofit as trombe wall.	234.6	28
New York	8	Double-glazed greenhouses with masonry storage walls.	299.5	46
Boston	6	Glazing retrofit for direct and indirect (water-storage) gain. Greenhouse additions on south walls.	152.5	45
Providence	3	Trombe wall replaces conventional siding.	141.3	54
Schenectady	3	Greenhouse addition on south wall of each floor.	91.5	26
Boston	14	Atrium sunspace with clerestory south-facing windows. Bay windows retrofit with double glazing. Greenhouse additions.	170.4	72
Bronx, N.Y.	10	Trombe wall retrofit.	277.5	53
Charlotte	4	Greenhouse retrofit with exterior walls used as storage.	89.11	34
Kansas City	8	New glazing and interior sheet rock and plaster storage. Trombe wall retrofit. Clerestory windows.	265	60
Roanoke	15	Retrofit greenhouse with water storage. Trombe wall retrofit.	303.1	33
Milwaukee	6	Two-story lean-to greenhouse. Thermosyphoning air heaters.	364.0	34
New Haven	7	Porches retrofit as greenhouses. Efficient glazing. Clerestory windows.	365.8	34
New Haven	6	Trombe wall retrofit.	335.2	42
St. Johnsbury (Vermont)	21	Fish tanks behind south-facing windows.	1520.0	11
New York	5	Trombe wall retrofit. Window box collectors.	185.0	58

Source: HUD Project Summaries (1980).

Tabe 8.15. Predicted Performance of Passive Designs for Commercial Building Retrofits.

LOCATION	BUILDING DESCRIPTION	SYSTEM DESCRIPTION	HEAT SPACE (FT²)	HEATING LOAD MMBTU/YR	SOLAR FRACTION %	COOLED SPACE (FT²)	COOLING LOAD MMBTU/YR.	SOLAR FRACTION %	LIGHTING LOAD MMBTU/YR.	SOLAR FRACTION %
Princeton, N.J.	Two-story university architecture design studio.	Heat and glare-free daylighting provided by a 2-story direct-gain thermosyphoning collector with 2 perforated, horizontal light shelves. Supplemented by 7 pairs of clerestories with insulating shutters. Window quilts on all windows.	13,700	1220.0	72	N/A	N/A	N/A	502.0	40
Philadelphia, PA	Four-story automobile rebuilding facility built in 1914.	A series of solar window furnaces retrofit to south wall provide both heat and daylight reflected by their light shelves. Fiberglass insulating panels replace old windows on northeast, northwest, and west walls. Improved ventilation.	57,000	5878.0	8	57,000	86.0	49	604.0	—

Table 8.15. Predicted Performance of Passive Designs for Commercial Building Retrofits (*Continued*)

LOCATION	BUILDING DESCRIPTION	SYSTEM DESCRIPTION	HEAT SPACE (FT²)	HEATING LOAD MMBTU/YR	SOLAR FRACTION %	COOLED SPACE (FT²)	COOLING LOAD MMBTU/YR.	SOLAR FRACTION %	LIGHTING LOAD MMBTU/YR.	SOLAR FRACTION %
Essex, MD	Retrofit and new addition to two existing schoolhouse-type buildings. Now a senior citizens center.	A courtyard placed between the old and new sections allows daylight penetration. A clerestory window faces south on new addition. Greenhouse is added to south facade, with storage in internal mass walls and floor. Sliding doors next to vestibule.	13,500	603.0	23	5,000	193.0	38	193.0	68

Location	Building	Notes							
Wassau, WI	Two-story retail/office building.	A sunspace is added to narrow south end of north-south elongated building, circulating warm air to both levels. Clerestories provide heat and light to upper level.	2,700	38.5	40	2,700	13.3	40	52.4	40
New Braunfels TX	Single-story mental health center workshop. Schoolhouse-type building.	South-southwest orientation allows considerable solar gain from double-hung windows in winter, and ventilation in summer. Most of retrofit work in the form of conservation.	4,800	128.3	31	4,800	55.5	16	7.2	26

143

Active Solar Heating. Active solar space-heating systems are neither as widely used nor as technically well-developed as active solar water heaters or some passive solar space-heating systems. But thousands of active solar space-heating systems—many of them government-funded—are in use across all areas of the country.

Active solar space-heating systems can be viewed as larger and more complicated versions of active solar water heaters. In all, solar energy is collected by panels and moved by pumps or fans either directly to the building or, more often, to a storage container. From storage, solar heat is transferred to the building's heating system as needed. Auxiliary energy from a conventional furnace, an electric-resistance unit, or a heat pump boosts distribution temperatures if the solar contribution alone cannot meet the building's heating needs.

Differences in active heating systems, all of which work this way, relate to the temperature that the collectors can reach, the transfer medium used, and the centralization or decentralization of the storage system.

Metal Flat-Plate Collectors. Most active solar space-heating systems sold today use active flat-plate collectors with metallic surfaces. These collectors can deliver temperatures exceeding 180°F to storage—more than adequate for space heating most buildings.

Two basic types of flat-plate systems are common: one uses a liquid solution for heat transfer; the other, air. Which collector, storage, transport, and distribution components these systems require is determined by the type of transfer fluid used. Liquid space-heating systems use the same collectors that active water-heating systems do. (See Chapter 7.) In air systems, metal absorbers (usually sheet metal or aluminum) are crossed with ductwork that allows the transfer medium to flow across the collector surface. Different levels of glazing and different selective absorber coatings can be used in both air and liquid collectors.

Both air and liquid flat-plate collectors can be purchased in modular, factory-built units assembled or built on-site. The metal absorber plates, the metallic or wooden framing and backing, the glazing, and the insulation that make up liquid collectors can be put together on-site. Most site-built air collectors are integrated with the roof structure: glazing is installed over wood plenums in which metal absorber plates are mounted.[71]

In liquid systems, pipes connect the collectors to a steel, concrete, or fiberglass storage tank. From storage, solar heat goes to the distribution system via other pipes that lead to a heat exchanger connected with ductwork in a forced-air system, or to piping in a hydronic system.

In cold climates, liquid solar space-heating systems need protection from freezing. Where city water or distilled water is used as the transfer fluid, freezing is prevented by allowing the fluid in the collectors to drain back into the storage tank when outside temperatures fall. The other method of freeze-protection uses an antifreeze solution as the transfer medium, circulating it through a "closed" pressurized piping loop and transferring heat to storage via a heat exchanger.

In air systems, collected solar heat is moved through ductwork either to the house itself or to a rock storage bin. In site-built systems, the plenum in the roof structure performs the same function as the collector ductwork in factory-built models. From storage, another line of ducting transfers solar heat to the ductwork in the building's conventional distribution system.

In both liquid and air systems, solar heat is transferred to the domestic hot-water tank by heat exchangers linked to the collectors or main storage tank.

Nonmetallic Flat-Plate Collectors.   If some of the lower-cost collectors used to heat swimming pools or water could be used for space heating, the cost of active solar heating could be reduced by 50 to 66 percent. These low-cost systems operate just like those using flat-plate metal collectors, though most nonmetallic active collectors use liquid rather than air as a transfer medium, so freeze protection is needed.

While some nonmetallic collectors have reportedly been used in space-heating applications, practical experience is so far limited.

Solar-Assisted Heat Pumps.   One anticipated problem with lower-cost collectors is that they may not deliver heat at temperatures high enough for space heating. One proposed solution is to combine lower-temperature, lower-cost collectors with heat pumps.

The solar-assisted heat pump offers some theoretical advantages over both stand-alone heat pumps and stand-alone active solar-energy systems. The coefficients of performance of conventional stand-alone air-source heat pumps decline as ambient temperatures fall. Active solar metal collectors are relatively expensive, and lower-cost solar collectors used alone may not do the job—hence, the appeal of a combination.[72]

In this hybrid system, the solar collectors can be connected to the heat pump in one of two ways: (1) making the heat pump either single-source or (2) dual-source. In a single-source arrangement, sometimes called a "series configuration," the heat pump uses only the solar-heated transfer medium as a heat source. In some single-source applications using higher-temperature solar collectors, solar heat is used alone whenever it is adequate to heat the building. In a dual-source arrangement, the heat pump can use solar energy, the ambient air, or a combination of the two as heat sources, depending on the amount of heat needed.

The efficiencies (amount of heat delivered divided by the amount of energy used to run the heat pump) of both single- and dual-source applications can be improved by reducing the amount of solar heat lost from storage. Some experts have proposed ground-coupling the storage system to link the solar collectors and the heat pumps. Since the temperature of the earth fluctuates less than that of the ambient air, systems that circulate solar-heated transfer fluid through buried water tanks that are pressurized and sealed, through buried rock bins, or through a network of pipes can provide warmer heat than ground-coupled heat pumps alone.[73]

The technical factors governing the optimal choice among the many combinations of systems available are numerous. The amount of sunshine available, the average ambient temperatures during the heating season, and the size and patterns of buildings' heating and cooling loads must all be taken into account. In cold but sunny regions, solar-assisted heat pumps and lower-cost collectors may be the optimal combination. Where there is less sunshine, stand-alone heat pumps may work best. For buildings whose peak heating demands occur at night, solar-assisted heat pumps and lower-cost collectors may be the optimal combination. Where there is less sunshine, stand-alone heat pumps are probably the optimal choice. Where heating loads are heaviest in the daytime and ambient temperatures are warm enough to allow a fairly high coefficient of performance, a stand-alone heat pump appears most appropriate. Where heating loads are high and cooling needs low, one should consider a stand-alone heat pump or a solar-assisted heat pump. But as cooling loads approach or exceed heating loads, using stand-alone solar systems for heating and energy-efficient window air conditioners or natural cooling becomes more appealing. (See Chapter 9.)

Trickle-Down Collectors.   Another comparatively inexpensive active solar-energy system that uses flat-plate metal collectors is the so-called "trickle-down" design, patented under the trade name of Solaris®. In this system, water trickles down a corrugated sheet-metal absorber plate mounted under glazing on the roof.[74] Heating up as it flows downward, the heated transfer fluid flows through pipes to a steel storage tank surrounded by a rock bed. From there, it is moved to the building through the home's conventional forced-air system.

High-Temperature Collectors.   The cost-effectiveness of solar space-heating systems can also be improved by upgrading collector performance. One option is to use collectors capable of delivering higher temperatures to storage. Few of the high-temperature active collectors marketed today are economical for space heating alone, though they are relatively more cost-effective in combined heating- and cooling-, industrial-process, or solar thermal-electric applications. Yet, at least one higher-temperature collector model has been used in active solar-heating systems. Instead of using flat-plate collectors, this system circulates a liquid through coated metal tubes encapsulated within larger glass tubes. The tubes can absorb more energy per unit of area than flat-plate metallic panels can, and their absorptive capacity is boosted even further by reflecting surfaces placed behind them to concentrate more light on the tubes' surface. Since the vacuum reduces heat losses, these "evacuated-tube" collectors do not have to be insulated or glazed.

Phase-Change Storage. While most active space-heating systems use centralized storage tanks or rock boxes, smaller systems that can be decentralized throughout the building are being developed. Now available are phase-change compounds that begin melting at relatively low temperatures (for example, 90°F) and begin recrystallizing when the temperature of the melted solution drops only a few degrees below the initial melting point. This means that recrystallization and heat releases can begin as the storage melts and loses heat naturally by conduction or as it is fan-cooled.[75]

Active System Performance. Monitoring the results so far indicate only broad trends. There is no typical active solar heating system since systems vary with the size of the building, the building's regional location, and the number of conservation measures and passive solar elements the building has. But while there are no standard performance results, solar system performance is commercially measured in terms of system efficiency (the number of Btu of solar heat delivered to a building divided by the number of Btu of solar heat striking the collector surface) and solar fraction (the proportion of the heating load provided by solar energy). As a rule of thumb, a well-designed and properly installed active heating system using flat-plate metal collectors should not have a system efficiency much below 20 to 30 percent or a solar fraction below 35 to 40 percent.

While the federal government, private engineering firms, and universities have monitored active heating systems, no performance pattern has emerged in terms of system type, system size, or regional locations. Indeed, system efficiencies and solar fractions vary with the building, system, and user. Collector efficiencies, transport efficiencies, storage efficiencies, and distribution efficiencies all depend on the quality of a system's design and installation. Solar fractions also tend to vary with the number of standard conservation features employed and with the occupant's energy-use habits.

Monitoring has revealed that a number of material, design, or installation problems keep many recently installed active space-heating systems from performing as well as expected.[76] Some collector materials deteriorate under high heat; nonpressurized piping systems corrode in the presence of water with a high oxygen content; and heat losses from improperly insulated or sealed pipes, ducts, or storage tanks are sometimes excessive. However, the active solar industry will resolve these problems as it improves its product designs, quality control, and installation techniques. Already, some government- and privately funded projects with apparently well-designed and properly installed systems have met over 50 percent of the user's annual heating needs with solar energy.[77]

Much less experience has been accumulated on either low- or high-temperature solar collectors. Whether low-temperature models can generate temperatures adequate for space heating remains an open question. In addition, the durability of these systems under prolonged exposure to high temperatures and their ability to provide freeze protection have yet to be demonstrated. (See Chapter 7.) For high-temperature collectors, the critical concerns are their cost-effectiveness in improving overall system efficiencies and their durability under prolonged exposure to high heat.

Experience with solar-assisted heat-pumps clearly shows that special care is needed in combining solar and heat-pump hardware and in controlling their separate and interacting operations. In some systems, solar storage has been oversized, so it cannot deliver high enough temperatures to the heat pump. Some engineers also claim that heat-pump designs will have to be modified for maximum effectiveness in solar heat-pump hybrids. Also, even with the use of lower temperature solar collectors, standard heat pumps may deteriorate.

Economics of Solar Heating

Active Systems.   Active solar space-heating systems can be quite economically attractive from a private home- or building-owner's point of view, especially where solar insolation levels are high during the heating season. Systems installed at construction time cost less than retrofits and can be more easily integrated into the building. They have added appeal where a high proportion of the combined water/space-heating load is for water and where electric-resistance heat or fuel oil is displaced. As Table 8.16 shows, a system designed to meet about half of a single-family home's space- and water-heating needs achieves positive cash flow in four years or less when electric-resistance heating is the auxiliary energy source and in six years or less with fuel-oil back-up.

To compute cash flows, weigh the annual principal, interest, and maintenance costs for the solar energy system against both annual utility bill savings and the tax savings. Consider that the federal tax credit allows 40 percent of the installed system costs to be deducted for the purchase year and that interest payments on the solar energy loan system are tax-deductible.

Table 8.16. Economics of Active Solar Space-Heating Systems in New Single-Family Homes Using Flat-Plate Metal Collectors ($1981).

CITY	COLLECTOR AREA (SQ FT)	AUXILIARY ENERGY	YEARS TO POSITIVE CASH FLOW	YEARS TO RECOVER DOWN PAYMENT	PAYBACK
Boston	510	Electric-resistance[1]	1	2	11
		Fuel oil #2 [2]	6	8	16
		Natural gas[3]	8	12	17
Wash, D.C.	350	Electric-resistance	4	2	14
		Fuel oil # 2	6	2	15
		Natural gas	10	15	19
Grand Junction, CO	215	Electric-resistance	1	2	10
		Fuel oil #2	2	2	10
		Natural gas	8	9	17
Los Angeles	80	Electric-resistance	2	2	11
		Fuel oil #2	N/A	N/A	N/A
		Natural gas	13	18	22

Asumptions:
1. Financing: Costs of solar included in 30-year mortgage at 13 percent.
2. Cost of solar system: $3,000 plus $25/ft^2; System sized to meet 50 percent of load.
3. Fuel prices (first year):
    Boston   Electricity 9.8¢/kwh
        Fuel oil $1.26/gal.
        Natural gas 62.5¢/therm
    Washington   Electricity 6.5¢/kwh
        Fuel oil $1.26/gal
        Natural gas 46.3¢/therm
    Grand Junction   Electricity 5.6¢/kwh
        Fuel oil $1.12/gal.
        Natural gas 34.4¢/therm
    Los Angeles   Electricity 6.9¢/kwh
        Natural gas 29.7¢/therm
4. Electricity prices expected to increase at 12 percent per year, fuel oil prices at 13 percent, and gas prices at 15 percent.

Tax provisions also shorten the time it takes to recover down payments for all systems using electric-resistance or fuel-oil back-up, even for systems with relatively long paybacks. Since homebuyers often care more about these two factors than about payback, most systems with electric-resistance or fuel-oil back-up as the auxiliary fuels are at least marginally appealing on a strictly economic basis. Different circumstances make active solar heating economically attractive in different locations. In Boston, the economics is favorable principally because electric rates are high (.098 cents/kwh). Years-to-positive cash flow for systems with electric-resistance back-up are longest in Washington D.C., which has relatively high heating loads and only moderate levels of solar insolation and moderately high electric rates (.065 cents/kwh).

Table 8.17. Economics of Active Solar Space-Heating Systems—New
Single-Family Homes—Variations (Washington, D.C.) (1981 $).

	COLLECTOR AREA (FT²)	YEARS TO + CASH FLOW	YEARS TO RECOVER DOWN PAYMENT	PAYBACK
Electricity				
Base case	350	4	2	14
Mini-load	270	6	2	15
Low-cost	380	1	2	10
Mini-load and low cost	290	1	2	10
Fuel oil #2				
Base case	350	6	2	15
Mini-load	270	7	2	16
Low cost	380	2	2	11
Mini-load and low cost	290	2	2	11
Natural gas				
Base case	350	10	15	19
Mini-load	270	11	16	20
Low-cost	380	7	2	15
Mini-load and low-cost	290	8	2	16

Source: Residential Solar Viability Program (1981).

In Table 8.17, the economic consequences of various system alterations are summarized. In the "mini-load" case, the building heat-loss factor has been cut to reflect strict energy-conservation measures. The "low-cost" case is predicated on the use of a system with lower-cost solar collectors, while the "low-cost plus mini-load" option involves using lower-cost collectors to meet a reduced load. Just as for active solar water heaters, the most dramatic economic improvements occur when solar system costs are reduced. (See Chapter 7.) The economics of both the standard and reduced loads are improved significantly if lower-cost collectors are used, even where natural gas is the auxiliary heat source.

Reducing heating loads by taking prior conservation measures has relatively small negative impacts on solar economics. The largest change in system appeal occurs in the shift from base case to mini-load with electric-resistance back-up; the achievement of positive cash flow is put off for two years by conservation improvements. Since all of the low-cost cases remain virtually unchanged by load reduction, low-cost solar energy and energy-conservation packages should hold broad appeal.

While solar systems in new homes may be marginally attractive, the economic barriers to active space-heating retrofits in single-family houses are significant—in part because conventional flat-plate collectors are expensive but also because interest on home improvement loans is high. (See Table 8.18 and compare with Table 8.19.) As Table 8.18 indicates, in no case does the system

Table 8.18. Economics of Active Solar Space-Heating Systems—Retrofit—Single-Family Homes—Variations (Washington, D.C.) (1981 $).

	COLL. AREA (FT^2)	YEARS TO POSITIVE CASH FLOW	YEARS TO RECOVER DOWN PAYMENT	PAYBACK
Electricity				
Base case	350	6	10	10
Mini-load	270	6	11	11
Low-cost	380	6	7	7
Low-cost and Mini-load	290	6	8	8
Fuel oil #2				
Base case		6	11	11
Mini-load		6	12	12
Low-cost		6	8	8
Low-cost and Mini-load		6	9	9
Natural gas				
Base case		6	15	15
Mini-load		6	15	15
Low-cost		6	12	12
Low-cost and Mini-load		6	12	12

Soruce: Residential Solar Viability Program (1981).

Table 8.19. Impacts of Special Financing on Economics of Active Solar Space-Heating Systems—Single-Family Homes—Retrofits Flat-Plate Metallic Collectors (Wash., D.C.) (1981 $).

	YEARS TO POSITIVE CASH FLOW	YEARS TO RECOVER DOWN PAYMENT	PAYBACK
Electricity			
Package one[1] (Base case)	6	10	10
Package two[2]	8	10	10
Package three[3]	6	2	15
Fuel Oil #2			
Package one (Base case)	6	11	11
Package two	9	11	11
Package three	7	8	16
Natural Gas			
Package one (Base case)	6	15	15
Package two	11	14	14
Package three	11	17	20

Code:
[1] Package one: 5-year loan at 16% interest.
[2] Package Two: 10-year loan at 8%.
[3] Package three: 30-year loan at 15½%.

Source: Residential Solar Viability Program (1981).

achieve positive cash flow within the life of the loan, and only low-cost collectors where electricity and fuel-oil are back-up heat sources pay for themselves within 10 years.

Table 8.19 summarizes the economics of three ways of financing a conventional active heating system installed in Washington, D.C.: (1) 5-year, 16-percent loans; (2) 10-year, 8-percent loans; and (3) 30-year, 16-percent loans. Even with lower interest rates, the longer terms usually mean longer paybacks. Since most home-owners use year-to-positive cash flow as their investment touchstone when inflation runs high, special financing programs that extend the loan terms make solar space-heating retrofits more economical.

The most promising situation occurs when low-cost collectors can be bought on favorable financing terms. (See Table 8.20.) Under both of the special financing programs considered here, systems with electricity and fuel-oil back-up achieve positive cash flow before the loan is due for repayment. Paybacks are, again, extended under the longer-term package. Load reductions affect the economics of these systems somewhat negatively, but prior conservation does not work against the otherwise economical solar retrofit.

Table 8.20. Impacts of Low-Cost Systems Plus Special Financing on Economics of Active Solar Space-Heating Systems—Single-Family-Retrofit—(Wash., D.C.) (1981 $).

	YEARS TO POSITIVE CASH FLOW		YEARS TO RECOVER DOWN PAYMENT		PAYBACK	
	STANDARD LOAD	MINI-LOAD	STANDARD LOAD	MINI-LOAD	STANDARD LOAD	MINI-LOAD
Electricity						
Package One[1] (Base Case)	6	6	7	8	7	8
Package Two[2]	2	4	2	2	7	8
Package Three[3]	2	2	2	2	11	11
Fuel oil #2						
Package One (Base Case)	6	6	8	9	8	9
Package Two	5	5	2	2	8	9
Package Three	2	4	2	2	12	12
Natural gas						
Package One (Base Case)	6	6	12	12	12	12
Package Two	10	11	12	12	12	12
Package Three	8	9	10	12	16	17

Codes:
[1] 5-year loan at 16% interest.
[2] 10-year loan at 8% interest.
[3] 30-year loan at 15½% interest.
Source: Residential Solar Viability Program (1981).

**Table 8.21. Economics of Active Solar Space-Heating Systems—
Multifamily Buildings—New Construction (Wash., D.C.) (1981 $).**

	COLL. AREA (FT²)	SYSTEM PRICE (DOLLARS)	YEAR NET PRESENT VALUE IS POSITIVE	IRR 5th YEAR (%)	AFTER-TAX CASH FLOW— (5th YEAR) (DOLLARS)
Base case	7,500	187,500			
Electricity			1	37.8	6,795
Fuel oil # 2			2	35.1	5,409
Natural gas			2	27.3	732
Mini-load	5,300	149,00			
Electricity			1	36.2	4,593
Fuel oil #2			2	33.7	3,584
Natural gas			2	26.5	177
Low-cost	8,000	160,00			
Electricity			1	41.4	7,419
Fuel oil #2			1	38.1	6,050
Natural gas			2	28.7	1,429
Mini-Load/Low-cost	6,200	124,000			
Electricity			1	39.6	5,096
Fuel oil #2			1	36.5	4,100
Natural gas			2	27.8	734

Notes:
1. The system's net present value is the sum of all fuel savings and tax savings minus all solar costs over the 20-year life of the system, all discounted at 15 percent. It also includes the (discounted) net after-tax proceeds of the sale of the solar energy system as part of the property in that year.
2. Internal rate of return (IRR), which treats the solar energy system as an investment, equals the after-tax return on the investment.
3. After-tax cash flow is defined as the annual utility bill savings attributable to the solar energy system minus annual principal and internal costs. It also includes tax benefits and tax losses (e.g., rates tax), but not solar energy equipment appreciation.

Active space-heating installations in new multifamily buildings hold considerable economic appeal if it is assumed that the solar equipment increases in value with the property. As Table 8.21 shows, all of the 12 systems analyzed reach positive net present values by at least the second year, and all systems with electricity or fuel-oil auxiliary have after-tax positive cash flows of at least $3,500 by the fifth year.

Assessing the economic feasibility of active space-heating in commercial structures is more difficult mostly because there are no "typical" commercial buildings, so there can be no "typical" active solar heating installation to service it. But it can be assumed that the positive factors that contribute to active solar system feasibility in residential buildings apply to commercial structures as well. In addition, some commercial buildings need less heat per square foot because they have larger surface-to-volume areas.

Solar-Assisted Heat Pumps. Few economic analyses cover all possible combinations of active solar collectors and heat pumps. But most research indicates that active solar heating and heat pumps can each save relatively large amounts of conventional energy, not that systems combining both save appreciably more energy or money than either operating alone. Ironically, while nearly all studies emphasize the need to analyze systems using lower-cost and lower-temperature collectors, few do.[78]

Passive Systems. Since no passive system or passive solar home is typical and many systems combine two or more passive technologies, research has resulted in various simulations or experiments instead of broad-based analyses. Obviously, however, the impacts of a passive heating system's energy savings on a building-owner's cash flow will be influenced by the initial cost of the system, the financing terms, the amount of conventional energy saved annually, and the price of conventional energy. While increases in interest rates and conventional fuel prices are hard to predict, both simulations and experience suggest that passive solar heating systems can deliver substantial amounts of energy to a building. Many passive solar energy systems also cost far less than active solar energy systems designed to deliver the same amount of heat. The costs of passive solar heating systems vary much more with the costs of the building than do the costs of active solar hardware, though they tend to fall between 5 and 10 percent of the total construction costs.[79,80] (See Table 8.22.) For passive retrofits, system costs can be calculated on the basis of square footage of the materials used. (See Table 8.23.)

Table 8.24 summarizes rough calculations that indicate the potential ranges of cost savings for a number of passive systems whose performance has been monitored. The figures are encouraging, especially for electrically heated homes. In six of nine cases, first-year savings in homes with electric back-up equal or exceed one third of the system's purchase price, which means that simple payback occurs in three years or less. While financing charges are not included, these systems appear relatively inexpensive, so additional interest payments may be largely offset by tax savings and increased utility bill savings as fuel prices escalate. Initial system costs could also be offset by lowered HVAC costs for smaller furnaces.

Combining Energy Conservation with Active or Passive Solar Heating. There are four generalizations so far supported by evidence. One is that the simpler conservation options are still the most cost-effective energy-saving strategies. Second, much depends on climate. Third, energy-conservation and solar-energy techniques can usually be combined most cheaply when a building is built or substantially rehabilitated, and it is always more economical to

Table 8.22. Construction Cost Comparison of New Conventional, Energy-Conserving and Passive Solar Homes (1979 $).

BUILDING CATEGORY	PASSIVE HOME	ENERGY CONS.	CONVENTIONAL
Block stemwalls	1,935	1,935	577
16-in Trombe wall	818	-0-	-0-
Rigid insulation	577	577	-0-
Exterior framing	1,868	2,278	2,100
Exterior sheathing	685	819	757
Interior sheathing and thincoat	2,104	2,317	2,317
Roof insulation	909	909	587
Wall insulation	496	872	582
Exterior stucco	3,320	3,658	3,658
Interior plaster	599	311	311
Venting windows	1,194	1,870	1,692
Fixed glass	2,029	-0-	-0-
	16,734	15,546	12,581
20% Labor fringe	645	604	515
	17,379	16,150	13,096
15% Overhead profit	2,697	2,422	1,964
	19,986	18,572	15,060
Total house costs	62,151	60,737	57,225
Cost per sq. ft. floor area:	45	44	42
% Cost increase over conventional house	8.6%	6.1%	

Source: Taylor 1979, Reproduced in *A New Prosperity: Building a Sustainable Energy Future,* 1981.

Table 8.23. Estimated Costs of Passive Solar Retrofit Techniques ($ 1977).

TECHNIQUE	ESTIMATED COST (DOLLAR/FT^2)
Movable insulation	.50– 5.00
Trombe wall	3.00–15.00
Glazing	5.00–25.00
Greenhouse	4.00–30.00
Thermosyphoning air heater	1.50– 9.50
Awnings	3.50– 6.50/ft^2 of shaded glazing
Porch/overhang	4.00–15.00/ft^2 of shaded glazing
Windbreak	1.50– 2.50/ft^2 of shaded glazing

Source: *A New Prosperity: Building a Sustainable Energy Future,* 1981.

Table 8.24. Estimated First and Fifth-Year Cost Savings of Selected Passive Solar Homes (1981 $).*

LOCATION	YEARLY HEAT LOAD (MMBTU)	SOLAR %	COSTS	YEAR ONE SAVINGS (ELECTRIC BACK-UP)[1]	YEAR ONE SAVINGS (GAS BACK-UP)[2]	YEAR FIVE SAVINGS (ELECTRIC BACK-UP)[3]	YEAR FIVE SAVINGS (GAS BACK-UP)[4]
Sommersworth, NM	56.0	66	1500	703.72	245.52	1,239.95	493.74
Ames, IA	44.4	32	6000	270.52	94.38	476.65	189.78
Prescott, AZ	72.0	81	3440	1,110.41	387.41	1,956.54	779.08
Royal Oak, MI	70.0	79	2500	1,052.91	367.35	1,855.22	738.74
Atascadero, CA	32.0	100	1815	609.28	212.57	1,073.55	427.47
Santa Fe, NM	69.0	96	N/A	1,261.21	440.02	2,222.25	884.88
Bozeman, MT	80.0	85	7920	1,294.72	451.71	2,281.29	908.39
Yakima, WA	32.0	63	1000	383.85	133.92	676.34	269.31
East Pepperell, MA	17.0	35	-0-	113.29	39.52	199.62	79.47

* Derived from Data Presented in *A New Prosperity: Building a Sustainable Energy Future*, 1981.
[1] Electric rates set at .065¢/kwh ($19.04)/MMBtu) — 100% furnace efficiency.
[2] Gas prices at .465/therm — ($4.65)/MMBtu) — 70% furnace efficiency.
[3] Electric rates at Year 5 reflect yearly increases of 12% to $33.55/MMBtu.
[4] Gas prices at Year 5 reflect yearly increases of 15% to $9.35/MMBtu.

implement an energy-conservation package than to add energy efficiency improvements one by one. Fourth, whether a "total-package" or "step-by-step" approach is used, many of the costs of active solar energy systems are fixed while for passive solar systems, few are. Thus, the potential for reducing systems costs by taking stricter energy-conservation measures is proportionately greater with passive than active systems. Finally, "solar-plus-conservation" packages are far more economical when compared with such expensive energy sources as fuel oil or electric-resistance heat than with, say, natural gas.[81]

City Applications

The most basic factor governing the technical and economic feasibility of on-site renewable-energy technologies is the availability of adequate south-facing roof, wall, or ground space unshaded from mid-morning to mid- or late-afternoon. With suitable collector space, orientation, and solar access, on-site solar heating is technically feasible in any building. Without any one of them, it is not. Other important factors include the building's size and structural characteristics and the size and pattern of its heating load.

The availability of solar access for active or passive solar energy collectors has been studied empirically in Boston, Philadelphia, Baltimore, Minneapolis, Denver, and Los Angeles.[82] Results in even the densest areas have been encouraging. For example, a city-wide study of a small randomly selected number of buildings in Boston indicated that about one third of the two- and three-story buildings had adequate solar access on south-facing walls and over one half had sun enough on south-facing roofs. While taller buildings had less wall access, many had significant roof access.[83] Yet, few studies have examined the actual structural space available for on-site solar collectors in buildings located in dense urban neighborhoods—a key factor even if solar access is good.[84]

Active Systems.   The space needed for active collectors is influenced by collector efficiencies, the size of the heating load, and regional locations. For example, an active solar collector array capable of supplying 50 percent of the heating energy used by a 1,500 ft^2 house would need 510 ft^2 of space in Boston but only 215 ft^2 in Grand Junction, Colorado.

The amount of space available on the roof of single-family detached homes varies by neighborhood and even by building. For example, houses in one single-family neighborhood in Denver have, on average, only about 377 ft^2 of available roof space, adequate there but not in other cities with less winter sunshine. But a study of Roxbury, Massachusetts, suggested that single-family homes could accommodate up to 545 ft^2 of active collectors.[85] Then too, single-family attached homes are also more likely than others to have on-site ground space available.

Dwelling units in attached multifamily buildings may not need collector areas as large as those that detached homes require. Average dwelling sizes may be smaller, and together with the use of common walls and floors, may cut heating needs. Moreover, the flat roofs often found in row-houses or low-rise apartment houses may accommodate a larger collector area than can the sloped roofs found on many single-family or semi-detached homes.

Even these advantages probably cannot compensate for the inadequacy of roof space on the top of mid- or high-rise buildings. And, although on-site ground-level space could be accessible to these buildings—on parking lots, for example—few high-rise urban apartments will have enough unshaded, south-facing ground to accommodate an array of collectors capable of heating the buildings.

Another remaining problem is that older row-houses may contain as much or more conditioned space as newer detached housing. In the Philadelphia neighborhoods under study, for example, typical two-story row-houses without garages contained less than 1,000 ft^2 of conditioned space, while both two-story row-houses with garages and three-story row-houses enclosed conditioned areas of between 1,500 and 1,800 ft^2.[86] In addition, even though they have common walls, older row-house or multifamily buildings that are not well insulated or weatherized may consume large amounts of space-heating energy.

The energy inefficiency of many city buildings can be gauged by comparing data on heat loads in existing buildings with projected heat loads of buildings similar to those assumed in the solar economics calculations above. When the effects of climate are taken into account, the "base case" load in the solar houses is less, sometimes substantially so, than those common in existing city buildings. (See Table 8.25.) Thus, either energy-inefficient city buildings need more collector area than the "base case" houses to meet 50 percent of their space-heating needs or extensive energy-conservation measures are needed before even the "base case" level of solar cost-effectiveness can be reached. As for conservation measures, retrofits do reduce heating loads, but the most extensive measures are relatively expensive and may compete economically with active solar heating.

Passive Systems—Residences. If orientation and solar access are correct, most single- and multifamily buildings with wood or masonry exteriors should be at least considered as potential passive solar retrofits. One or two structural features are especially well-suited for retrofits: relatively large areas of south-facing glass or masonry walls. To be sure, a large south-facing glass area does not a passive solar energy system make. On the contrary, large areas of single-

Table 8.25. Monitored Heat Loads in Existing Buildings Compared with Heat Loads Derived from Assumptions Used in Solar Economics Calculations.

HOUSE TYPES	BASIC FEATURES	CONDITIONED SPACE (FT^2)	ANNUAL HEATING LOAD (MMBTU)	ANNUAL HEATING LOAD (MMBTU/FT^2)
1. BALTIMORE				
Solar base case				
Single-family detached	8.42 heat loss factor × degree days × sq ft conditioned space	1500	59.7	39.8
Existing houses				
Single-family detached[1]	Two stories/ basement	1695	71.0	41.9
Single-family attached[1]	Two stories/no basement	1300	63.2	48.6
Rowhouse[1]	Three stories/ basement	2255	44.2	19.6
OTA "1973" House[2]	One story/three bedrooms	1000	55.4	46.1
2. PHILADELPHIA				
Solar base case				
Single-family detached	8.42 heat loss factor × degree days × sq ft conditioned space	1500	60.8	40.5
Existing houses				
Row-house[3]	Brick/two stories/ no basement	1210	88.9	73.5
Row-house[3]	Brick/two stories/ basement/ garage	1315	71.8	54.6
Row-house[3]	Brick/three stories/ basement/ garage	1906	88.3	46.3
Twin[3]	Brick/two stories/ basement/ garage	1436	94.9	66.1

Sources: 1. Twiss (1979), derived from Hittman (1975).
2. *Residential Energy Conservation,* Office of Technology Assessment, 1979.
3. Philadelphia Solar Planning Project (1980).

paned glass are net heat losers in winter. However, residential buildings with large south-facing window areas do have substantial passive solar potential, and glazing can be added to buildings. A simple strategy is to enclose an existing wooden or stone porch with glass, in effect turning a porch into a greenhouse.[87] More expensive alternatives are to "build out" a sunspace from the building's exterior surface or to replace part of a wall with glazing.[88]

Buildings with masonry surfaces have combined collectors and storage components potentially in place. Exterior masonry surfaces can be glazed, creating a "trombe wall" retrofit. Given the number of single-family, multifamily, and commercial buildings with masonry exteriors, the potential for this application is massive in some cities.[89] However, the availability of potential collector area is not the sole prerequisite for a passive retrofit. Interior building materials and design are critical, so rehabilitation may be needed, especially in wood-floored and wood-walled buildings.

Passive Systems—Commercial Buildings. How well a commercial building takes to passive solar retrofitting depends on how it uses energy, as well as on how it is built. Small retail stores and offices or warehouses and factories with unshaded south-facing masonry walls probably have the greatest retrofit potential. Retrofitting trombe walls onto these buildings is not much more complicated than residential retrofits, and many such buildings have energy-use patterns well suited for passive heating retrofits. Since smaller buildings have larger surface-to-volume areas than bigger structures, they need relatively more heat than larger commercial buildings but (because of the benefits of relatively greater internal heat gains) less heat per square foot than residences. Thus, a relatively small passive contribution could satisfy a relatively large portion of the heating demands.

In large buildings with large glass areas, enhancing heat collection and retention on the south side and reducing heat loss on the north can cut heating loads dramatically. But since many larger commercial buildings use much of their energy for lighting and cooling, care is needed to make sure that steps taken to improve winter heat gain do not add to cooling loads. Similarly, efforts to reduce winter heat loss should not be allowed to create a need for more artificial lighting. All glass areas will probably need double glazing and night insulation, while most south-facing areas probably require some overhangs or movable blinds.[90]

In any large building with relatively heavy heating, lighting, and cooling needs, replacing some of a south-facing wall with glazing makes sense. This move could directly reduce heating and lighting needs and indirectly cut demands for cooling. But using the wrong kind of glass or the wrong kind of structural controls could lead to both nocturnal heat loss (increasing heating demand) and daytime over-heating (increasing cooling demand).

Since passive systems can be adapted to diverse building types, urban locations, and climates, most will fit into the cityscape. Buffalo, New York, has 100 double-envelope houses, and 12 earth-sheltered townhouses have been built in downtown Minneapolis next to an interstate highway. In a new Tennessee Valley Authority building in Chatanooga, passive technology will provide lighting, heating, and cooling.

Institutional Constraints and Scoio-economic Considerations

Solar space-heating applications require more space and solar access than solar water heaters do. How much can be allotted depends on how much shade-free south-facing area is already available and what future land development holds in store. In a single-family, low-density area where vegetation is a major source of shading, a shade-control ordinance could be effective. Where many taller buildings already have shaded walls, protecting roof access is relatively more important. (See Chapter 3.)

The widespread use of solar heating is not likely to prompt an adverse electric utility reaction unless a large number of homes in the city are electrically heated. Even then, solar heating equipment could be combined with utility load-management techniques to reduce peak demands for electricity. (See Chapter 5.)

Low-income city residents cannot afford most solar energy equipment without some form of financial assistance. Nor will solar heating appeal to many owners of multifamily buildings. However, some solar installations implemented by city governments and community groups have helped low-income consumers cut utility bills. A Community Action Agency in Arkansas, for instance, installed low-cost air heaters in the homes of low-income elderly people, saving them up to 25 percent on heating bills, even though their homes had already been weatherized.[91]

9. Cooling Buildings

While demands for cooling energy do not begin to approach demands for heating energy in the U.S., reducing cooling loads and finding renewable sources of energy for cooling are vital for a number of reasons. First, in a small but growing number of cities, cooling represents the largest building energy drain. For example, the South Florida Regional Planning Council found that air conditioning constituted the highest building energy demand (45 percent, overall) for a tri-county area that includes Miami. According to OTA, a centrally air conditioned single-family home in Houston demands about seven times the amount of cooling as heating energy. Of course, since not all homes are centrally air conditioned, actual cooling energy-uses per house are commensurately less. Still, the South Florida Energy Audit predicted that fully 75 percent of all new construction would incorporate central air conditioning, while DOE predicts that 90 percent of all new construction in the Southwest will be centrally cooled.[1,2] Even in the somewhat cooler climate of Greenville, North Carolina, newer multifamily low-rise buildings use almost 20 percent of their energy for air conditioning, as compared with 7 percent for older single-family detached houses.[3]

Second, numerous factors combine to create higher cooling demands in cities than in surrounding suburbs. One is urban form. The same features that help to cut winter heating needs (dense housing patterns, taller buildings that function as wind breaks, and so on) often add to summer cooling loads. And lack of trees and other vegetation in many urban neighborhoods also makes people and buildings hotter. Moreover, in some large commercial buildings, annual cooling needs exceed demands for heating, and high cooling demands can indirectly add to heating energy use. (See Table 9.1.)

Still another factor is that, insofar as cities contain relatively more poorly insulated buildings than surrounding suburbs do, urban demands for cooling per square foot of mechanically cooled space may well be greater. And because cities contain proportionately more lower-income residents than do suburbs, many of the window air-conditioning units in place in cities may be older and less efficient than the norm. A final incentive for reducing the use of cooling energy is that most mechanical cooling is supplied by electricity—usually the most costly source of energy.

Many of the space-heating technologies described in the last chapter can also

Table 9.1. Relative Annual Heating and Cooling Demand in Selected Cities.

CITY	PERCENT OF SPACE-CONDITIONING DEMANDS USED FOR HEATING	PERCENT OF SPACE-CONDITIONING DEMANDS USED FOR COOLING
Minneapolis	45	55
New York	31	69
St. Louis	29	71
Atlanta	21	79
Los Angeles	12	88

Source: S. Robert Hastings, Passive Solar Design for Urban Commercial Environments, Presented at the 1979 National Passive Solar Conference.

be used in cooling. Improving the building shell and making cooling systems more efficient can help reduce conventional energy use for cooling. On-site passive and active solar energy systems can cool as well as heat air. In addition, renewable sources of electrical energy (such as wind, hydropower, and photovoltaics) can be used to power electrical cooling devices. (See Chapter 11.)

COOLING WITH CONSERVATION

Technical Approaches and Issues

Shell Improvements. In smaller residential and commercial buildings with relatively high surface-to-volume ratios, building-shell improvements that save heating energy save cooling energy, too. Reducing infiltration keeps hot air outside the building and mechanically cooled air inside, and better insulating most parts of the shell helps reduce conductive heat gains in summer. Using efficient window designs and materials can also reduce overheating.

The effects that these improvements have on cooling loads vary by climate. In general, their overall impacts will probably be proportionately less on cooling than on heating demands. According to the National Association of Home Builders Research Foundation:

First, even in very hot climates, where summer design temperatures are near 100 degrees F, the temperature difference between the outside and inside air is smaller than for most winter temperature differences. All other things being equal, the amount of energy used is directly related to the temperature difference. Second, the effect of double glazing for summer conditions is considerably less than for winter conditions. Third, the amount and effect of wind infiltration under summer design conditions is substantially less than for winter conditions.[4]

Nonetheless, shell upgrading can significantly reduce heat gain. For instance, technically simple shell retrofits could reduce total heat gains in a 1,500 ft^2 house in Baltimore by 38 percent.[5]

Most simple shell-conservation measures—adding attic insulation, double-glazing, plugging leaks, and so on—require no special modifications to reduce cooling and heating loads. However, such relatively elaborate shell retrofits as adding large quantities of insulation or extensive south-facing window areas may. For example, although adding floor insulation to prototypical single-family houses in Baltimore and Chicago cuts heating needs enough to make the net benefits worthwhile, cooling demand will increase slightly because such houses do not lose heat to the cooler ground in summer. But in Houston, adding floor insulation does not increase mechanical cooling energy needs because floor heat losses are negligible in hot, humid climates. (See Table 9.2.)

In buildings with high internal heat gains, the relative impacts of conservation measures on heating and cooling demands require even more analysis. Consider, for example, the results of Energy Ltd.'s analysis of a prototypical supermarket in Seattle. Adding attic insulation to R-19 levels was predicted to reduce heating needs by over 70 percent but to increase cooling loads by over 30 percent. While the net benefit (assuming the building is both electrically heated and electrically cooled) was still positive, increasing ventilation and recovering waste heat would realize equally great amounts of heating-energy without increasing cooling loads. Clearly, trade-offs between heating and cool-

Table 9.2. Relative Impacts of Shell Improvements on Heating and Cooling Demands.

CITY	CONSERVATION LEVEL	HEATING LOAD (MMBTU)	COOLING LOAD (MMBTU)	TOTAL LOAD (MMBTU)	TOTAL LOAD DIFFERENCE FROM BASE CASE (MMBTU)
Baltimore	Base Case	55.4	18.0	73.4	
	Moderate	43.7	15.8	59.5	13.9
	Heavy	9.9	20.4	30.3	43.1
Chicago	Base Case	81.1	14.0	95.1	
	Moderate	64.7	12.3	77.0	18.1
	Heavy	20.2	17.9	38.1	57.0
Houston	Base Case	8.4	56.9	65.3	
	Moderate	5.2	51.3	56.5	8.8
	Heavy	0.0	43.4	43.4	21.9

Source: OTA, Residential Energy Conservation (1979, p. 36).

ing demand fluctuations have to be assessed in detail for different types of commercial buildings.

Since moisture may accumulate in heavily insulated buildings, latent heat gain resulting from additional humidity is another potential drawback to insulating to cut down cooling loads. In practice, though, it is so difficult to add large amounts of insulation to all portions of the building shell that the problem is hypothetical.

South-facing glazing, too, can add to cooling loads in summer, as the previous chapter makes clear. Overheating from south-facing glazing can be prevented by properly sizing the glass area, using overhangs that block summer and early fall sunshine, and using reflective blinds or shades. In a few cases, adding south-facing glass areas should be discouraged altogether. (Florida's model building code encourages less glass on unshaded south-facing walls.[6])

In all-glass office buildings, the challenge is more complicated. Unwanted heat gain should be stopped but not at the sacrifice of daylighting, which complements artificial lighting—a high energy use in its own right and a chief source of internal heat gain. In such cases, an integrated retrofit strategy may be necessary: reducing some glass areas on easily-accessible floors, installing reflective blinds or insulated shades, and eliminating unnecessary artificial lighting.

Reducing Internal Heat Gains. Much of the cooling demand in smaller buildings is climate-related. But in larger ones, cooling needs are dictated mainly by the heat generated inside the buildings. Body heat, constantly running machines, and high levels of artificial lighting all contribute to internal heat gains.

There are three main strategies for reducing internal heat: (1) using more efficient appliances and machines, (2) improving the design of building interiors, and (3)—often most important—reducing lighting levels. Reducing lighting by using less bright bulbs or by removing some fixtures cuts both direct energy use and space-cooling demands dramatically.[7] For example, if buildings designed to use 5.0 watts of lighting per square foot of work space were improved to require only 1.5 watts per square foot,[8] some 17.5 kw of cooling capacity could be saved in each 10,000-ft^2 office building.

Of course, the impacts of lighting improvements on cooling and heating demands must be assessed carefully. In colder climates, internal gains can lessen heating loads. So the impacts of different lighting-efficiency strategies on different energy demands, the types of energy used to meet each of those demands, and the benefits of alternative approaches that combine selected lighting-efficiency improvements and shell-conservation measures must all be considered.

Table 9.3. Energy Demand Impacts of Lighting Efficiency Strategies on Heating, Cooling, and Electric Loads in Prototypical Commercial Buildings—Seattle (in MMBtu).

SMALL OFFICE	HEATING	COOLING	ELEC.	TOTAL
Base Case	127.6	50.3	331.0	508.9
Delamping	180.1	37.1	252.1	469.3
Efficient Bulbs	127.6	50.3	292.2	470.1
Efficient Bulbs and Reduced Infiltration	100.1	49.8	292.2	442.1

MIXED OFFICE/RETAIL	HEATING	COOLING	ELEC.	TOTAL
Base Case	165.9	42.9	318.0	526.8
Delamping	290.2	32.1	253.1	575.4
Efficient Bulbs	165.9	42.9	276.4	485.2
Efficient Bulbs and Reduced Infiltration	134.9	42.3	276.4	453.6

SUPERMARKET	HEATING	COOLING	ELEC.	TOTAL
Base Case	992.1	509.9	4,138.6	5,640.6
Delamping	1,385.1	411.7	3,549.2	5,346.0
Efficient Bulbs	992.1	509.9	3,874.9	5,376.9
Efficient Bulbs and Reduced Infiltration	975.3	512.5	3,874.9	5,362.7

Source: Energy, Ltd.

Table 9.3 indicates some lighting-conservation options open to building designers or owners. In each building category, delamping provides the greatest combined cooling and service-electricity energy savings. But in all cases, delamping also increases heating loads significantly. Where heating energy is provided by natural gas, the delamping strategy may be preferable. But where space heating is provided by electric-resistance heat, delamping may be less cost effective.[9] It increases total energy demands in mixed office/retail buildings. In both the small office and mixed office/retail buildings a combination of more efficient light bulbs and basic shell-conservation techniques reduces total energy demands, and the additional savings in expensive electric energy can probably justify the additional costs.

Types of Systems.   Most air conditioners in single-family homes use electricity. Window units operate on a vapor-compression cycle, while central air conditioning systems use vapor-compression chillers or heat pumps in the cooling mode. At least one company also markets gas-fired absorption chillers for residential applications.[10]

In larger buildings, most cooling systems chill water and then circulate it throughout the building. With a forced-air distribution system, the chilled water circulates through a liquid-to-air heat exchanger in the ductwork. The cooling modes of the more complicated systems that combine heating and cooling work on the same general principles as the complex heating system described in the previous chapter: the chiller simply replaces the furnace.

In the chiller, a working fluid (refrigerant) under low pressure absorbs the heat contained in the water returning from the conditioned space, cooling the water so that it can be recirculated. Heat is released by the refrigerant as it is pressurized by an electric compressor or as it comes into contact with an absorbent solution. Heat can then be discharged from the chiller by air-, water-, or evaporative cooling. Some air-cooled systems (such as reciprocal electric compressors) employ a remote air-cooling/condensation unit, while a cooling tower is needed in water-cooled systems.

Standard window units generally have measured capacities of from 7,000 to 14,000 Btu, which means that they can remove that amount of heat from conditioned space in one hour. Larger central air conditioner capacities are generally given in tons, with one ton representing a capacity of 12,000 Btu per hour. Smaller single-family central air-conditioning units are in the 3- to 5-ton range. Moderate compression and absorption systems can vary from 10 to 25 tons, while larger buildings can use units of 100 tons or over. Single absorption units can reach over 1,500 tons of capacity, and the very largest compression units can exceed 10,000 tons.[11]

System Efficiencies.   The efficiency of an air-conditioning system or component is generally given in coefficients of performance (the Btu of heat removed divided by the Btu of energy used to operate) or in energy efficiency ratios (the Btu of heat removed divided by the watt-hours of operating energy). Currently, the coefficients of performance (COPs) of new room and central compression air conditioners fall between 1.6 and 3.1 (5.46–10.58 Energy Efficieny Rating—EER).[12] COPs of larger electric vapor-compression chillers vary more, depending upon the type and capacity of the compressor used and whether the system is water- or air-cooled. As a range, COPs of compression chillers run from about 2.85 to 3.6 for reciprocating units, 4.1 to 4.9 for centrifugal units, and 3.3 to 4.5 for screw-type units.[13] COPs of absorption units are generally much lower, from 0.45 to 0.70 for "single-effect" and from about 1.0 to 1.15 for "double-effect" chillers.[14] However, primary-fuel COPs of compression and absorption units appear much more comparable if the penalties imposed by electric generation and transmission losses are taken into account. (For example, if two thirds of the primary fuel input to electricity production is lost, the primary fuel COP of an electric-compression chiller with a nominal COP of 3.0 is approximately 1.0.)

Improving Chiller Efficiencies.   Since both heat pumps and electric air conditioners operate on vapor-compression cycles, many of the techniques used for improving heat-pump efficiencies also work to improve the performance of room or central air conditioners as well. (See Chapter 8.) At least one residential-sized gas-fueled absorption air conditioner is now on the market, and gas-fired heat pumps that operate on the absorption cycle in both heating and cooling modes are also being researched. Either could cost less to use in residential buildings than electricity.

Absorption-chiller efficiencies are expected to improve as designs using the "double-effect" absorption principle come into wider use. Because such units can accommodate higher temperatures, they can separate refrigerants and absorbents more completely, thus allowing for more effective heat transfer in the condensation, refrigeration, and absorption stages. Double-effect absorption chillers can easily reach COPs of over 1.0, and thus compete with some of the larger electric-compression units on a cost/Btu basis. Double-effect absorption chillers are available only in relatively large models (385 tons or over) and cost about 30 percent more than a single-effect unit of equal capacity.[15]

Both absorption- and compression-chiller COPs could also be improved with the use of on-site chilled water storage, which would allow the absorption unit to operate longer at steady-state conditions, shaving electric demand peaks and cutting cycling losses. Another potential economy would be to use waste heat to drive chiller combustion in compression processes.[16]

Newer Approaches.   An innovative cooling system now operating in a house in Knoxville uses a water-source heat pump in winter to make ice that is then stored and used for cooling in summer. The heat pump of this system (called the Annual Cycle Energy System—ACES) operates in winter using a large (10,000 ft^2) water tank in the basement as the heat source.[17] Its annual COP is 12.7. Variants of this kind of system are being studied carefully for their potential as community heating and cooling systems. (See Chapter 10.)

Advanced thermodynamic technologies could also be used in cooling systems. Perhaps closest to market readiness is an on-site heat engine fired by natural gas; it generates the electricity needed to run a vapor-compression chiller or heat pump that can be operated in the cooling mode. Prototype chillers operating on Rankine thermodynamic cycles can be obtained from such firms as Carrier, General Electric, and Honeywell on special order, though heat engines have been coupled with heat pumps or air conditioners only rarely.[18] Such heat engines could generate electricity for sale under the provisions of PURPA 210, allowing their owners to defray systems costs.[19] It can also be economical to use the heat engine to run a heat pump that both heats and cools.

Improving System Efficiencies.   Improvements to heating and cooling systems should be undertaken together, especially in large buildings. Heat-recovery systems can remove heat from buildings, as well as help run absorption chillers.[20] Enthalpy heat-exchange systems can cool and dehumidify incoming air in summer, much as they can heat and humidify incoming cold air in winter.[21] Sophisticated control systems that monitor conditions in different building locations and climatic conditions can often reduce the amount of artificial cooling needed. For example, a system known as an air-economizer cycle can substitute outside air for chilled air when outdoor air temperatures are cooler and drier than those indoors.[22] Control strategies governing the more efficient use of a building's total energy usually reduce cooling energy needs the most because loads in many large buildings are cooling-dominated.

For perspective, Florida's Model Energy Efficiency Building Code, developed for an area where cooling loads certainly dominate, contains several provisions for reducing cooling energy-use in large buildings. For example, it encourages the use of automatic temperature resets to raise the temperature of cold air or lower the temperature of heated air in order to control interior temperatures with minimal energy expenditures. For buildings that need simultaneous heating and cooling, the code requires sequential temperature controls that raise or lower air temperatures to delivery levels instead of mixing hot and cold air streams.[23]

A newer cooling approach is through dessicant systems that enhance the enthalpy systems described above. Dessicant cooling systems use some high surface-to-volume area solid such as silica gel, a molecular sieve, or zeolite to dry return air from conditioned space before the air is evaporatively cooled by a humidifier. Unlike enthalpy systems, dessicant systems have a drying agent, so they work relatively better in humid climates.[24]

Economics of Conservation Techniques for Cooling

Building Shells.   Few economic analyses of shell-conservation techniques distinguish between heating savings and cooling savings. Most, in fact, concentrate on heating-energy savings because their magnitude is so much greater. Since most basic shell improvements cut both heating and cooling energy needs, however, it does make sense to count cooling savings simply as a dividend of economical retrofit strategies. (See Chapter 8.) To the extent that shell-conservation improvements do increase cooling energy demands, such impacts will have to be calculated case by case.

Air Conditioning Equipment.   The economics of replacing less efficient with more efficient window units or with small central air-conditioning systems var-

Table 9.4. Incremental Costs of Improved Performance in New Central or Room Air Conditioners.

A.C. TYPE/ CAPACITY (BTU/ H)	BASELINE EFFICIENCY RATING	NEW UNIT EFFICIENCY RATING	ADDED COSTS FOR IMPROVED EFFICIENCY RATING ($)	COSTS/WATT IMPROVEMENT	HOURS OPERATED (YEARLY)	FIRST-YEAR SAVINGS ($)*	LIFETIME SAVINGS ($) (10 YEARS)*	SIMPLE PAYBACK OF ADDED COSTS (YEARS)*
Central (19,750)	7.5	8	33	20	800	8–60	151	3–4
Central (19,750)	7.5	13	446	40	800	57.90	1,016.1	5–6
Room (7,000)	6.5	7	15.40	20	700	3.50	61.50	3–4
Room (7,000)	6.5	11.6	190.00	40	700	21.60	378.50	6–7

*At electricity costs of 6.5¢/kwh in year one, with 12% price escalations in succeeding years.
Source: Derived from data collected by the Natural Resources Defense Council.

ies with many factors. If new air-conditioning systems are to be purchased, the (sometimes) higher costs of more efficient systems are more than made up for in utility bill savings. (See Table 9.4.) In most cases, however, replacing air conditioners that are still in good condition is not economically worthwhile.

The economics of the experimental ACES system have been calculated by researchers at Princeton University and Oak Ridge National Laboratories.[25] At baseline costs of 6¢/kwh for the electric-resistance system, it is cheaper to install the ACES system than to purchase additional electricity for the resistance system. However, the air-to-air heat pump is the better investment still on a cents-per-kilowatt-hour basis, though the pump saves far less energy than the ACES system (2,600 versus 4,300 to 6,900 kwh).

The economics of fossil-fueled equipment that can replace electrical vapor-compression units varies greatly. For example, Allied Chemical Corporation and Phillips Engineering are developing a gas-fired absorption heat pump with a predicted COP of 1.2 for heating and 0.5 for cooling.[26] Since natural gas currently costs about one third the price of electricity on a dollar per Btu basis, such a heat pump could deliver energy at costs equivalent to an electric heat pump with a heating COP of 3.6 and a cooling COP of 1.5. However, the relative advantages of gas-fired heat pumps may decline as gas prices are deregulated.

Costs for newer technologies are far less competitive. Rankine-cycle engines alone cost between $900 and $1,600/kw in 1978 dollars. Even if the costs of smaller (5–10 kw) Rankine units decline as predicted to $200–$300/kw (.20–.30 cents/watt) with mass production, the costs for compression-cooling (the other "half" of the Rankine chiller) must still be added.[27,28]

City Applications

Nearly all city buildings can economically accommodate some improvements to both shells and cooling systems. While shell improvements are likely to be proportionally more effective in smaller buildings, they still tend to affect heating-energy savings more than cooling-energy savings. Opportunities for improving cooling systems are increasing. As noted earlier, over 75 percent of new residences in warmer areas will most likely be mechanically cooled. Even in cooler areas, new townhouse units or multifamily low-rise buildings will probably incorporate proportionately more mechanical cooling than will single-family homes. This is true for cities like both Greenville and Boulder, although the size of their respective cooling loads certainly differs.[29] Then too, growing or redeveloping downtown areas can expect increases in office building construction with consequent increases in total cooling loads.

Institutional Constraints and Socio-economic Considerations

Unlike reductions in heating-energy demands, reductions in cooling-energy needs directly affect the public utility serving the city. Summer-peaking utilities will experience a seasonal production reprieve, and cooling-load reductions and system efficiency improvements will also help shave daily peaks. Consequently, utility reactions will, most likely, be determined by the utility's overall fuel supply and capacity picture. (See Chapter 5.) Where utilities operate with little excess capacity, the conservation of cooling energy should dovetail well with utility efforts at peak-load management. But where excess capacity is ample or plant construction under way, conservation efforts may not be well received.

Relatively few low-income persons live in buildings that are centrally cooled.[30] Most of those that do probably live in apartment buildings where landlords can make energy-conservation retrofits. Most mechanically cooled low-income housing probably has window units, often purchased second-hand.[31] Consequently, a city government's options for reducing the impacts of cooling costs on low-income people are extremely limited. Weatherizing shells could help cut cooling demands somewhat, but few other cooling-conservation programs directly benefit poor people. Similarly, the most direct job benefits for chronically unemployed persons are likely to arise from weatherization programs, rather than from programs for improving cooling systems.

SOLAR COOLING

Technical Approaches and Issues

In general, solar cooling techniques and technologies are not as well-developed as those for heating. But numerous exciting passive solar designs can potentially replace some mechanical cooling energy in new and retrofitted buildings. Active solar collectors can be used to run absorption chillers or, in more experimental applications, to run heat engines in compression-cooling systems.

Passive Systems.   As noted earlier, all passive solar systems must include some means of preventing heat gain. (See Chapter 1.) In hot climates, designers must pay special attention to using overhangs, reflective blinds and shades, vents in collector glazing, strategically placed vegetation, and earth-berming. (See Chapter 8.)

Perhaps the simplest technique for removing heat is through ventilation. To maximize ventilation in new buildings, the structure should be designed to take advantage of prevailing wind patterns. In addition, solar heat can be collected, absorbed, and transferred in ways that enhance air convection. For example, one multifamily retrofit project in Boston uses part of a Trombe wall to trap heat on summer days. At night, vents from the glazing to the outside and from the building interior to the sun wall are opened, so hot air flows to the outside. Opening windows on the building's north side draws cooler night air through the entire structure. Average temperature swings between daytime and nighttime are 16°F—adequate for natural ventilation. A building's designers should expect room remperatures to stay within 4°F or so of those that window air conditioners would supply.[32] Natural ventilation can also be facilitated by interior atriums that can be opened to the outside by roof temperatures. An enhancement of the "whole-house" fan concept, the design facilitates the natural upward movement of air—the "chimney effect."

Another way to remove heat from buildings via passive solar design is to absorb hot interior air and then discharge it outdoors—the passive heating cycle in reverse. Hot interior air is collected and absorbed in storage materials and either discharged immediately or else stored. Whether discharge is immediate or delayed depends on the temperature difference between the storage materials and the heat sink. For example, "ground-coupled" mass-storage could discharge heat immediately because ground temperatures are generally lower than room and storage temperatures. However, designs that release heat to the outside air will usually need to delay heat dumping until nighttime, when outside temperatures fall.

Ground-coupling can be accomplished in a number of ways. In slab-on-grade construction, the mass floor-slab can simply function as the storage-dissipation material. In buildings with basements, the mass storage in walls or floors will need to be thermally connected to the foundation. Not surprisingly, this technique works best where ground temperatures are lower than ambient temperatures.

A radically different way to use the ground as a heat sink is to employ so-called "cooling tubes." One basic tube design draws outside air through a buried run of tubing made of aluminum or some other metal. As air moves through the tubes, it cools by losing heat to the ground. Air-flow through the tubes can be induced by natural ventilation or by a small attic fan. In either case, cooled air exiting the tubes replaces heated air, since ground-coupling and ventilation techniques are used in tandem. Other tube designs could feature "closed" tubular loops, which could circulate return air from the building, or a transfer fluid that could absorb heat from the building and discharge it to the earth.

The most common method of "air-coupling" uses the roof's materials to store and dissipate ambient heat. Sometimes the approach is to use a "roof pond" or water-storage technique. In such air-coupled cooling systems, bags of water are placed atop a reinforced roof and covered with a movable shading device that prevents the water from absorbing too much outside heat. Heat from the building's interior flows naturally to the bags by convection and conduction. At night, the movable shade is removed and the absorbed heat is reradiated to the outside air. In winter, the system works in reverse, discharging absorbed heat into the house at night.

Since the roof's heat-storage and -dissipating functions often occur when differences between inside ambient temperature, the outside ambient temperature, and the storage temperature are relatively small, heat absorption areas must be as large as possible. Accordingly, some designs increase the surface area by allowing water or air to circulate around the building shell, absorbing heat from walls, as well as from the interior roof surface.[33]

In multistory buildings, using the roof to dissipate heat will not work because only the top floor would be cooled. Yet, some experimental designs can make air-coupled heat dissipation possible in large buildings. In one such design, a heat-dissipating sill would be installed above each floor. As hot air in the room rises, it is vented to sills where heat is lost to the air by radiation and evaporative cooling (from a built-in water pipe). As air in the sill cools, it convects to the bottom of the sill, and from there flows through ductwork under the floor to an air vent that opens to the room. This process continues until a sensor closes the sill vent.

Another approach is to use innovative thermal mass construction to maximize heat absorption. For instance, precast hollowcore concrete can be used to absorb building heat during the day and to release it at night. The next day, thermal absorption begins again.[34]

Both absorption and compression-chillers cool water that in turn cools a building through either a hydronic or forced-air distribution system. In some climates, passive heat transfer can be used to help make ice in winter that can then be stored until summer, when it can be used to chill water. One such chiller design calls for the use of heat pipes that will allow natural heat-transfer from a heavily insulated water bin under or near a building until the water freezes. In summer, the ice serves as a capacious heat sink to chill a transfer fluid that circulates through the building.[35] A more ambitious "hybrid" approach, developed by the Princeton Energy Group, is to use snow-making equipment or atomizing spray nozzles to produce ice by exposing water to cold winter air. Stored in an insulated reservoir until summer, the ice can then help meet cooling loads: as it melts, it is circulated through pipes to a liquid-to-air heat exchanger that chills the air in a conventional forced-air distribution system. The ice supply perpetuates itself as returning water is sprayed over the remaining ice mass. Some prototypical ice-making cooling systems have predicted COPs of up to 23.1.[36]

In general, techniques for reducing heat gain are effective if properly designed and built. True, overheating in summer is more than an occasional problem even in well-designed passive buildings, as attested to by technical evaluations from the HUD Residential Solar Demonstration Program and building surveys of passive solar home occupants done by the Memphremagog Group in Vermont.[37,38] Yet, most such problems can be easily remedied or controlled in a well-designed system.[39]

Of the passive designs for removing heat, those relying on ventilation and direct dissipation by ground- or air-coupling seem most promising. Direct ground-coupling or air-dissipation systems seem most appropriate for new buildings, while ventilation techniques also hold appeal for use in retrofits.

The performance of all of these techniques in hot, humid climates is a subject of some disagreement. Both ground-coupling and air-coupling seem to hold less promise where diurnal temperature swings are slight, and ventilation is less effective if incoming air contains large amounts of latent heat. Yet, reducing infiltration to allow for less moisture build-up, using dessicant material to help dry incoming air, or using energy-efficient dehumidifiers are all solutions to the humidity problem.

Active Systems. Active solar cooling systems can be used effectively in far fewer settings than passive cooling systems. Confined mostly to larger buildings, active cooling may nonetheless work better than passive cooling in hot, humid climates.

Most active cooling systems use solar heat to help drive conventional cooling processes that remove both sensible and latent heat from conditioned areas, while at least one more experimental collector is designed to produce chilled water directly in the collector itself.

The active solar cooling applications that are closest to commercial readiness are solar absorption chilling systems. In these systems, high-temperature collectors (either troughs or evacuated tubes) supply some or all of the thermal energy needed to raise the absorbent/refrigerant solution to working temperatures—200 to 300° F.

At least two major HVAC equipment companies, ARKLA and Carrier, have made commitments to producing solar-powered absorption chillers.[40] Also, numerous absorption systems have already been installed, a few in residential buildings and more in smaller or larger commercial facilities. In colder climates (such as Minneapolis and Keystone, S.D.), these systems also provide space heating.

Active high-temperature collectors can also provide some of the thermal energy needed to run heat engines that, in turn, generate electricity to run compression chillers. The heat collectors transform the pressurized working fluid into a vapor that can be used to produce mechanical power for electric generation. (See Chapter 11.) Some innovative Rankine chillers may be able to efficiently use lower-temperature heat in this process.[41]

A final approach to active solar cooling is to use active collectors to help dry dessicants in dessicant cooling systems. (After the dessicants absorb moisture from return air, they must be dried before the cycle can begin again.) Currently, a gas- or oil-furnace is used to dry dessicants, but solar heat could be used instead.

Since most dessicant systems for residential and commercial uses are still being pioneered, solar-assisted dessicant systems will not be commercially ready for some time.[42]

All these active approaches combine active solar collectors with conventional absorption, compression, or dessicant-cooling equipment. A much more innovative design still in the demonstration stage produces chilled water in the collector itself. The Zeropower® system makes use of a fine-grained mineral called zeolite, which gives off refrigerant vapor when it is heated and absorbs it when cooled and can thus heat water in a collector during the day and cool it at night.[43] This unique system requires no additional chillers or cooling towers. Also, because the system does not need to produce the high temperatures that absorption chillers or heat engines need, the collector area can be smaller (as much as 75 percent, the manufacturer claims).

More generally, active cooling systems are not as developed or as reliable as active heating systems. Because all cooling applications require higher temperatures than solar heating, problems with heat loss from collectors, transport, and storage are more serious in cooling systems. Worries about the durability of collector materials under high temperatures also abound. In the case of Rankine cooling, the nonsolar chiller system (engine plus compression-chiller) needs further development to bring down costs and help compensate for the relatively high cost of the solar energy system. (Another improvement would be the addition of a "cold storage tank" to solar cooling systems.) Yet, active solar cooling applications have met with some success. A solar-assisted absorption-chiller is meeting the cooling energy needs of a 1,000 ft^2 college bookstore complex at the University of Minnesota.

Economics of Solar Cooling

Relatively little information on the economics of solar cooling systems exist, partly because separating cooling-related costs and performance from the system's overall costs and performance can be difficult. A few passive systems, however, have been monitored specifically for cooling performance. One new house in Starkville, Mississippi, uses special window designs, overhangs, and extra insulation to cut mechanical cooling energy needs by 30 to 50 percent.

The solar features added 3 to 5 percent (or $1,200) to the total costs of the home and stand to save between $1,400 and $2,100 over the 10-year life of the air-conditioning system, with simple payback occurring in 5 to 7 years.[44]

More strictly "solar cooling" systems are those using active collectors to run absorption-chillers. In some applications, these systems provide both space cooling and space heating. However, in warmer areas some such systems are used primarily for cooling, so their costs and performance levels can, in theory, be isolated. Currently, small (3- to 5-ton) absorption systems that provide both space cooling and heating cost an estimated $30,000 to $35,000 installed. A typical 25-ton absorption unit costs nearly $100,000 installed.[45]

If any generalization is possible, it is that passive cooling features are likely to be much more economical than active cooling systems for the foreseeable future. Costs for more elaborate ice-making systems remain unknown, although some reasonable cost data will soon be available.

City Applications

Passive cooling features can be included in almost any passive solar heating system, though the need for extensive structural adjustments will limit the feasibility of many passive cooling designs in retrofits. For example, a "roof pond" or even some of the innovative structural materials discussed above will probably be confined to new buildings. Most passive cooling retrofits will probably rely on techniques for reducing heat gain, although some can include fairly sophisticated mechanisms for removing heat. While ice ponds could be integrated with existing building cooling systems, land availability may limit retrofit potential in dense areas.

While active absorption of compression-cooling systems can theoretically be used in single-family applications, for the near future, these systems will be installed in larger buildings. However, even in smaller residential structures, more collector area will usually be necessary because of the higher temperatures needed. Limits on space and solar access may restrict use of these active solar cooling technologies to relatively low-density areas or to community-scale systems.

Institutional Constraints and Socio-economic Considerations

Passive cooling applications will likely raise the same land-use issues as passive solar heating systems do. (See Chapter 8.) As for active cooling systems, they will likely raise more concerns than active heating systems because they require larger collector areas.

Public utility reactions to solar cooling will depend largely upon how much back-up power is needed and when. If solar cooling equipment or designs reduce the total need for electrical energy for cooling but not the demand for back-up cooling electricity at peak periods, utilities may levy additional demand charges on solar users. (See Chapter 5.) Owners of passive systems may be slightly more vulnerable to this risk than owners of active systems, since active systems deliver the bulk of their output in the daytime or early evening, when peak electrical demands are highest. However, passive systems relying on such principles as nighttime ventilation may not be able to reduce peak-load consumption much.

Some building owners or occupants may be willing to pay higher unit prices for infrequently used back-up power. But this arrangement could work against the adoption of more expensive conservation or solar techniques that would reduce the need for off-peak electric cooling. A plausible solution here is simply not to rely on electricity for back-up. In some passive applications, no mechanical air conditioning may be needed. In active solar-absorption combinations, auxiliary energy can be obtained simply from the back-up gas boiler.

Potential conflicts with utilities could be minimized if off-peak storage devices were included in solar-cooling applications. For example, in active absorption systems, off-peak electricity could be used to charge the solar storage tank at night to keep storage temperatures hot enough to run absorption units during the day. The appeal of this approach would rest on the costs of off-peak electricity. Economy could also be improved by using a heat pump to heat the solar storage. Conversely, electric chillers could be run in off-peak hours to chill water that could then be stored for use in tandem with solar cooling the next day. In passive systems, the back-up cooling system could be operated at night, allowing the thermal mass to absorb the cool air and reradiate it the next day.

Passive solar cooling features can help to reduce electricity bills or (where mechanical cooling is not used) to help low-income persons stay comfortable and healthy. A HUD-funded passive retrofit in Charlotte, North Carolina, eliminated the need for mechanical cooling and thus helped low-income tenants realize substantial utility bill savings.[46] Even where the passive cooling output is not large, poor people, especially older folk, can benefit palpably by adding passive cooling additions to their homes.

10. Community Heating and Cooling Systems

District-heating systems produce heat centrally, then distribute it to individual buildings through a piping network. Sources of district heating include cogenerated heat from power plants, waste heat from industrial processes, heat produced expressly for district heating by heat-only boilers, heat from municipal solid waste-to-energy systems, small-scale cogeneration systems, renewable sources of district heat, or geothermal energy.

Most of these "community-scale" systems serve large-scale central business districts. Yet, some smaller-scale heating systems are also in use, and some hospitals, campuses, and shopping centers are using community heating or combined heating-and-power plants.[1]

Community-scale heating systems are being examined with growing optimism for several reasons. First, because urban heating and cooling demands in cities are highly concentrated, costs per delivered Btu could be relatively low. It may be more energy efficient and more economical to provide heat from a highly efficient centralized source than to rely on individual on-site heat-only systems. Second, district heating systems using cogeneration can provide a combined heat-and-power package that is more energy efficient overall than centralized power-only plants and on-site fossil-fueled heat-only systems. (See Table 10.1) Finally, district-heating plants can lessen dependence on oil and gas by substituting heat from coal-fired power generating plants or centralized boilers.

Table 10.1. Relative Fuel-Input Efficiencies of Electric-Only vs. Cogeneration Power Plants.

POWER PLANT	FUEL INPUT FOR ELECTRICITY PRODUCTION	% FUEL INPUT FOR THERMAL ENERGY	% FUEL INPUT EXPELLED AS WASTE HEAT
Electric-only	33	0	67
Cogeneration	25	55	20

Source: Oak Ridge National Labs, *District Heating/Cogeneration Application Studies for Minneapolis/St. Paul Area.*

For all their potential flexibility and economy, however, few district-heating systems in the United States are currently operating successfully. Some of the newer and smaller-scale integrated energy systems installed in institutional or business complexes show promise of success. But most large-scale urban district-heating projects are losing both money and customers.

Most of these systems are old—built before World War II. Although initially economical, they lost out as oil and gas prices declined, as more sophisticated on-site heating systems were developed, as population density fell, and as power plants became larger and less likely to be built within city limits. While interest in upgrading, expanding, and building systems has increased along with oil and gas prices, cities with existing systems have an inefficient and outmoded technology on their hands. These old systems all distribute district heat with steam. Most district-heating systems in Northern Europe— where the technology is advanced and widely used—distribute heat through water. Such systems can move heat more efficiently and over longer distances, and the costs of water piping per foot of trench are lower, so systems are more cost-effective and appeal to more customers.[2]

In Minnesota, a proposed retrofit of St. Paul's district-heating system illustrates the technical state of the art in this country.[3] The system will use cogenerated heat from a power plant to provide 250°F water to buildings in the city's central business district. Using 250°F water instead of steam will allow the system to cogenerate heat at less cost because the lower distribution temperatures will mean less reduction in the plant's electrical generating capacity. Moreover, fitted steam pipes can be replaced by lower-cost prefabricated, polyurethane insulated pipes. Buildings with steam-distribution systems will need to be retrofitted for hydronic operation, but costs can be reduced if the steam piping already in place is used and if heat loads are reduced through end-use conservation. Buildings using hot-water or hot-air distribution systems can be retrofit somewhat more cheaply, simply by adding heat exchangers that connect the building's system with district heating piping. Yet, the total price tag for St. Paul's system is estimated at $77 million (1980), less than $15 million of which is for heat sources. Piping costs of the St. Paul system range from $95 to $930 per foot of trench, and average $200.[4]

A final problem with conventional district heating concerns its impact on end-use conservation efforts. A district-heating system financed and built without prior end-use conservation could damage both the attractiveness of conservation initiatives undertaken after the installation of district heating and the long-term stability of the district-heating system itself. With such a system, customers who make subsequent energy conservation improvements could end up paying higher unit costs, just as some electric utility customers who conserve must pay higher unit costs if they diminish the utility's load factor without contributing to shaved peaks. (See Chapter 11.) OTA calculated that if 40 percent of a district-heating system's customers took strict end-use conservation measures, total capital costs per delivered Btu would increase by about 40 percent.[5]

As the experiences of many European cities attest, district-heating systems can nevertheless be both energy-efficient and profitable. Europe's successes can be attributed partly to municipal ownership.[6] Another factor is strict building energy-conservation codes, which minimize potential conflicts between district-heating system requirements and on-site energy conservation. Last and perhaps most important, European district-heating technology is making great strides in efficiency and economy. The use of innovative distribution materials and designs could cut costs from $700 to $1000 per foot of trench to less than $100 per foot. Smaller-scale cogeneration systems can make the production of heat and power on a block- or neighborhood-scale a real option, and techniques are being developed to store collected heat or cold over whole seasons.

Of course, how these various technical innovations affect the economics of any particular system will vary with the heat sources available, the heat distribution systems in place in buildings to be served, and the ease of laying distribution pipes of varying sizes. Using low-temperature heat decreases some distribution and building connection costs, but increases others. Where net distribution and connection costs are higher, they may in some cases be offset by cheaper heat. In others, the use of a slightly higher-temperature heat source may allow somewhat lower distribution and connection costs. And in some instances, extensive end-use conservation may make any district-heating system uneconomical.

The evaluation of proposed district-heating systems requires analyzing their three principal components: (1) end-use systems, (2) distribution systems, and (3) heat sources—both separately and together. Most analyses of district heating systems consider heat sources first, then examine distribution and end-use systems. But, the efficiency and economy of most systems depend on making end-use and distribution systems as cost-effective as possible.

THE EFFICIENT USE OF DISTRICT HEAT

Although many Swedish district-heating systems distribute hot water at 250°F, properly built or retrofitted buildings can use much lower temperatures: 100 to 120°F in hydronic distribution systems or 90°F in a hot-air system. In existing buildings that retain their internal heating systems, distribution temperatures will have to be higher, 120 to 140°F in hydronic systems and 110 to 120°F in forced-air systems. While actual minimum levels will depend on a number of building-specific factors, virtually all retrofitted buildings will be able to use lower temperatures than those that conventional hot-water district-heating systems provide.

Equipping buildings to use district heat efficiently requires three actions. First, end-use conservation is necessary to reduce heating loads. Second, internal heat distribution systems need to be adapted to use lower temperature heat to maximum advantage. Third, the interconnection between the building and its district-heating system must be designed so that hot water is returned at the lowest temperature technically and economically practical.

Adapting Internal Distribution Systems

Technically and economically, it is quite easy to design new buildings to make effective use of low-temperature hot water. Newer types of hot-water radiators have surface areas and heat transfer capacities far above conventional baseboard hot-water radiators, and cast-iron radiators from older buildings can be used very efficiently in low-temperature hot water district-heating systems.

The use of such energy-conservation measures has important implications for the retrofit of internal building distribution systems for low-temperature district heating. After conservation measures have been implemented, building "internals" sized to meet preconservation loads will be larger than needed to meet post-conservation demands. In such oversized systems, lower temperatures can be used once conservation measures have been taken.

If existing buildings use hydronic distribution, conversion to low-temperature district heating is relatively simple technically. But in the U.S., many types of heat distribution systems are in use: hydronic, forced-hot-air, one- and two-way steam, and electric-resistance. (See Chapter 8.) Of these, forced-hot-air and two-way steam are most easily retrofitted for hot-water district heating.

To retrofit a forced-hot-air system is to do nothing more than add a liquid-to-air heat exchanger to connect the district-heating system's distribution pipes with the forced-air ductwork. However, the size of the heat exchanger needed will increase as district heating supply temperatures drop.

The retrofit of two-way steam systems to hot water is not technically difficult or costly if existing piping and radiators are sound. Usually, the only additions needed are a hot-water valve and a key air vent to allow air to bleed from the system, though steam vents must be eliminated in some cases.[7] Using hot water instead of steam in an existing system lowers maintenance costs and allows a more even flow of heat through the system.

Converting one-way steam heating systems to hot water poses much greater technical and economic problems. Radiators must be replaced, hot-water return pipes installed, and both hot-water valves and air vents added.

Relatively few city buildings use electric-resistance heat. But those that do can be retrofitted for hot-water district heating only at great difficulty and pro-hibitive costs. The electric-resistance system must be replaced with one compatible with the hot water supplied by district heating. In most cases, this means a hydronic system since it requires no heat exchanger and allows for direct connection. This extensive retrofit is justified only in those rare instances where the delivered costs of low-temperature district heat (including costs of end-use conservation) are very low as compared to electric-resistance heating.

Interconnecting Buildings and District Heat

Numerous techniques have been developed for interconnecting buildings and district-heating systems. Which is best depends on the kind of internal distribution system in place and on the "match" between the quality (temperature and pressure) of heat that the building needs and the quantity of heat that the district heating system supplies.

Conventional district-heating systems in Sweden use heat exchangers for interconnections because of the great differences between the temperature and pressure of district-heating hot water (250°F) and the distribution temperatures and pressures required in the building (164°F).[8] Heat exchangers have also been necessary where land elevation varies wildly or where high-rise buildings are plentiful (since both increase static pressure).

Relatively efficient, heat-exchange systems still lose heat as they transfer it from the district-heating system to the building. Also, as hot-water distribution temperatures decrease, heat-exchange surface area (and costs) go up. This is a problem particularly when district heat from hot-water distribution must be transferred to a hot-air distribution system inside the building. According to one study, the required area for water-to-air heat exchangers increases 51 percent if the water temperature is reduced from 160 to 120°F and 21 percent with a temperature drop to 140°F.

One alternative to heat exchangers is the direct connection of the district-heating system to building distribution systems. Thermodynamic efficiency is greater, capital costs are less, and heat losses in the distribution system are less. The direct-connection system also needs significantly less water over time than systems employing heat exchangers, so less electricity is needed for pumping power. Two other advantages are that corrosion is easier to control in water pipes in direct connection systems, and the speeds of distribution pumps can fluctuate with demand.

Compared with retrofits involving systems using high supply-and-return temperatures, indirect connection, and heat exchangers, a low-temperature distribution system retrofit interconnected directly can save 47 to 53 percent in capital costs for internal building distribution systems for district heating.[9]

However, direct connection can only be used with hydronic systems and in district-heating service areas where the water and pressure in the distribution system are relatively low.

One interconnection method currently under development would make low-temperature, hot-water district heat and existing forced-air building distribution systems more compatible. A three-stage heat pump would boost temperatures of 120°F hot water to levels that can be used in hydronic, modified steam, or forced-air systems. Automatic controls would regulate the temperature of the water or air in the building's distribution system: all of the relatively little heat needed on mild days would be supplied from the district heating system's hot water via the heat exchanger, while on colder days the control system would turn the heat pump on little by little starting at a pre-set minimum temperature.

A district-heating system combining the use of cogeneration, the production of low-temperature hot water at a maximum of 120°F, and multistage heat pumps at the point of end-use can be ultra-efficient. Depending on building efficiency and the heat distribution system's ability to use low-temperature hot water or hot air efficiently, buildings could achieve COPs (co-efficients of performance) of over 20. Of course, the overall district-heating system COP would be much lower since many buildings require higher supply temperatures than 120°F and thus use more operational energy to run heat pumps.[10]

Heat pumps can be used to solve some particularly difficult problems encountered in adapting low-temperature hot water district-heating technology to American buildings. If older buildings with hydronic, forced-hot-air, or two-way steam systems are connected to hot water district-heating systems without significant end-use conservation or heating-system modifications, heat pumps are the only alternative to higher distribution temperatures. Studies of English buildings that retained their internal distribution systems after directly connecting to hot-water district heating indicated that it would be prohibitively expensive to reduce supply temperatures below 158°F, even if energy prices tripled over their 1980 level.[11]

Solar District Heating—End-Use Implications

End-uses optimized for low-temperature heat are hallmarks of most new Swedish solar district-heating systems. Why is plain: solar collectors perform best at low collection temperatures; interseasonal heat stores hold heat most efficiently when lower-temperature hot water is stored; and heat losses in distribution systems are minimal when end-users need only low-temperature hot water.

Most buildings serviced by new Swedish solar district heating systems are designed to use hot water at very low temperatures. The first prototype Swedish solar district-heating system at Studsvik supplies an office building that uses hot water at 86°F via a water-to-air heat exchanger. The heat exchanger meets all the building's requirements, even though the differences between the distribution supply and return temperatures are only slightly greater than the temperatures required by the building. Within the building, warm air is distributed through hollow-joist flooring with inlets to rooms along the external walls. About half the heat is supplied from convection through the floor. The rest is supplied by ducts along the external wall.[12] More recent Swedish solar district-heating systems supply warmer water to end-users.

Domestic Hot Water From District Heat—End-Use Implications

Many district-heating systems that employ direct connections or heat exchangers utilize instantaneous electric domestic hot-water heaters. (See Chapter 7.) Systems with electric heaters do not have to provide domestic hot water when space-heating demands are negligible. Nor is there need for a separate distribution pipe for domestic hot-water distribution.

Of the district-heating systems that do heat domestic hot water, many use an instantaneous water heater connected via a heat exchanger to the district-heating system or to the building's space-heat distribution system. This method is effective, but large quantities of water from the main distribution system are needed to meet maximum demand. An alternative is to endow domestic hot-water heaters with enough hot-water storage capacity to level out the extreme peaks and valleys in domestic hot-water demand curves.

Numerous types of storage water heaters can be employed in a low-temperature, hot-water district-heating system using direct connection for space heating. Domestic water heaters are not, however, directly connected to the district heating system since the corrosion inhibitors in district heating water are toxic. Domestic water heaters with storage have a heat exchanger through which district heating water passes, a thermostatic valve that controls the flow of the district heating water, and a return temperature limitor that ensures that the district heating water deposits as much heat as possible before being returned.[13]

Domestic water can also be heated via a heat exchanger interfacing with the space-heat return pipe. The advantage here is that return temperatures can be greatly reduced when demand for domestic hot water is high. Since peak demand for domestic hot water is so much higher than peak space-heat demand, ample hot water storage allows a decrease in the peak demand and an increase in the base-load demand for domestic hot water. In systems where domestic hot-water requirements are met by return space-heating water, storage also contributes to relatively even return temperatures, though in summer most such systems are shut down.

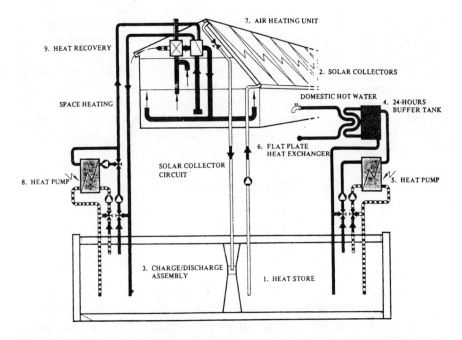

SCHEMATIC OF SOLAR DISTRICT HEATING FACILITY in Lambohov, Sweden. Water
is pumped from the heat store (1) to the solar collectors (2), where it is heated to about 7° to 8°
C before returning to the charge/discharge assembly (3) in the heat store. The charge/discharge
assembly ensures that the returning heated water enters the tank at the correct level for its temperature. The temperature difference between the top and the bottom of the tank is greatest in
November, falling to a minimum in February, while the actual temperature at the top of the
tank varies from 20° C in March to 70° C in August. The solar collectors are drained whenever
there is a risk of freezing. Water for heating the houses and for producing domestic hot water is
always taken from the top of the heat store. Water for heating domestic hot water is stored in
two 16-m³ buffer tanks (4) intended to even out demand peaks during the day. When the temperature of the water obtained from the heat store drops below 55° C, a heat pump (5) automatically cuts in to raise the temperature of the water passing from the heat store to the buffer
tanks. The temperature at the top of the buffer tanks must never fall below 55° C. Water from
the buffer tanks transfers its heat to the domestic hot water circuit in a flat plate heat exchanger
(6), through the secondary side of which the domestic cold water supply passes to be heated. The
heat exchanger has a rating of 465 kW. For space heating, hot water from the heat store heats
a circulating warm-air system in each house through a local heat exchanger. When the temperature at the top of the heat store exceeds the minimum temperature necessary for space heating,
the water is pumped directly to the warm air units (7) in the house. When the water temperature
falls below this level, a heat pump system (8) is engaged so that the supply temperature is raised
to the necessary level. Each house also contains a heat recovery unit (9), where heat from the
exhaust air preheats the incoming fresh air. Ventilation can be controlled as required by the
occupants. *(Courtesy of Solvarmecentral Lambohov, Sweden)*

During summer, instantaneous demand gas-fired or electric-resistance domestic water heaters can be used. Since domestic hot-water demand declines appreciably in summer, energy consumption for seasonal off-peak domestic water heating is relatively small. Still, the ideal would be to meet this demand via the district-heating distribution system, which is idle during summer if it is not also supplying space cooling.

District Cooling—End-Use Implications

District-cooling systems cannot compete economically with ultra-efficient on-site chillers unless building conversion costs are kept as low as possible. Most cost-reduction techniques are still developing. Yet, direct connection of district cooling offers economic and thermodynamic advantages. The system needs no heat exchanger, which is expensive and carries a thermodynamic penalty. Also, the capital, operation, and maintenance costs of absorption- or compressor-chillers are high, so if a building can use cold water directly, no capital need be expended on on-site air-conditioning equipment.

Differences in heating and cooling distribution system capacity requirements can be minimized with the use of on-site storage. In a two-pipe system, cooling can use the same storage unit that holds domestic hot water during the winter, though additional storage capacity may have to be added since storage requirements for chilled water are likely to exceed domestic hot-water storage capacities. On-site thermal storage would be particularly important if a full rnage of services was supplied from a four-pipe or pulsing distribution system.

DISTRICT-HEATING DISTRIBUTION SYSTEMS

The most expensive component of any district heating system is the distribution network. In older steam systems and in higher-temperature hot-water systems, distribution costs can account for up to 75 percent of total system costs. (In the St. Paul system, estimated capital costs for distribution are $22.2 million for the first phase.[14]) Currently, most analysts assume that high "heat densities" alone are the key to reducing distribution costs. But, lowering district heating supply temperatures can also reduce the costs of distribution system materials and installation. Adopting innovative routing practices and installation techniques can significantly lower capital costs and the costs to consumers.

Types of Distribution Systems

Steam Distribution Systems.   Although the first district-heating systems used steam, steam-based systems are inferior to contemporary alternatives because the distribution range of steam is usually three miles or less. In cogeneration, steam is usually extracted from a high-pressure turbine or from the cross-over point between a high- and low-pressure turbine, so the turbine does not generate all the electricity it could and steam heat thus costs commensurately more. Steam distribution systems also require elaborate, heavily insulated piping systems made of stainless steel and other expensive materials. Steam pipes require steam traps, inclined piping, manholes, and many expensive expansion joints to absorb the relatively high thermal stresses that the low-pressure steam imposes.

Two other problems are corrosion and inefficiency. Steam ruins pipe insulation much faster than either moderate or low-temperature water systems do—a factor that contributes to heat losses equalling 15 to 45 percent, as well as to maintenance costs. Steam also corrodes both steel and stainless steel pipes. Although piping, boiler, and maintenance personnel codes are more stringent for high-temperature steam systems, even high standards do not prevent equipment failure or alleviate the problems of operating inefficiencies under part-load conditions.[15]

Particularly inefficient are low-pressure steam systems that neither cogenerate nor return condensate. Cogeneration and condensate return can boost efficiencies (from lows of 30 percent) to 60 percent. But electrical efficiency is still low compared to cogeneration systems that produce hot water.

Steam district heating should be used only where steam is the only medium that can provide the required services: in some industrial-process heating applications, or in light industrial and commercial applications where switching from steam to hot-water distribution is prohibitively difficult and costly.[16]

Hot-Water Distribution Systems.   Hot-water district-heating technology has been developing steadily since World War II ended. The last decade saw breakthroughs in distribution systems, and even more exciting changes are in the offing. Research has focused on making the most expensive component of district-heating systems less costly, more efficient, and longer lasting.

The pace of innovation in hot-water district heating increased considerably when fabricated pipe-in-pipe was developed in the late 1940s. Since the mid-1960s, pipe-in-pipes with plastic/polyurethane external waterproof jackets have been extensively marketed. Typically, a conventional hot-water district-heating system has one pipe for supplying hot water to buildings and another for returning water to the heat source, where it is reheated and then recirculated.[17] In the two-pipe configuration, the system supplies domestic hot water inefficiently in summer because the pipes are sized to provide space heating in the winter. One way around this problem is to install hot-water heaters in individual buildings and to shut down the district-heating system when space-heating demand falls off. But even this method is less than efficient.

In a three-pipe system, two pipes are supply pipes—one for space, the other for domestic water heating. The temperature in the space-heating pipe varies with space-heating needs. In the other pipe, water is at a constant temperature year-round. This configuration allows a relatively low volume of moderate-temperature hot water to be supplied throughout the year in pipes sized to meet annual peak demand for domestic hot water. As for the larger volumes needed for space heating, the system can meet them and then shut down when there is no demand. The drawbacks of three-pipe systems are that they waste hot water because they have no return pipe, and they have had performance problems when hot-water demand falls.

The four-pipe system represented a further refinement. Here, two pipes each are used for space heating and domestic hot-water supply and return. The two-, three-, and four-pipe configurations can each be combined in a single large prefabricated insulated pipe-in-pipe for higher pipe quality and lower installation costs. While multiple pipe systems have been too expensive to use on a large scale, they might become more attractive as fuel costs escalate, especially in multithermal (heating, cooling, refrigeration) district systems.

Low-Temperature Hot-Water Systems. The costs of district-heating distribution systems can also be cut by reducing hot-water supply temperatures. For example, although many Swedish hot-water district-heating systems supply water at maximum temperatures of 250°F, end-use conservation in conjunction with adjustments in internal building-heat distribution systems can reduce required supply temperatures to 120°F in new or extensively rehabilitated buildings and to 140°F (for hydronic systems and even lower for hot-air systems) in existing buildings.

Most new Swedish solar district-heating systems also supply hot water at low temperatures to optimize solar collector efficiency. The prototype Swedish solar district-heating system at Studsvik supplies 86°F hot water to an office block equipped with a water-to-air heat exchanger and a forced-hot-air heating system able to use very low temperature heat.

Conventional Swedish district-heating systems operate with a maximum supply temperature of 248°F and a minimum return temperature of 112°F. In contrast, some Danish district-heating systems have much lower supply temperatures so as to obtain higher levels of electricity output in cogeneration. In Danish systems, the district-heating system and individual building (hydronic) heat distribution systems are directly connected, so the thermodynamic and economic penalties associated with the use of heat exchangers are eliminated. Heat from the Odense (Denmark) district-heating system costs about one third as much as space heat from on-site oil-fired, heat-only boilers.

Innovative Distribution Techniques

New Piping Technologies.   Of the numerous recent advances in distribution piping technologies and techniques, most involve the substitution of nonmetallic for carbon steel pipe. Fiberglass-reinforced plastics (FRP), cross-linked polyethylene, polybutylene, and concrete pipes have all been used as replacements. While FRP carrier pipes cost more than the equivalent in carbon steel, the installed costs are 10 to 50 percent lower due to other factors: savings in labor for handling and assembling the lightweight FRP piping; reduction in the insulation thickness required; reduction in the number of expansion devices needed; elimination of the need for a groundwater drainage system; and some reduction in trenching cost.[18]

Polybutylene pipes are flexible and up to 90 percent lighter than steel. They also resist most organic solvents, salts, acids, and alkalines even at elevated temperatures. But since polybutylene is partially soluble in aromatic and chlorinated hydrocarbon solvents above 140°F, polybutylene pipe is recommended for use only in some secondary low-temperature district-heating systems, pending further testing.

A number of pipes made of stressed concrete, polymer concrete, and asbestos cement have also been developed. The first two, relatively large, are used mainly in long distance heat-transmission systems. They are particularly effective in transmitting fluids, such as geothermal brine, that would erode steel pipes. They can also withstand relatively high pressures and temperatures, though concrete is particularly sensitive to temperature shock. While both types of piping cost less and perform better than the carbon steel alternative, both need further testing. The asbestos cement pipe also resists corrosion and chemical attack, but it is brittle and subject to rupture from severe temperature shock. Its cost is only about three-quarters that of the carbon steel alternative, but it too requires additional testing before it can be employed widely.

Along with the FRP pipe, perhaps the most promising recent innovation in district-heating piping technology has been the cross-linked polyethylene pipe used for years to distribute heat inside buildings. This piping has also recently been used underground in district-heating networks. Cross-linked polyethylene pipes are corrosion-resistant and strong, and less pumping power and no expansion devices are needed when these lightweight pipes are used. As for major disadvantages, operating temperatures cannot surpass 200°F; the pipes deteriorate when exposed to direct sunlight; and oxygen diffusion in the pipe could corrode the fittings. Although cost reductions of up to 50 percent over conventional steel conduits may be expected with cross-linked polyethylene piping, further development and tests are needed.

Low-Cost Adaptations of Existing Technology. Some conventional technologies could also be adapted to use low-temperature district heat. A unique approach to district heating involves a low-cost method of retrofitting power plants so that they can become "cogenerators" and thus provide 120°F hot water with low fuel costs. Sewer pipe, costing one fourth of what insulated steel pipe-in-pipe costs, would be laid in sand trenches below the frost line. More sand—a good, low-cost insulator—would cover the pipe. The system would carry water at temperatures no higher than 120°F, so heat losses would be minimal.

Lower-temperature hot water also requires a higher flow rate than a higher-temperature system does. Without some system modifications, the pumping power requirements of very low temperature systems might thus increase electricity use unacceptably. However, larger pipes would allow higher flow rates per unit of pumping energy expended.

Reducing Heat Losses. Combining lower-temperature distribution with improved piping designs reduces heat losses. Annual heat loss declines approximately 16 percent when the supply water temperature is reduced from 160° to 140°F, approximately 30 percent if to 120°F. Heat loss increases about 11 percent if 160°F water is raised to 180°F and about 24 percent if to 200°F.[19]

Reducing heat loss both saves fuel dollars and improves the viability of district heating in lower-density areas. Although the annual distribution system losses amount to no more than about 5 percent in a conventional hot-water district-heating system serving high-density areas, distribution heat losses increase as densities decrease and as buildings become more efficient.

Reducing heat losses in these situations might require adopting an alternative piping system. If two feeder pipe trenches were extended under sidewalks on the edges of a central area, the feeder-to-house connection distance would be reduced from 80 to 30 ft. Besides saving capital, this technique could also reduce distribution system heat losses to 5 percent in thermally typical buildings and to 10.5 percent in moderately conserving buildings.[20]

Another way to reduce heat loss is to super-insulate pipes. The key to knowing how much to insulate is knowing the value of the heat being transported: high-cost heat requires more insulation than low-cost heat, and supply pipes distributing higher-temperature heat to end-users require more insulation than return pipes do.

Maximizing Differences Between Supply and Return Temperatures.
Increasing the temperature differences between supply and return pipes also
optimizes system efficiency. With lower-temperature return water, pumping
energy needs are reduced and heat losses are cut. Dropping return-water temperatures can also make it possible to use much less expensive piping materials
than those needed to transport higher-temperature return water and much less
insulation. Since overall system efficiency is also enhanced by lower supply
temperatures, achieving the maximum temperature differences between supply
and return water hinges on minimizing return water temperature—basically,
a matter of end-use efficiency.

Another way to lower return temperatures would be to use the return water
to meet low-level heating needs. For example, a heat pump connected to secondary distribution circuits could be used to supply heat to greenhouses in a
district-heating system, as proposed in Denmark.[21] Other useful applications
for low-temperature return water include aquaculture, cropdrying, or the
drying of biomass or municipal solid-waste fuels.

Planning District-Heating Distribution

Thermal Load Density.   In planning a district-heating system, one of the first
steps is to find out whether heat demands are sufficient to pay back investments
in the system. An area's thermal load density is determined by the thermal
load per unit of building floor space, the number of building stories, and the
number of buildings in the area to be served. When heat demands in all potential district-heating service areas are mapped, these areas are then classified
according to their heat density—the maximum heat demand the district-heating system can meet. Even residential areas with relatively low thermal
demands per unit of floor space can have high thermal load densities if buildings are multistoried and close together. In contrast, some industrial districts
have high thermal load densities, even though many buildings are single-story
and far apart. In cities, heat density is usually highest in the central business
district.

In engineering, heat density is expressed in terms of Mw/square km or
MMBtu/h/acre, which indicates the gross quantity of heat demand for a given
area. But energy quality or the temperature of the heat required is not measured this way. Therefore, heat mapping should also include assessments of the
minimum temperature required by each building after all economic investments in increased efficiency have been assumed.

District-heating systems should be planned and designed to account for the
amount of heat needed, the types of energy systems currently in place, and the
temperature levels of existing demands. At the same time, additional maps

should indicate the impact of various energy-conservation measures on both the quantity and quality of heat demand. Conservation measures should be ranked (least costly first), and the combined effects of several measures should be gauged. Such analyses will help planners determine optimal levels of investment in end-use efficiency: the most cost-effective quantity and temperature for supply and return water, the right size pipes, the most appropriate piping materials, the best transmission and distribution system routes, the required capacity of heat sources, the best location for the heat sources, and the temperature of the hot-water output from the heat sources.

Although only areas with the highest heat requirements can make economic use of district heating, different assumptions and analytical techniques yield different conclusions. As Table 10.2 indicates, downtown highrise and multistory commercial and residential areas are prime candidates. Even so, the costs of installing district-heating distribution systems are high. A veritable tangle of pipes and wires rests beneath an urban street. Lines for gas, water, sanitary and storm sewers, telephones, electricity, subways, abandoned utilities, and so on pose very difficult challenges for routing and installing district-heating pipes. Equipment could be relocated, but only at great cost, and excavation can take place only when traffic is light—the very times workers must be paid premium wages. Indeed, some studies demonstrate that piping costs double when such interferences are included.[22]

Table 10.2. Heat Density as an Indicator of District Heating Potential.

HEAT DENSITY	AREA	CATEGORY
70 (Mw/km^2) over 0.97 (MMBtu/h/acre)	Downtown high rises	Very Favorable
70–50 0.97–0.70	Downtown multi-storied buildings	Favorable
80–20 0.70–0.28	City core— commercial buildings and multi-family apartment buildings	Possible
Less than: 12 0.17	One-family houses	Not Possible

Source: Wahlman (1978).

Distribution Systems Design Alternatives.   Although such extreme difficulties with installation of district-heating systems are less typical of areas outside central business districts, excavation should always be avoided if possible. One less disruptive alternative is tunneling through the bedrock underneath the city and driving shafts vertically to connect urban blocks or groups of blocks. Pipes are then run through the basements of attached buildings to supply buildings with space- and water-heating.

Alternatively, block-, multiblock-, or neighborhood-scale district-heating systems with basement-pipe routing might be designed to interconnect with a large district-heating system later. Supplying a group of "small-time" heat consumers with a single connection in effect makes the group a large user of heat: it can reduce distribution system costs enough to offset the higher costs of district heating in areas with low-heat density or difficult distribution-system installation conditions.

Another alternative is to locate distribution pipes under the sidewalks rather than in the middle of the street. (Work time can thus be cut anywhere from 33 to 42 percent.)

District Heating in Areas of Low-Heat Density.   If distribution-system capital and installation costs are reduced, distribution systems can tolerate greatly decreased heat densities and still be economically attractive. Even areas that are not densely developed may be feasible for some form of district heating if the population is concentrated along major traffic arteries and utility corridors.[23] A bonus of using district heating in such pockets of heat density is that substreets are less likely to be cluttered and commerce less likely to be disrupted. A drawback, on the other hand, is that in low-density areas, each delivery unit may have to be connected to the district-heating system.

"Greenfield" Sites.   Another potentially excellent site for a district-heating and cooling system is the "greenfield" site—any site for which new, large-scale residential, commercial or industrial development or thoroughgoing redevelopment is planned. At such sites, district-heating mains can be installed at the same time as other services, so installation costs are low. In addition, both new and rehabilitated buildings can be designed to use energy efficiently and to tap into the district-heating system from the outset. Further, if energy cascading techniques can be employed in industrial developments, investments in efficiency and heat supply systems are optimal and affordable.

Greenfield sites are also attractive as sites for solar district-heating and cooling systems. Besides the advantages already noted, the relatively large land areas required by solar collectors and by heat- and cold-storage facilities are easier to acquire in developing than in developed areas.

Facilitating High Rates of Interconnection. One factor crucial to the financial success of a district-heating system is interconnecting a high proportion of consumers to the system. (In Sweden, law requires consumers to interconnect with the system, though district-heating costs are notably lower than any alternative anyway.)

Where interconnection rates are low, the utility will have to charge connected parties relatively more to cover the underutilized system's high fixed costs. In the worst case, declines in interconnection as a result of rising heat prices could cause a district-heating system to fail financially.

In the United States, planners of district-heating systems have sought ways to ensure high rates of interconnection. Long-term "take-or-pay" contractual relationships between the district-heating utility and consumers could ensure sufficiently high rates, though take-or-pay contracts hold little appeal unless appreciable investments have been undertaken in increased building efficiency, delivered thermal services cost less than conventional forms of energy, and project cost-overruns appear unlikely.[24] Thus, rather than compelling consumption, district-heating planners should encourage the application of technical innovations that could significantly reduce district-heating capital and fuel costs.

One factor that both encourages high initial rates of interconnection and jeopardizes those rates over the long term is inefficient end-use. Ignoring end-use efficiency will intensify heat density and perhaps even lower initial consumer costs. But, unless long-term take-or-pay contractual agreements are in effect, consumers may later invest in more efficient on-site conservation devices and disconnect or reduce demand. If, in turn, district-heating rates of return decline, utilities could raise rates and bring about additional disconnections or cutbacks in demand—essentially, a "no win" situation. The way around this problem is to determine the optimal level of investment in building conservation and to provide consumers with access to affordable financing for these measures—in effect, considering building conservation, modification of internal building distribution systems, interconnection to the district-heating system, and the district-heating system itself as components of a single energy strategy.

Increasing District-Heating Load Factors. A district-heating system's load factor is the ratio of the amount of energy demanded annually to the system's peak capacity. (See Table 10.3.) Obviously, the ideal city from a district-heating perspective has long cold winters and cool summers. But nearly all northerly U.S. cities have space-cooling requirements, and most southern U.S. cities spend more energy on cooling than on heating. Thus, nearly every urban electric utility in the U.S. experiences a summer electrical-demand peak because of air conditioning.

Table 10.3.   Climatic Determinants of Heating Loads for Selected Cities.

REGION CITY	ANNUAL HEATING DEGREE DAYS	HEATING DESIGN DAY (97.5%) (°F)	DESIGN DAY	"LOAD FACTOR"	LATITUDE
Northeast					
Boston	5621	9°	56	.28	42°20′
New York	4828	15°	50	.27	40°45′
Philadelphia	4865	14°	51	.26	40°00′
Averages	5105	13°	52	.27	41°02′
North Central					
Chicago	6497	−4°	69	.26	41°55′
Cleveland	6154	5°	60	.23	41°30′
Detroit	6228	6°	59	.29	42°20′
Milwaukee	7444	−4°	69	.30	43°00′
Minn.-St. Paul	8159	−12°	77	.29	45°00′
St. Louis	4750	6°	59	.22	38°30′
Averages	6539	−0.5°	65.5	.27	42°02′
West					
Denver	6016	1°	64	.26	39°45′
Los Angeles	1819	40°	25	.20	34°03′
Phoenix	1352	31°	34	.14	33°30′
Seattle	5183	26°	39	.36	47°35′
Averages	3643	24.5°	40.5	.24	38°44′
South					
Baltimore	4729	13°	52	.23	39°20′
Dallas	2382	22°	43	.13	32°50′
Houston	1434	33°	32	.12	29°45′
Memphis	3227	18°	47	.19	33°05′
Miami	206	47°	18	.03	25°55′
Washington	4211	14°	51	.23	38°50′
Averages	2698	24.5°	40.5	.16	33°37′

Also worth noting is that in most northern urban areas, heating and cooling loads with end-use conservation are higher together than heating loads alone prior to conservation. Thus, providing cooling services can both increase the distribution system's post-conservation load factor and cool buildings efficiently.

One method of supplying district cooling is to supply hot water to on-site absorption chillers. However, this would require a relatively high-temperature distribution system since most conventional (single-effect, lithium-bromide) absorption chillers operate at heat-source tempeatures above 185°F. Another alternative would be to produce chilled water at a central location and distribute it to buildings for direct use. But unless the differences between supply and return temperatures were relatively even for heating and cooling distribution temperatures, the pipes would have to be oversized for the heating season in order to provide enough peak cooling.

If on-site chillers become extremely efficient, conventional district cooling may not be able to compete with them on its own. Therefore, district-cooling systems may have to make use of inexpensive cooling sources or provide additional services such as refrigeration. In a "full service" approach, district systems could employ a four-pipe system to iron out peak supply and return temperature differences and to equalize demand levels. Cold and hot water would seasonally alternate in the same supply and return pipes, while smaller supply and return pipes would distribute hot and cold water in alternate seasons for domestic hot water and refrigeration. In summer, space cooling and refrigeration could be supplied from the larger pipe and hot water from the smaller one. In winter, space and water heating could be provided by the larger pipe and refrigeration by the smaller. Another possibility would be using a two-pipe system that would alternate the supply of district heating in winter with district cooling in summer. Domestic hot water could then be efficiently supplied in the summer from on-site equipment such as solar collectors.

In any attempt to deliver additional service, one overriding fact needs attention: the fewer thermal services supplied, the lower the distribution-system capacity factor. If one of the primary reasons for providing additional thermal services is to increase the distribution system's capacity factor so as to maximize investments, the capital costs of adding each service will have to be carefully analyzed.

SOURCES OF DISTRICT HEAT

Given the relatively high cost of distribution systems and building conversion in conventional district-heating systems, the costs of the heat-sources must be kept as low as possible by using existing sources of waste heat, new heat sources with low capital costs, and new heat sources that will use low-cost fuels. Of course, the type of heat source used will affect not only costs related to heat sources, but also the capital and operating costs of the entire system. Using low-temperature heat sources can be particularly helpful because many lower-temperature heat sources have lower capital costs than higher-temperature options and because lower-temperature heat can often be supplied by waste heat from power plants or industries. Even when newly constructed heat sources are used, lower-temperature options are often less expensive. Then, too, the use of lower temperatures can facilitate many distribution system economies. Finally, where cogeneration is used to provide heat, providing lower-temperature heat can make it possible to produce relatively more electricity.

Cogeneration

Both in this country and in Europe, one of the most common sources of district heat is cogeneration—the production of both heat and power in a single energy-conversion system. Cogenerating both power and heat is significantly more efficient and economical than producing each separately.

Cogeneration systems can be topping cycles or bottoming cycles. In the more commonly employed topping cycle, electricity is generated first, and then heat in the exhaust stream is made available at various pressures and temperatures. In a bottoming cycle, thermal energy is produced for process uses, and relatively high-temperature and high-pressure waste heat is then recovered and used to generate electricity.[25]

The most commonly employed topping-cycle technology used to supply district heat is the steam turbine. Although steam turbines have lower electricity-to-thermal ratios than other heat engines such as the combined-cycle gas turbine or diesel, they can utilize a wide variety of fuels, including oil, gas, coal, wood, and wastes. Since most gas turbines and diesels now available run on natural gas or refined oils, steam turbines are thus likely to continue to be extensively used in cogeneration.

Two main types of steam turbines are used in cogeneration. Backpressure steam turbines expand steam through all stages of the turbine, exhausting steam at the temperatures and pressures required by an industrial process or district-heating system. The amount of heat produced in a backpressure turbine is inversely related to the amount of electricity generated. Therefore, to obtain exhaust temperatures useful for district heating, some power output is sacrificed. Still, the combined thermal and electrical efficiency of the backpressure turbine can be very high, especially if the heat outputs are closely matched to industrial or district-heating loads. The electrical and thermal output of backpressure turbines cannot fluctuate with varying loads, so these devices are usually used to meet the base heat-load requirements of either an industry or a district-heating system. In such a set-up, electricity production cannot be increased as heating loads drop, but both heat supply and electrical capacity can be increased significantly if sufficient thermal storage is provided (for instance, about five hours in a large urban district-heating system in Northern Europe.)[26]

The inflexibility of these units during periods of low heating or cooling requirements (spring and fall) does limit their potential as base-load electrical generators. Indeed, backpressure turbines are most appropriate for use in industrial environments where heat and electric demands are daily and seasonally constant. However, since backpressure steam turbines supply both electricity and heat efficiently and economically when operated at full loads, modifying rather than rejecting the technology might be worthwhile.

The second type of turbine commonly used in cogeneration is the extraction-condensing turbine (also called the condensing or intermediate take-off turbine). In this turbine, steam is bled off during the intermediate stages of the turbine cycle and is then either passed through heat exchangers to produce hot water or depressurized to supply consumer demands. Compared with a backpressure set, this type of turbine is easier to control and more flexible. If extra power is required for electricity generation during hours or seasons when heat requirements are minimal, an extraction-condensing unit can be precisely regulated to produce little or no heat while generating electricity at either partial or full capacity. If thermal storage is adequate, the heat supply can be reduced and electrical capacity increased while still meeting the requirements of the district-heating network. An extraction-condensing plant with thermal storage can meet up to 85 percent of a district-heating system's peak heating load and also generate electricity year round.

Although a new extraction-condensing cogeneration plant is significantly more expensive than a plant with the equivalent backpressure capacity, its much higher diurnal and seasonal electricity capacity factors compensate. In retrofits, however, condensing turbines for cogeneration are less economical because it is difficult to vary the steam-extraction, and retrofitting permanently derates electrical capacity to some extent.

Efficiencies of Cogenerating Steam Turbines. In terms of cogeneration/district-heating systems, efficiency is a relative concept. Since the efficiency of a single operation can be examined only as one component of an overall system with a number of interacting operations, it means little to say that a cogeneration unit has an "efficiency" of 85 percent. The key factors are the relative sizes of power and heat outputs.

In a cogenerating steam turbine, the cost of thermal energy is directly related to the cost of electrical energy: both utilize the same fuel for production, and both are outputs of a single process. Since electrical output decreases in inverse proportion to the temperature of the heat needed, the cost of thermal energy includes the cost of "sacrificed" electricity.[46] As for the overall efficiency of a steam turbine, it does not change appreciably as long as low-temperature hot water (under 212°F) is being produced. But the amount of heat obtained for each unit of electricity lost varies considerably with the water's output temperature—from 6.12 units of heat for every unit of electricity sacrificed if the water is at 216°F to 27.80 units if the water temperature is 120°F.

So if the goal is to increase electrical-generation efficiencies and reduce heat-energy costs, district-heating output temperatures should be kept as low as possible.

Retrofitting for Cogeneration.   Since many American cities with a potential for district heating have older, small-to-medium-sized straight-condensing turbines located relatively near high urban heating loads, these units have attracted attention as sources of district heat. Few such units are among the most important in an electric utility's capacity, so using them in cogeneration need not jeopardize electrical output.[27]

The appeal of retrofits stems partly from an analysis of utility capacity. Many utilities in the Northeastern and North Central states face the prospect of excess capacity while electricity growth rates remain relatively constant. In these regions, building new cogenerating facilities makes little sense. (See Chapter 5.) Then too, conversion typically costs only 35 to 40 percent of new construction.

The most common conversion is changing a turbine from a straight condensing unit to a backpressure unit. When several low-pressure turbine stages are removed, the exhaust pressure increases enough to heat the district heating water. This involves the loss of some electric-generating capacity during periods of low heat demand, but simplicity and low cost make this approach appealing nevertheless.

Converting Turbines to Low-Temperature Cogenerators.   Another method of converting steam turbines to cogenerators appears both simple and inexpensive. In order to regulate the power plant's flow of cooling water to produce a supply of low-temperature ($104°$ to $120°F$) water for district heating, a shutoff valve in the water discharge system is used to divert the entire flow through a central valve and forwarding pump. A forwarding-pump control is used to vary water flow so that changes in the electrical load on the plant will not change the outlet temperature of the cooling water. The cost of converting a conventional steam turbine to this type of low-temperature cogeneration has been estimated at about $25 per kW. However, using this system for district heating would require some low-temperature distribution and end-use modifications.

Coal-Fired Generation.   Since steam-turbine cogeneration units can burn alternative fuels and oil is no longer a bargain, steam turbines are economical sources of base-load district heat only when converted to cogenerators that run on alternative fuels. Converting oil-fired power stations to coal can be difficult. Many urban power stations that used to burn coal converted to oil in order to comply with environmental regulations. Switching back would require extensive modification of air-pollution control equipment. Even for new coal-fired plants on open sites, the capital costs of pollution-control equipment can be high, and trouble-free operation is by no means assured.

One alternative fuel that an oil-fired power station may adapt relatively easily is coal-oil slurry—a mixture of about 40 percent pulverized coal and about 50 percent heavy fuel oil. Many oil-fired boilers can burn it without extensive boiler retrofits, especially if such boilers were initially designed to burn coal. Technical, economic, and environmental considerations permitting, the use of slurry-burning boilers could enable small-to-medium-sized urban cogenerating power stations to supply district heat at affordable prices.

If the oil in the slurry mix later becomes unacceptably expensive, a later conversion to fluidized-bed boilers may be in order. In fluidized-beds, finely ground particles of inert matter (such as residual ash from coal combustion, sand, graded limestone, or dolomite) absorb sulfur dioxide during combustion on a base surface impervious to air. Air at high pressure and velocity is blown upward through the particle bed so that the particles remain suspended, behaving much like violently boiling liquid. Fuel is injected into this turbulent inert matter, which is preheated by gas. Combustion is rapid and efficient, and heat transfer (through tubes immersed in the bed) is much more efficient than that in conventional boilers, and these compact systems are relatively easy to integrate with steam-turbine cogeneration capacity. Also important, sulfur dioxide emissions can be reduced by 80 to 90 percent if fluidized-beds replace coal-fired boilers.

So far, coal-burning, fluidized-bed units are currently operating only at Georgetown University in Washington, D.C., and in Enkoping, Sweden. But small-to-medium-sized units should be commercially available within the next few years, and pressurized fluidized-beds may be available within a decade.

Small-to-Medium Cogeneration Technologies. Cogeneration technologies using oil or natural gas will become ever less attractive as fuel prices increase. But their use may be economical over the short term. Relatively small-sized units—which are highly efficient—could supply both heat and power to commercial, institutional, or larger residential complexes. (See Table 10.4.)

Although only natural gas or highly refined petroleum fuels can be used in gas turbines, gasifiers can be placed in combination with gas turbines or dual fuel engines so that coal, wood, peat, agricultural waste, and urban waste can be used. While gasification entails some loss of efficiency, most losses can be recovered if the low-temperature heat is used in a district-heating system, or if it is used to dry the solid fuel prior to gasification.

Only small gasifiers are easy to buy in the United States. But small-scale gasifiers are being developed for diesel power plants so that these plants can

Table 10.4.   Characteristics of Small-to Medium-Sized Cogeneration Plants.

TURBINE	OUTPUT RANGE (Mwe)	PEAK ELECTRICAL EFFICIENCY (%)	PEAK OVERALL EFFICIENCY (%)	POWER-TO-HEAT RATIO AT 80°C (176°F)	POWER-TO-HEAT RATIO AT 160°C (288°F)	PEAK-LOAD EFFICIENCY
Backpressure	1–125		83–92	0.4–0.6	0.27–0.40	Good
Uncontrolled Extraction	5–300	33–40	up to 89	0.4–0.6	0.27–0.40	Good
Controlled Extraction	60–700	28–41	up to 85	0.4–0.6	0.27–0.40	Good
Open-Cycle Gas Turbines	1–100	22–32	65–83	0.4–0.7	0.4–0.7	Poor
Combined-Cycle Gas-Steam	6–300	33–45	65–83	0.6–1.1	0.6–1.1	Good
Diesel or Gas Engines	0.1–30	38–40	60–80	Diesel Oil 0.9–1.1 / Gas 0.6–0.8	Diesel Oil 0.9–1.1 / Gas 0.6–0.8	Very Good

Source: Robinson, Peter J., Transmission and Distribution Networks and The Consumer—The Potential for Development in W. R. H. Orchard and A. F. C. Sherratt, Combined Heat and Power: Whole City Heating—Planning Tomorrow's Energy Economy, Halsted Press, Division of John Wiley and Sons, N.Y., 1980, p. 188.

use gasified peat, wood, agricultural waste, and coal. In cogeneration, these diesels can obtain efficiency levels over 80 percent if used to supply a low-temperature district-heating system. In electricity production, they frequently achieve over 40 percent electric generation efficiency prior to any waste-heat recovery. At the same time, the capital cost of diesels is relatively low—about $500 to $800 per kw(e).

Overall, diesels are very reliable engines (particularly those of the large, heavy, slow-speed, marine variety) and easy both to buy and maintain. Nitrogen oxide emissions resulting from fuel combustion do continue to pose problems, however. Yet, in district heating salvation may consist in the fact that while NOx levels would rise greatly near the cogeneration site, total air pollution levels in the area served by the diesel cogeneration district-heating system would fall if individual, heat-only boilers were replaced by this system.

Thermal Storage. Using heat-only boilers to meet peak heating loads over the long term is inefficient and expensive in most applications. An increasingly plausible alternative is using thermal storage—especially with low-temperature district-heating systems—to increase the base-load unit's contribution to peak heating requirements and to increase its electricity capacity rating. Storage could also increase the proportion of heat supplied by the base-load cogeneration unit, thus decreasing the requirements for peak-load boilers.

If hot-water storage is to contribute significantly to the peak-load contribution from heat-only boilers, storage facilities must be located near heat loads. One option is to use large water-storage tanks at atmospheric pressure that would be charged by the base-load cogeneration unit when demand is low. When demands peak, some of the hot water could be supplied to the system.

Another option for meeting peak daily heating demands mostly through base-load capacity would be to combine various scales of thermal storage with variable-speed pumping capacity. Instead of merely increasing hot-water temperatures in response to peak demand, water-flow velocities could be controlled to deliver hot water at relatively constant temperatures and at appreciably greater flow rates. Heat pumps could also be an economical source of peak heat supply.

Given the alternatives, oil-fired, heat-only boilers should be used only where they have already been installed in buildings slated for district heating hookup. There, they could meet system peak-load requirements for individual buildings while the system is developing. But if larger-scale boilers are needed to provide transitional hot-water supplies during system development, small fluidized-bed boilers capable of using coal or urban waste could make a more economical and efficient choice.

District Heating Using Industrial Waste Heat

Residential and commercial areas near industrial facilities could use the excess heat from industrial processes as a source of district heat. Industrial cogeneration and industrial-process heating can be "cascaded" so that energy of progressively lower quality is matched with industrial processes requiring progressively lower temperatures.

This strategy can be extended beyond the boundaries of the industrial plant to the community, with the low-temperature district-heating system at the bottom of the energy cascade. While the efficiency and economy of this system depends largely on the distribution system's ability to utilize lower-temperature hot water effectively, industrial waste heat is at least potentially a very inexpensive source of district heat.

The contribution of low-temperature industrial waste heat to a district-heating system can be optimized in various ways. Since light industrial, commercial, and residential users' temperature demands vary both diurnally and seasonally, thermal storage facilities can maximize the contribution of industrial surplus heat. Another option is using heat pumps to boost low-temperature heat to levels adequate for space- and domestic water-heating. Here, the closer the temperature of the surplus heat source and the temperature required by the district heating system, the better that system efficiency and economy will be.[28]

Using low-temperature industrial surplus heat for district heating also increases the cost-effectiveness of more intensive and costly industrial energy-conservation measures. The possibility of earning substantial profits from selling low-temperature surplus heat to a district-heating system should prompt industry to invest in process-energy conservation.

Implementing extensive industrial-conservation measures before providing waste heat for district heating is also essential to assuring the long-term availability of a heat source. Reducing either the quality or the quantity of available waste heat once the system is in place could seriously imperil the technical and economic viability of the district-heating system. Unless industrial processes employ cost-effective conservation at the outset, energy price increases could stimulate conservation investments that might reduce the level of waste heat supplied to the district-heating system. Thus, industrial energy conservation and cogeneration dovetails with commercial and residential district heating. (See Chapter 13.)

To help use excess heat to meet heating needs, energy-intensive industries should locate close to other industrial producers or consumers of waste heat.[29] Cogeneration capacity can then be pooled, and energy of varying quality can be cascaded among different industries and processes. In Denmark, energy-intensive industries have already been encouraged and subsidized to locate near communities with district-heating systems that can effectively utilize industrial surplus heat.[30] In general, the availability of lower-cost energy could be a powerful advantage for communities that want to attract new plants or encourage local industries to remain in the community or to expand production.

Municipal Solid-Waste Systems as a Source of District Heat

With landfill areas becoming scarcer and environmental controls more costly to meet, some cities have built plants to recover resources—and marketable energy—from municipal solid wastes. While nearly all existing or planned municipal solid-waste-to-energy systems in the United States produce either steam or fuel for sale to utilities or industries, only a few supply urban areas with district heat.

Some "resource-recovery" systems have as their first order of business the reclamation of metals or glass; others, energy production; and still others, both. Most energy resource-recovery plants coming on-line recover at least some metal and glass at the plant, but none has been built in conjunction with a good source-separation program that would both maximize recyclables and boost the waste stream's homogeneity.

If recycling is not made part of every resource-recovery plan and plant, future recycling efforts will be stifled because the energy value of garbage will serve as an incentive to waste. Alternatively, oversized resource-recovery plants could become noncompetitive when recycling diminishes the waste stream. Indeed, successful "source-separation" programs can make "garbage-to-energy" projects uneconomical.[31]

Of the 57 municipal solid-waste-to-energy facilities operating or under construction in 1981, only seven were connected to district-heating or cooling systems. In part, this is because in most cities the waste stream cannot meet a large base-heating load. As OTA notes, only 23 cities produce over 1,000 tons of refuse per day, and even 1,000 tons of daily refuse would generate only about 700,000 MMBtu of heat in a heating season.[32] Then too, many cities prefer to sell the products of resource-recovery facilities to selected local industries that can help to maintain the tax base and provide jobs.[33]

These limitations aside, energy from municipal solid-waste facilities could meet a portion of a large city's need for baseload heat and supplement the contributions of cogeneration or industrial waste-heat systems. They could also be used to help supply base-heating loads for clusters of industries or businesses whose access to cheaper or more secure heat sources would help stabilize the local economy. Currently, over 250 plants in Europe are supplying district heat from incinerated urban wastes.[34] In this country, a resource-recovery plant processing 140,000 tons of refuse a year provides about 25 percent of the total heating and cooling needs of 28 commercial buildings in downtown Nashville.[35]

Energy resource-recovery systems can produce different forms of energy from municipal solid waste via different technologies. Watcrwall combustion systems use furnaces jacketed with water-filled tubes. Heat is recovered as either low- or high-temperature steam, depending on whether or not electricity is to be generated.

Widely used combustion and RDF (refuse-derived fuel) production technologies have had substantial technical problems. The pipes and boilers of incineration systems have corroded in the presence of hot flue gases produced in combustion, a problem that could be reduced if plastics were removed from the waste before incineration. Refuse-derived fuel can be difficult to store and handle. Recyclable materials can foul production or combustion equipment. And the relatively low energy content of RDF limits economies in transportation and combustion.

Still in the development stages are pyrolysis and bioconversion. Pyrolysis exposes organic matter to heat in the absence of oxygen. The process yields either liquid or gaseous fuels, depending on the specific process used. It is a potentially attractive technology because the liquid or gaseous fuels produced can be easily stored, transported, or used on-site to generate electricity or steam. Its drawbacks are high costs, complicated technology, and operating problems.[36] Bioconversion involves extracting gaseous fuel from municipal solid waste through a process similar to that now commonly used to produce sludge in many sewage plants.

Technical Concerns and System Efficiency.   One of the most significant technical problems surrounding a municipal solid-waste processing plant is oversizing or undersizing the plant relative to the size of the waste stream so that the plant is either overloaded or operating at less than full capacity. To avoid this problem, cities must compare a resource-recovery facility with all plausible alternatives: continued landfilling, incineration, and vigorous source-separation and recycling.[37]

Another design problem is redundancy of equipment, which oversizes a plant but which is needed as protection against breakdowns. Some redundancy is needed, especially with very large-scale plants, but it could be minimized by correcting other design and operational problems.

The reliability and durability of many major system components also remains questionable. Because many waste-to-energy plants use relatively new technologies or new combinations of existing technologies, equipment mismatches or control problems cause short to extensive shutdowns. In 1981, a

Table 10.5. Comparison of Energy-Recovery Efficiencies for
Various Solid-Waste Energy-Recovery Processes.

PROCESS	NET FUEL PRODUCED	TOTAL AMOUNT AVAILABLE AS STEAM
	(EXPRESSED AS PERCENT OF HEAT VALUE OF INCOMING SOLID WASTE)	
Water-Wall Combustion	—	59
Fluff RDF	70	49
Dust RDF	80	63
Wet RDF	76	48
Purox Gasifier	64	58
Monsanto Gasifier	78	42
Torrax Gasifier	84	58
Oxy Pyrolysis	28	23
Biological Gasification*		
With use of residue	29	42
Without use of residue	16	14
Brayton Cycle/combined cycle	19 plus	
Waste-Fired Gas Turbine	12 directly as electricity	

*Includes energy recovered from sewage sludge.

survey of plants indicated that 11 of 45 operational facilities were "temporarily closed."[38]

Nationally, technical problems affect system efficiency. Typically, the heat value of unprocessed municipal solid waste ranges from 3,500 to 6,500 Btu/lb. Overall, the efficiency of a waste-to-energy system is the percentage of this value ultimately converted to useful energy (generally steam). For the broad range of energy-recovery processes listed in Table 10.5, from 14 to 63 percent of the energy contained in combustible refuse can be recovered. The overall efficiency of most systems actually falls between 40 and 60 percent.

Economics. While resource-recovery plants require major capital investments, the range of capital costs per input ton per day of capacity is considerable. (See Table 10.6.) Assessing the economics of a waste-to-energy system requires considering both how much money is saved on the purchase of conventional fuels and how much money is saved on alternative means of waste disposal.

Many factors influence the costs of steam or electricity produced by a resource-recovery facility. These include the reliability of markets for metal and glass, the size of tipping fees for unloading refuse, and the cost increases

Table 10.6. Capital Costs per Input Daily Ton—Municipal Solid-Waste Plants.

Steam Systems

LOCATION	CAPACITY (TONS PER DAY)	CAPITAL COSTS (MILLIONS)	CAPITAL COSTS/ CAPACITY ($/TPD)
Batesville, Arkansas	50	1.2	24000.
North Little Rock, Arkansas	100	1.45	14500.
Osceola, Arkansas	50	1.1	22000.
Bridgeport, Connecticut	1800	53.	29500.
Windham, Connecticut†	108	4.125	38195.
Dade County, Florida†*	3000	165.	55000.
Jacksonville, Florida	40	2.8	70000.
Burley, Idaho†	50	1.5	30000.
Chicago, Illinois	1600	23.	14375.
Ames, Iowa	200	6.3	31500.
Auburn, Maine	200	3.97	19850.
Baltimore, Maryland	600	30.1	50167.
Baltimore County, Maryland	1200	8.4	70000.
Braintree, Massachusetts	250	2.8	11200.
East Bridgewater, Massachusetts	550	10–12	20000.
Saugus, Massachusetts*	1200	50.	41667.
Genesee Township, Michigan	100	2.	20000.
Duluth, Minnesota†	740	19.	25676.
Ft. Leonard Wood, Missouri	75	3.3	44000.
Durham, New Hampshire	108	3.3	30556.
Glen Cove, New York†*	225	34.	66176
Monroe County, New York	2000	62.2	31100.
Oceanside, New York	750	9	12000.
Harrisburg, Pennsylvania	720	8.3	11528.
Nashville, Tennessee	530	24.5	46226.
Hampton, Virginia (NASA)	200	10.3	51500.
Ft. Eustis, Newport News Virginia	40	1.4	35000.
Norfolk (VA) Naval Station	360	2.2 (1967)	6111.
Norfolk Naval Ship (Portsmouth)	160	4.5	28125.
Salem, Virginia	100	1.9	19000.
Tacoma, Washington	500	2.5	5000.
Madison, Wisconsin	400	2.5	6250.

†Soon to be operational.
*Steam for producing electricity.

for plant operation relative to increases in oil and gas prices.[39] Another economic variable is the possibility of offering lower rates to customers with long-term contracts to enhance the market's stability. For example, Ohio's 1,000-ton-per-day facility charges from $1.87 to $4.37 per 1,000 lb of steam, depending in part on the reliability and duration of the customer's demand.[40]

A final economic consideration is that new municipal energy-recovery projects can be economically competitive today only if they receive "special" financing. Sometimes, this translates into public ownership, especially in cities that own their public utilities. Sometimes, it means developing some creative public-private arrangements. In Saugus, Massachusetts, a privately-owned resource-recovery system was designated as real property to be taxed at full value, so the city was willing to help finance the project through tax-exempt revenue bonds. Now, the facility pays the town $1.4 million in taxes each year.[41]

Institutional Constraints. Even relatively small resource-recovery facilities require fairly large tracts of land. Small plants processing from 200 to 500 tons input per day require three to five acres of land just for plant operations, and larger plants (1,000 tons or more input per day) can require as many as 10 acres. Yet, where landfilling sites are becoming scarce, municipal solid-waste processing could reduce the net flow of material into existing landfills. Moreover, resource-recovery plants can be well integrated with centralized industrial operations, with ready markets for recovered energy and materials.[42]

Lack of utility cooperation in resource-recovery projects can be a major stumbling block. Utilities have little economic incentive to participate in resource-recovery projects since even large facilities will be unable to produce more than a small fraction of the steam that a utility needs.[43] But if a district-heating system (or processing plant) can provide an attractive primary market for energy from resource-recovery plants, utilities' unwillingness to cooperate may not be a serious problem.

From an environmental perspective, resource recovery is usually preferable to landfilling and simple incineration. If materials are reused, less land is altered by landfilling or mining. On the other hand, resource recovery is a more centralized and visible process than "source separation," so it might help institutionalize environmentalism.

The production and use of energy from garbage also entails environmental problems. Burning plastics and hazardous wastes can create unacceptable emission levels of toxins and organic hydrocarbons.[44] Supplementing coal with refuse-derived fuel in boilers can actually increase stack emissions, and corrosion can also reduce operating efficiency of pollution-control equipment.

Smaller-Scale Sources of District Heat

Most district-heating applications have been used in central business districts or large cities. But with the rise of fuel costs and the maturation of efficient small-scale cogeneration technologies, neighborhood- or block-scale district-heating systems are becoming increasingly practical.

The U.S. Department of Energy has recently supported research on the potential of relatively small cogenerating engines.[45] (See Table 10.7.) All the engines studied have combined heat and power efficiencies of 60 percent or greater, and the generation costs (cents/kwh) associated with such engines are becoming competitive with those in central generating plants. Although the advanced closed-cycle Brayton and Stirling engines cost more than currently available technologies, their potential efficiencies and lower fuel costs (for coal) will make these developing technologies increasingly competitive.

Assuming that the coal-fueled technologies utilize fluidized-bed combustion units, most give off fewer emissions than centralized electric-generating stations and on-site fossil-fueled boilers. Compared to an all-coal utility, all total energy technologies except diesels and steam turbines showed a significant reduction in overall emissions.

Other Potential Advantages.   Relatively small district-heating systems also have the potential to distribute lower-temperature heat and thus make use of relatively inexpensive pipe. Routing such pipes through the basements of build-

Table 10.7. Fuels, Efficiencies and Capital Costs of Various Cogeneration Engine Types.

		EFFICIENCY‡ (%)			CAPITAL COST[1] ($/KW)	
PRIME-MOVER	FUEL	ELECTRICAL*	RECOVERABLE HEAT	REJECT HEAT	1000 KW	3000 KW
Diesel	Oil	36	42	22	375	336
Gas Turbine (Simple)	Oil	25	45	30	519	372
Gas Turbine (Regen.)	Oil	38	22	40	735	513
Adiabatic TC Diesel	Oil	47	36	17	397	354
Stirling	Oil	34–46	41–54	12–13	420–487	373–429
Stirling	Coal	34–46	41–54	12–13	537–604	479–535
Open-Cycle Brayton	Coal	34	36†	30	814	713
Closed-Cycle Brayton	Coal	34	41†	25	574	443

Notes: 1. 1977 Prices.
*Includes Generator Efficiency of 95%.
†At 150°F exit temperature from heat recovery exchanger.
‡At lower temperatures (120°F) greater percentages of recoverable heat are possible.

ings served by block- or multi-block systems further reduces capital costs. Overall, using smaller distribution networks greatly diminishes overall thermal losses in the distribution system and facilitates the direct connection of hydronic distribution systems in buildings to district-heating systems.

Renewable-Energy Sources of District Heat

Community-scale solar energy systems have several potential advantages over on-site solar energy applications. One is economy of scale. Smaller collector and storage areas per dwelling unit are needed in larger systems, and economical and efficient storage is possible only in large-scale systems. Since storage tanks have smaller surface-to-volume areas as volume increases, larger-volume storage tanks can store more heat (a function of volume) at reduced unit costs (a function of surface area). A second advantage is that using a community-scale system eliminates the needs for properly oriented and unshaded space for collectors on each building and for installing individual storage tanks in retrofits.

Renewable-energy-based community systems can also be attractive long-term alternatives to conventional district heating. A "solar district-heating" system entails no pollution control problems: it can provide lower-temperature heat than conventional systems can, so materials and installation costs for the distribution are commensurately less; and its use is more compatible with extensive end-use conservation measures.

For all their theoretical advantages, however, serious practical concerns surround the use of solar district heat. First, economies of scale may not be great enough to offset the high purchase price of active collectors. Second, numerous technical questions need answers. Will some of the materials and installation problems experienced with on-site active solar systems recur in community-scale applications? (See Chapter 8.) Do community-scale systems that distribute heat over relatively long distances lose too much heat or use too much operating energy to be economical? Finally, can adequate space and solar access for collectors be secured?

Active Solar District Heating.   Several district-heating systems using active solar collectors are being operated in Sweden. The largest is designed to meet about 90 percent of the space-heating and water-heating requirements in a 56-house project now under construction. Theoretically, these systems work the same way that on-site active solar heating systems do. However, the collector, storage, and distribution components of community-scale systems do have special design requirements.

With only limited experience to draw on, engineers cannot yet identify a typical, let alone an optimal, active solar district-heating system. While all systems will doubtless use water (rather than air) as the transfer and storage medium, other materials and design details will vary from project to project. Some will use decentralized collectors with centralized storage, while others will use centralized collectors and decentralized storage, and still others will use both centralized collectors and storage.

In one Swedish project, all the collectors have been dispersed onto the roofs of attached buildings. But this "landless" approach is feasible only in a new community or in a neighborhood with excellent solar access. Elsewhere, collector locations will have to be centralized—a drawback in some urban areas. The amount of land area needed, however, can be minimized by reducing heating loads through end-use conservation, by using long-term storage (see below), or by using such centralized auxiliary systems as heat pumps.

Whether flat-plate metal collectors or concentrating collectors constitute the optimal choice in district-heating systems depends mostly on the design delivery temperature of the system. Where design delivery temperatures are less than, say, 130°F, it makes sense to use lower-temperature flat-plate panels. In colder climates, to avoid dissipating heat from storage water to the outside in winter and to take advantage of the 30-fold increase in sunlight in summer, interseasonal storage is needed to keep the amount of collector area reasonably low.[46] For example, a space-heating system in Seattle without long-term storage would require almost nine times the collector area as one with storage.

In Sweden, various long-term storage designs have been employed. In one project (a prototype demonstration of the interseasonal storage concept), storage consists of a pit in the earth in the shape of a truncated cone. The upper edge consists of concrete blocks added above grade level, and the pit's interior is sealed with a plastic liner and then insulated. Research conducted in both Sweden and the United States indicates that this design—which includes no separate tank or costly structural support—is probably the most cost-effective. But it is feasible only where the existing geological structure is highly stable.

An interseasonal storage system's effectiveness is also a function of the internal tank configuration. Because hot water rises to the top of the tank and cold water sinks to the bottom, the supply and return piping must enable the coldest water to circulate to collectors and the warmest water to the heating load.

One way of keeping storage and expenses to a minimum is to use solar collectors that provide hot water at the highest temperature that collectors can tolerate without performance losses. However, higher-temperature concentrating collectors cost significantly more than flat-plate models. Both storage and distribution components also become more expensive as a system's "design temperatures" increase. (An alternative to this approach is to use lower-temperature collectors and to boost storage temperatures with heat pumps.)

The Swedish systems developed so far should perform well. But their capital costs are very high. For example, the Lyckebo project costs about twice as much to build as a conventional district-heating project in Sweden—an estimated $93.00 per MMBtu, if amortized at .15 per year.[47] Costs can also be reduced by using lower-cost or higher-efficiency collectors or (primarily in new developments) by optimizing the entire solar district-heating system for low-temperature collection, storage, distribution, and end-use.

Solar Ponds. Solar district-heating systems using solar ponds as heat sources can be far less expensive than those using active collectors, partly because ponds combine collection and storage functions. A solar pond could supply district heat by circulating a transfer fluid between the pond and end-uses via pipes and heat exchangers. Solar ponds capable of reaching temperatures sufficient for electricity generation will probably be located in more southerly regions. (See Chapter 11.) But solar ponds capable of providing heat energy can operate in colder climates as well.

Numerous types of solar ponds exist. The salt-gradient pond utilizes high concentrations of salt near the bottom and a black plastic liner to collect solar radiation and store heat in the pond's lower regions. This increased density prevents the continuous convection of the warmed water from the bottom of the pond to the cooler layers near the top, so heat is transferred to the surface primarily by conduction, which is slow enough to enable the lower regions of the pond to maintain high stable temperatures of up to 212°F.

The so-called "membrane pond" contains a horizontal partition to separate the lower convecting zone from the middle nonconvecting zone. Some variations also feature a second partition slightly below the surface of the pond to minimize the depth of the surface-convecting layer.

Solar ponds require large land areas and ensured solar access. Consequently, solar ponds are not likely sources of district heat for neighborhoods in larger cities. But in smaller communities, the story is different. For example, in a Hampshire College study, a modified salt-gradient pond was selected for a proposed district-heating system in Northampton, Massachusetts, even though almost 2 percent of the city's land would be needed for the project.[48]

While solar ponds of this type can collect and store thermal energy at very low temperatures, their exposure to sunlight is restricted in winter, and they can lose large amounts of heat to the cold air if left thermally unprotected. Potentially, the large amounts of salt needed create environmental problems as well. As a result, some solar ponds have been modified in view of these problems. For example, a layer of fresh water under an insulating layer supports a shallow black-bottom pool that functions like a flat-plate collector. As water in the pool heats, it could be circulated to the pond and be replaced by cooler water that would, in turn, be heated. This pond would work better than the salt-gradient pond in colder months because both the floating lid and the plastic liner would guard against heat losses.[49]

Modified solar ponds could probably provide district heat at far less cost than active solar collectors. But salt-gradient ponds are not cheap. The proposed Northampton pond, for example, would cost an estimated $100 million—excluding land and distribution costs for some 445 acres of ponds.[50]

Natural and Waste-Water Sources. Bodies of surface, ground, or waste water could also supply district heat. Technically, they would work much the same way as solar ponds, though in most cases the temperatures of natural water resources will have to be boosted via a heat pump because the temperatures available from these sources alone would be too low.

Diesel-driven heat pumps combined with natural or waste-water heat sources have been designed to supply the base-heating load to some small Danish towns. By one calculation, a fossil-fueled heat pump could boost the temperatures of a seawater heat source enough to supply district heat to 2,000 homes using about half the fossil fuel a heat-only boiler system would require. In another design, heat from sewage water combined with recaptured waste heat from the diesel-powered heat pump supplies about 35 percent of a small town's heat-energy demand.[51]

Since more than 60 percent of the land area (and 75 percent of the population) in the continental United States is reasonably close to aquifers, these underground stores of water could also be used for district heat. Indeed, a University of Alabama research team has proposed a community heating and cooling system combining an aquifer heat source with temperature-boosting heat pumps in each building in the system.[52]

Ice Ponds for District Cooling. The ice pond concept being demonstrated in Princeton, New Jersey, could be used to provide community-scale or on-site cooling. (See Chapter 9.) Probably the most promising immediate candidates for community-scale ice pond systems are relatively small clusters of residential or commercial buildings adjacent to land areas large enough to accomodate a reservoir. Obviously, multiple reservoirs could also be interspersed among clusters of buildings. In the future, very large ice ponds could be developed to serve large developments. At any size, ice ponds are probably most feasible in new developments or in existing ones near adequate open lands.

Biomass. Most of the known biomass-burning plants in the United States produce electricity or cogenerate electricity and process steam. However, these plants could also provide district heat if they were suitably retrofitted. (For a

Table 10.8. Western Cities of 100,000†
Population Close to Proven Hydrothermal
Resources Over 150°F.

POPULATION LEVEL	CITIES
Over 200,000	Albuquerque
	Long Beach
	Los Angeles
Circa 100,000	Boise
	Concord, California
	Fremont, California
	Huntington Beach, California
	Las Vegas
	Mesa, Arizona
	Riverside, California
	Salt Lake City
	San Bernadino
	Tacoma

Sources: Allen and Shreve (1980); SAI (1980); Struhsacker and Blackett (1981).

discussion of biomass resources in the production of electricity and industrial process steam, see Chapters 11 and 12.)

Geothermal.   Few cities are close to rich high-temperature geothermal resources. Those that are, however, could use geothermal energy for district heat, at least on a neighborhood scale. (See Table 10.8.) Boise, Idaho, is currently rehabilitating its pre-1900 district-heating system to use local hydrothermal resources. When completed, the system could serve up to 500 homes and 40 commercial buildings, as well as state office facilities—replacing 1,750 billion Btu of fossil fuels (natural gas) a year.[53] Klamath Falls, Oregon, will use hydrothermal resources in a hot-water district-heating system that will eventually heat its entire 54-block central business district.[54]

Of course, geothermal resources are ultimately exhaustible. Accordingly, geothermal-based district-heating systems could circulate the hot water back to the wells after it is passed through the heat exchangers and thus recharge the store. How effective this strategy is only experience will tell.

Socio-economic Considerations.   Some form of public support may be necessary to keep financing costs down. According to the Office of Technology Assessment, under private financing arrangements, conventional district heating is economically risky and marginal.[55]

Subsidizing a district-heating project is a substantial undertaking for cities acting alone. Usually, the sale of revenue bonds is unwise unless many customers are committed to hooking up to the system.

Customer availability may. be limited, especially when existing sources of heat (such as natural gas) cost less. Building conversion costs tend to be high, and without subsidies, they will be unthinkably so for owners in marginal financial shape.

District heating projects have often been praised as sources of jobs. But how many of those jobs would be locally based depends largely on the nature of the local labor pool. Extensive on-site conservation programs may provide more and longer-term jobs for the local population.

III. POWER AND FUELS

11. Electricity

Electricity is a versatile form of energy that can heat, cool, light, and run appliances. But in most cities, electricity costs more per Btu than either fuel oil or natural gas. (See Table 11.1.). Inherently inefficient to produce, electricity from thermal power plants is likely to remain the most expensive fuel source, too.

Fortunately, numerous technologies can be used to improve electrical efficiency on-site, thus offsetting some of the inefficiencies of centralized power stations. For example, an electric heat pump with a coefficient of performance of 2.0 is as fuel efficient as a gas furnace: it can, in other words, deliver about the same number of Btu for an equal Btu input. The efficiencies of lighting and other appliances are improving. And numerous technologies for producing power on-site are available—advanced cogeneration, fuel cells, wind-energy conversion systems, photovoltaics, and others.

Cities need to take an increasing interest in producing electricity more efficiently and economically. A recent International City Managers Association study indicated that electricity for street lighting was second only to gasoline among most cities' energy purchases.[1] Electricity costs can also affect a city's prospects for economic development because commercial and industrial sectors tend to consume proportionately more electricity than residences. Cities with large numbers of office buildings or large industrial areas may find that improving electrical efficiency will help to stabilize local consumer prices and job opportunities. In cities that are growing rapidly, improved electrical efficiency can help to shape utility peaks, thus stabilizing electricity prices and helping attract new businesses.

A third factor is that cities must now cushion the impact of rising electricity costs on ordinary citizens. Although relatively few older cities in the Northeast and Midwest make much use of electric heating, in many cities in the West and South substantial growth in electric heating in new construction is projected. For example, Portland predicts that 6,000 of 8,000 new single-family homes and 30,000 of 32,000 new apartments built between 1975 and 1995 will use electric heat.

Electricity usage for residential cooling is also likely to grow. Although the amount of energy used for cooling is still small in most cities (see Table 11.2), it is increasing with the popularity of window air-conditioning units and central

Table 11.1. Current Energy Costs by Fuel Type (Selected Cities).

CITY	NATURAL GAS		FUEL OIL		ELECTRICITY	
	¢/THERM	$/MMBTU	$/GALLON	$/MMBTU	¢/kwh	$/MMBTU
Boston	.61	6.1	1.278	9.128	.083	24.444
New York City	.74	7.4	1.28	9.143	.127	37.077
Buffalo	.51	5.1	1.291	9.222	.067	19.631
Newark	.74	7.4	1.28	9.143	.127	37.077
Pittsburgh	.41	4.1	n/a	n/a	.063	18.318
Philadelphia	.60	6.0	1.235	8.822	.077	22.631
Baltimore	.53	5.3	1.235	8.822	.068	19.906
Washington, D.C./VA	.57	5.7	1.277	9.122	.068	19.866
Atlanta	.48	4.8	n/a	n/a	.051	14.886
Miami	.55	5.5	n/a	n/a	.075	22.095
Cincinnati/KY	.43	4.3	1.226	8.757	.054	15.758
Cleveland	.41	4.1	n/a	n/a	.082	24.150
Detroit	.50	5.0	1.267	9.050	.068	19.830
Chicago/N.W. IN	.46	4.6	1.236	8.829	.082	24.168
Milwaukee	.55	5.5	1.187	8.479	.055	16.109
Minneapolis/St. Paul	.47	4.7	1.181	8.436	.059	17.294
St. Louis	.49	4.9	1.187	8.479	.056	16.365
Kansas City, MO	.38	3.8	n/a	n/a	.069	20.173
Dallas/Fort Worth	.46	4.6	n/a	n/a	.065	19.043
Houston	.47	4.7	n/a	n/a	.064	18.805
Denver	.45	4.5	n/a	n/a	.065	19.149
Los Angeles	.37	4.1	n/a	n/a	.071	20.760
San Diego	.41	4.1	n/a	n/a	.098	28.584
San Francisco/Oakland	.39	3.9	n/a	n/a	.053	19.433
Portland, OR	.59	5.9	1.170	8.357	.042	12.399
Seattle	.68	6.8	1.269	9.064	.024	6.986
Anchorage	.25	2.5	1.232	8.800	.054	15.723

Source: Bureau of Labor Statistics (1981).

Table 11.2. Residential Electricity Use (Selected Cities).

CITY	ELECTRIC ENERGY FOR SPACE HEATING		ELECTRIC ENERGY FOR WATER HEATING		COOLING		LIGHTING OR OTHER ELECTRIC	
	BBTU	% OF TOTAL RES HEATING	BBTU	% OF TOTAL DHW HEATING	BBTU	% OF RES USE	BBTU	% RES USE
Boulder	260.0	7.6	60.0	6.8	54.0	1.0	614.0	12.0
Bridgeport	513.65	1.1						
Dayton	560.67	7.0	196.31	8.6	363.2	3.0	1127.5	9.3
Dearborn	19.75	.43	22.0	3.4	55.28	.87	838.87	13.2
Greenville	59.12	6.0	118.35	58.6	116.01	7.3	240.51	15.0
Knoxville	4616.56	59.8	1487.55	87.4	365.0	3.0	1714.0	14.3
Los Angeles					2500.0	.7	5000.0	1.4
Miami	3390.0	56.3	9960.0	80.0	3247.0	45.3	15900	22.2
Portland (1971)	4704.03	15.4	4378.78	72.7	—	—	1720.9	4.0

Sources: Portland: *Energy Conservation Choices for the City of Portland (1977).*
Other Cities: *CCEMP Energy Audits (1979).*

221

air conditioning systems in new construction. (See Chapter 2.) Since many cities are served by public utilities with summer peaks, electricity for cooling usually costs more per kilowatt hour than electric heat. (See Chapter 5.)

Even in cities that expect to see relatively little new residential construction, lighting and appliances already consume relatively large amounts of electricity in homes. In both Dayton and Boulder, for example, residential lighting and appliance operations account for more than twice the electricity use that residential space- and water-heating do. Philadelphia's nonheating residential electric use is over 70 times that of its electric consumption for home heating.

Finally, cities that own their electric utilities have considerable incentive to reduce on-site electrical consumption and to develop cheaper ways to produce electricity. Municipal utilities such as Jacksonville's need to decrease their dependence on imported oil, while those like Seattle's need to consider how they can meet increasing demands without building or buying into new generating facilities. End-use electrical conservation or smaller-scale electrical production initiatives are particularly needed for municipal utilities that distribute only power generated elsewhere.

City governments can take action. Oakland cut its annual street lighting bill by over $1 million simply by replacing its less efficient light bulbs.[2] Other cities have developed and enforced building- and lighting-efficiency standards, provided incentives or information to developers, and made new subdivision ordinances and zoning provisions reflect the need for energy conservation. Efficiency standards for older buildings could also be employed, though with greater difficulty. As for appliance efficiency, information programs are the answer.

Cities can also take steps to encourage the use of on-site renewable-energy technologies that produce electricity. If electric rates are high and sunshine or wind resources abundant, incentive programs to encourage the use of photovoltaic or small wind-energy conversion systems make sense.

Of course, most cities exercise power over utility supply decisions as advocates and petitioners before the state Public Utility Commissions—not as utility owners. Cities can scrutinize utility plans for expanding conventional power production and recommend and support PUC consideration of alternative electrical conservation production programs or technologies.[3] They can also forestall or block utility discrimination against residents who become small power producers under the provisions of Section 210 of the Public Utilities Regulatory and Policy Act (PURPA), which allows owners of small cogeneration or renewable-energy electrical systems to sell power to public utilities. Most significantly, cities themselves could become power producers by developing existing hydropower sites or constructing city-owned wind-power systems or cogeneration facilities.

Some recent "supply-side" developments may also prompt utilities to welcome city participation in electrical conservation programs. Many oil-dependent utilities serving older cities must find substitute fuels or reduce demand. In high-growth areas, utilities are straining to meet increasing demands even as the costs of new thermal power plants skyrocket.

Most utility conservation programs now in effect either encourage builders (or owners) to cut electricity use in buildings or help local governments undertake larger-scale load-reduction efforts (See Table 11.3.) A few investor-owned utilities have investigated renewable technologies capable of supplying electricity to the grid. A Southern California Edison study shows that (on a life-cycle basis) wind, solar thermal-electric, hydroelectric, and geothermal systems using 1981 technology are cost-competitive with combustion turbines. Wind and hydropower can compete with coal in utility applications, and both solar thermal-electric and photovoltaic systems are expected to compete economically with combustion turbines and coal by 1981. (See Table 11.4.)

Partially as a result of its analysis, Southern California Edison (SCE) plans to include about 2,000 Mw of "renewable" electrical supply—a third of the capacity increases projected for the next decade.[4] Also in California, PG&E,

Table 11.3. Selected Utility On-Site Conservation Initiatives.

UTILITY	PROGRAMS
Pacific Gas and Electric	• Rate incentives for energy-efficient new construction. • Zero-interest loans for weatherization of existing homes in service area. • Incentive payments to communities for demand control (demonstration).
Southern California Edison	• Street-light conversion assistance. • Field tests of heat-pump water heaters. • Load management devices installed in demonstration homes.
New England Electric System	• Distribution of high-efficiency light bulbs. • Distribution of guide to energy labelling. • Demonstration field tests of interactive load-management devices. • Demonstration of off-peak heat storage in ceramic tiles. • Partial grant for attic insulation in electrically heated homes.
Arkansas Power and Light	• Encouraging load management of air conditioning.
Tennessee Valley Authority	• Rate incentives for energy-efficient new construction. • Load management of water heaters and air conditioners via demand controls. • Low-interest loans for heat pumps.

Source: William Snyder, Report on Survey of Utility Investments in Conservation and Renewable Energy Resources, Center for Renewable Resources, 1981 (unpublished).

Table 11.4. Summary of Southern California Edison Study of Current and Future Costs of Renewable Sources of Electricity.

| | APPROXIMATE ESTIMATED LIFECYCLE COSTS (1980 CENTS PER KWH) | | | |
	1981 TECHNOLOGY	1990 TECHNOLOGY	FUEL COSTS	OPERATION AND MAINTENANCE COSTS
Nuclear	8.0	8.0	1.5	1.25
Coal	13.5	13.5	4.5	2.0
Combustion Turbines	15.5	15.5	12.5	.5
Wind	14.0	12.0–10.0	-0-	.5
Solar Photovoltaic	160.0	31.5–14.5	-0-	.5
Solar Thermal	26.0	20.0–16.0	-0-	1.0
Geothermal	20.0	20.0	8	1.5
Fuel Cells	35.5	26.0–15.5	11.5	1.25
Hydro	15.0	15.0	-0-	N.A.

Source: Energy Issues . . . and Answers, Southern California Edison, Congressional Briefing, July, 1981.

owner of the Geyser Geothermal plant, is reportedly planning to retire about 1,000 Mw in the next decade and is considering using a large (151 Mw) new hydroelectric plant, a 50-Mw biomass-fueled power plant, and a pilot 2.5-Mw wind program. On the other side of the country, the New England Electric System plans to get 60 Mw of its capacity from hydro, wood, and wind power by 1996 and to purchase power from the developer of a small hydro project in Lawrence, Massachusetts. It also intends to study the feasibility of other hydro and wood-fired plants as well. Cooperating with Wind Power, Inc., New England Electric System is also installing up to 60 small wind machines that will eventually generate up to 13 million kilowatt hours annually.

In the South, Arkansas Power and Light is developing advanced cogeneration technologies.[5] Surveying its service area, the utility had identified 23 potential industrial customers, for whom it is developing a wide range of financing options, including third-party arrangements and the creation of a nonregulated subsidiary. The company will conduct a preliminary economic and engineering feasibility study on its own, and encourage the potential cogenerator to jointly finance the final engineering study if preliminary results are encouraging.

ON-SITE CONSERVATION OPTIONS

Electricity is generally considered the most elastic (price sensitive) of energy demands.[13] Electricity consumption can be reduced easily by using fewer appliances, using appliances less often, or turning off appliances and lights not in use. Consumers can also cut their electrical bills by trying to use off-peak power. (See Chapter 5.) Information programs will encourage consumers to achieve these savings, but consumer-oriented information programs need to be supplemented by equipment and system-related changes—the subject of this discussion.

Energy-Efficient Appliances

From an energy-efficiency standpoint, appliances can be compared either in terms of how much energy each needs to supply its rated output or in terms of the average amount of electricity (kilowatt hours) each will use per year under normal operating conditions.

Like air conditioning systems, a new generation of light fixtures and appliances is becoming more efficient. (See Chapter 2.) For example, new high-efficiency, long-lived fluorescent bulbs for indoor use and high-pressure sodium outdoor lamps can deliver many times the light output (lumens) per electrical input (watt) of incandescent lamps.

Making any single appliance more efficient may produce relatively small savings. But household energy consumption can be greatly affected by cumulative appliance efficiency gains. In the mid-1970s, an average household used some 9,000 kwh annually, assuming no electric heating or cooling. Substituting new top-of-the-line, appliances could reduce that figure by one third or more. (See Table 11.5.)

Table 11.5. Projected Improvements in Appliance Efficiencies.

APPLIANCE	ELECTRICITY CONSUMPTION 1980 STOCK (KWH/YR)	ELECTRICITY CONSUMPTION 1980 NEW (KWH/YR)	ELECTRICITY CONSUMPTION 1985–2000 NEW (KWH/YR)
Freezer	1285	1285	1000–500 (1985–1995)
Refrigerator	1700	1700	750–500 (1990–2000)
Electric Ranges	970	900	600–100 (1991–2000)
Televisions	350	250	100 (2000)
Dishwashers	250	250	150 (1990)
Room Air Conditioners	1440	1170	936–720 (1990–2000)

Source: A New Prosperity: Building a Sustainable Energy Future, Solar Energy Research Institute, 1981.

In larger commercial or institutional buildings where electrical use is often close to half of total energy consumption, efficiency-related savings can be greater still.[6] In many buildings, simply using more efficient lamps can save up to 25 percent of lighting electricity usage.[7] Seattle calculated that "relamping" could cut service electricity use by up to 15 percent in four prototypical commercial buildings.

Industrial electricity conservation also starts with the use of more efficient appliances. A Britsh research team has optimistically concluded that fully half of the electricity used in industry could be saved by downsizing motors, improving maintenance, using better controls, and the like.[8]

Planning and Managing Electrical Systems Better

The other side of increasing the efficiency of lighting and appliances is planning and managing buildings' electrical systems better. In many larger residential or commercial buildings, systems are not integrated with daylighting techniques, and illumination levels are higher than needed. Researchers say that lighting standards could be cut in half without impairing visual acuity.[9]

Lighting system overdesign can be combatted. In new buildings, the daylit areas could either be used for work needing more light, or artificial lighting levels in daylit areas could be reduced. Since the extensive use of daylighting also affects heating and cooling loads, daylighting features need to be evaluated in terms of their impacts on total building energy use. (See Chapters 8 and 9.) For example, using many north-facing windows for daylighting purposes could add to heating loads in cold climates. In most cases, though, a balance can be struck.

Homes, offices, and commercial buildings can be "delamped" without loss of utility or comfort. The simplest way to delamp is to remove or disconnect fixtures. An alternative is to replace higher wattage bulbs or tubes with lower ones. Still another is adopting such new devices as dummy tubes (which replace one lamp in a two-lamp fluorescent fixture) and light-polarizing acrylic panels (which improve the quality and distribution of reduced lighting levels).[10] All of these strategies, of course, need to be evaluated in terms of their impacts on lighting, heating, and cooling demand.

Electrical System Management

Even a well-designed electrical system using efficient lighting fixtures and appliances can achieve its potential savings only if it is properly managed. In smaller residences, businesses, or industries, simple, accessible controls and occupant education can help ensure better energy management. For example, switches installed in small areas of office buildings can allow an employee to turn the lights out when they are not needed. Electrical machines that shut off automatically can also save energy. Some industrial machines can be designed so that they will not run unless a worker is present.[11]

Larger buildings or complexes can profitably use more centralized control systems, some of which involve computers. Many of the heating-system control strategies discussed in Part Two, Chapter 9, can be used to improve electrical system efficiencies as well. Area-wide lighting levels can, for instance, be controlled automatically either by timers or by light-sensitive cells (photocells). More sophisticated electrical control systems can monitor a building's overall electrical consumption to protect owners or occupants from high electricity costs or power outages. Computer-operated load-shedding programs can even predict peak demands and reduce or shift consumption to shave peaks and save money.[12]

Economics.   Replacing a relatively new appliance with a more energy-efficient model does not make sense until the appliance begins to wear out, unless the appliance is highly inefficient and local electricity costs high. But replacing an inefficient refrigerator (205 yearly kwh consumption) with a more efficient $600 model (90 yearly kwh) is an investment that will pay for itself in six years where electricity rates are over 10 cents per kwh.[13]

Larger equipment purchases or systems cost more, but savings on energy can be greater too. Oakland's new street lighting system will pay back its capital costs ($4 million) in less than three years. An apartment building owner in Santa Monica, California, saves at least $700 annually in lighting costs as a result of a $1,500 investment in more efficient lamps and better design.[14]

Institutional Constraints and Socio-economic Considerations.   Electricity conservation directly affects a public utility's financial situation. A utility straining to meet capacity or to reduce oil consumption may welcome cities' efforts to reduce electricity consumption. But those with excess capacity or relatively stable primary fuel sources may fear that conservation efforts will cut revenues without shaving peaks. Where utilities are uncooperative, the city government has numerous strategies at its disposal. (See Chapter 5.)

Current federal and state subsidies, tax policies, and regulatory practices encourage utilities to build power plants at the expense of on-site conservation and smaller-scale supply technologies.[15] Utilities can receive tax benefits for large-scale construction programs, but not for conservation programs. Subsidies for conventional power plants are, by one count, 12 times those available for conservation.[16] Some regulatory provisions also penalize utilities if sales fall below a certain level: utilities aren't allowed to raise rates to compensate.

Finally, relatively few PUCs help utilities to consider alternative conservation and small-scale supply options. Cities can, however, address these larger-scale problems by lobbying state and federal representatives to make constructive changes.

A primary social concern is that the electricity that poor people purchase is usually used inefficiently because their appliances are so old.[17] Moreover, many residential electrical uses and costs are beyond the individual consumer's direct control. Even in apartment buildings with individual metering, electricity costs for such communal uses as hall lighting and elevators are apportioned among tenants.

How best to reduce the electric bills of low-income tenants depends on the nature of the housing stock. Where most low-income people own their homes or where much of the rental stock is in single-family houses, replacing inefficient appliances should be a top priority. Alone or in a public-private partnership, cities could purchase newer appliances in bulk and sell them at cost or at subsidized below-cost rates to low-income persons. In multifamily buildings, electrical conservation programs—including a basic, low-cost audit that would cover the electrical system—could be targetted at landlords.

SMALL-SCALE (DISPERSED) ON-SITE POWER PRODUCTION

Opportunities for individual home or building owners to produce their own electricity are expanding. Technical advances make on-site power production feasible on a smaller scale: the cost of electricity purchased from the grid makes small electrical facilities more economical, and new federal and state laws encourage the development of small power facilities by requiring regulated electric utilities to purchase power from them.

This section provides a general overview of small power production technologies—on-site cogeneration and renewable sources of electricity—and a preliminary assessment of their potential contribution in cities.

On-Site Cogeneration

Many industries, businesses, and apartment complexes already produce their own power by using on-site heat engines to run electric generators. They can then recapture the excess heat energy and use it to heat space or water. Now, with the development of smaller cogeneration units, combined power and heat production may be possible in some individual homes, too.

Commercial and industrial cogeneration units can make use of topping or bottoming cogeneration cycles and a number of different fuel sources. In the topping cycle, a steam turbine or a fuel-fired heat engine runs an electric generator, while the waste heat is used for water or space heating. The bottoming cycle uses waste heat from an industrial process to run a turbine or heat engine that, in turn, operates a generator. Oil, gas, coal, wood, or municipal wastes can be used as fuel for either cycle. Smaller topping-cycle cogeneration units using gas turbines, diesels, or advanced heat engines can deliver heat and electricity to single businesses and residences or clusters of each.

Economics.   The economic appeal of on-site cogeneration technologies is based on two related assumptions. First, as it was early in the century, it may now be cheaper to use lower-quality heat energy to generate electricity on-site than to purchase large quantities of electric energy from the grid. Second, it is cheaper still if on-site process heat can be used to generate power or if the excess heat from power-production can be used or sold.

Accessing the economics of a typical cogeneration system is difficult because few typical systems exist. Cycles, output levels, and power-to-heat ratios vary. In addition, sometimes end-use conservation options are economically preferable to cogeneration. (See Table 11.6.) Then too, the economics of a particular cogeneration system will depend not only on financing terms but also on the differences between the price of the cogeneration fuel and the price of electricity, the power-to-heat ratio of the cogeneration facility, and the nature of other

Table 11.6. Typical Capital Costs for Various Cogeneration Technologies.

TECHNOLOGY	CAPITAL COSTS (S/KW)
Gas turbine (topping cycle—5 Mw)	500[1]
Diesel engine (topping cycle—5 Mw)	550[1]
Steam turbine oil-fired (topping cycle—5 Mw)	875[1]
Steam turbine coal-fired (topping cycle—5 Mw)	1250[1]
Steam turbine coal-fired (topping cycle—27.5 Mw)	319[2]
Oil- and gas-fired turbine (topping cycle—16.6 Mw)	322[2]
Gas- and oil-fired turbine (topping-cycle—13 Mw)	445[2]
Coal-fired steam turbine (topping cycle—29 Mw)	486[2]
Oil-fired steam turbine (topping cycle—12 Mw)	416[2]
Steam and organic bottoming cycle up to 0.5 Mw	400– 700[3]
0.5 to 10.0 Mw	700–1000

Sources:
1. Eric Bazques, Brief Overview of Cogeneration: Technological and Economic Factors, Office of Technology Assessment—internal draft, September 25, 1980.
2. H. M. Bushby, Industrial Cogeneration: Five Case Studies—Synthesis and Observation, (unpublished draft) (1981).
3. Resource Planning Associates (1979).

Table 11.7. Summary of Economic Analysis of Five Cogeneration Case Studies.

INDUSTRY TYPE	FUEL SOURCES	ELECTRICAL DEMAND (PEAK CAPACITY— KW)	PROCESS STEAM DEMAND (LB/HR)	COGENERATION CAPITAL COSTS ($/KW) (1980 $)	FUEL USED IN COGENERATION EQUIPMENT	CAPACITY FACTOR (%)	% ELECTRICAL DEMAND FROM COGENERATION	AFTER TAX IRR (%)
Paper/pulp	• Black liquor (coal by-product) • Electricity	26,000	60,000	319	coal	51	65	39.4
Chemical	• Residual oil • Natural Gas • Electricity	14,200	48,000	322	oil and gas	48	51	38.0
Food processing	• Residual oil • Natural Gas • Electricity	11,000–13,000	35,000	445	gas and oil	58	98	33.9
Textile	• Coal • Electricity	35,000	70,000	486	coal	57	55	29.3
Oil refining	• Natural Gas • Fuel oil • Fuel by-products	27,000	33,000	416	oil	93	54	29.0

Source: Bushby, 1981; (see Table 11.6).

energy conservation measures employed at the site. Finally, a favorable utility sell-back rate may make cogeneration systems profitable investments no matter how much competing fuels cost.

Although few studies of the economics of generic cogeneration systems exist, the case studies summarized in Table 11.7 do reveal the kinds of situations in which cogeneration systems are economically viable. For example, two of the plants (paper/pulp and textile) use relatively low-cost coal as a cogeneration fuel. Although the chemical plant uses oil for 61 percent of its cogeneration input fuel, that oil is replacing very high-priced electricity. The importance of system design features is illustrated by the food processing and oil refining cases. The food processing plant uses oil and gas as cogeneration fuel sources to meet all of its relatively small electrical demand. Finally, the oil-refining plant uses oil to cogenerate, but achieves an acceptable rate of return because the cogeneration unit has a high utilization factor (93 percent).

City Applications. Any large user of heat and power is a good candidate for an on-site cogeneration system. Along with industry, commercial facilities with high thermal loads (such as laundries, hotels, schools, or hospitals) should have their cogeneration potential investigated. So should larger residential complexes located where prices for both heat and power are high.

Assuming high prices for competing fuels or good sell-back rates, more specific factors will determine whether a facility is suitable for cogeneration. (See Table 11.8.) One key factor is consistently high year-round demand for both thermal energy and electricity. Potentially even more important is a favorable

Table 11.8. Factors Affecting the Economics of Cogeneration.

Helpful	*Harmful*
• High purchased electric rates.	• Low purchased electric rates.
• Available low-cost cogeneration fuel.	• Available cogeneration fuels high in cost.
• High year-round demands for heat and power.	• Heat and power demands seasonal.
• Stable power/heat demand ratio.	• Power/heat demand ratio fluctuates.
• Favorable sell-back rate and no penalty for interconnecting.	• Low sell-back rates and/or back-up connection rate penalty.
• No violation of new-source review requirements under Clean Air Act.	• New-source review violation—offsets needed or cleanup of existing sources.

sell-back rate. Also important—primarily in equipment choice—is how heating and electrical demands relate to each other over the year. Since most cogeneration units that can "track" fluctuating heating and electrical demands are more expensive than those delivering a constant thermal/electric output, a facility whose proportional demands for heat and electricity are relatively constant may find cogeneration more economically attractive than one that requires much more heat than electricity in one season, much less in another.

Finally, environmental considerations may prevent some urban facilities from installing new cogeneration equipment if the systems would violate air quality standards. Especially controversial are new cogenerating facilities using diesel fuel, oil, coal, or waste for fuel.

Institutional Constraints and Socio-economic Considerations.   One institutional concern surrounding the use of on-site cogeneration facilities is regulating the development and operation of such facilities. A second concern is relating cogenerating facilities to the problems and needs of public utilities. While the Public Utility Regulatory Policies Act (PURPA) of 1978 addresses both concerns, some institutional barriers remain, and numerous implementation decisions still rest with the state public utility commissions.

As noted, the impacts of state and federal clean air standards on cogeneration facilities are both complex and site-specific. Owners of new cogeneration facilities must show that they have complied with clean air requirements. When they violate these standards, they must either offset emissions from other sources or clean up existing sources so that no net increase in emissions results.[18,19] Finally, even if no significant pollution-control investments prove necessary, most cogeneration facilities have to go through a long licensing process. New York State, for example, requires a construction permit and operating certificate for any owner of an emissions source with a net input greater than one million Btu per hour if residual oil, coal, and other solid fuels are used.[20]

The reactions of public utilities to cogeneration facilities are also critical to cogeneration's economic success since most potential cogenerating facilities are large electricity users. Utilities with growing demand will probably support or encourage the development of cogeneration facilities to offset the need for new power plants. For example, Pacific Gas and Electric recently invited 12,000 potential industrial suppliers of cogenerated electricity to submit proposals to deliver electricity to the company's grid. At the other end of the spectrum, utilities that use expensive primary fuels or have excess capacity may react adversely. Consolidated Edison in New York, for example, has discouraged large customers from using on-site cogeneration, contending that other ratepayers would be harmed if the utility's load factor were reduced.[21]

Utilities affect the economics of cogeneration systems through the rates they charge for backup power and by their willingness to purchase excess power from cogeneration systems. PURPA 210 both prevents local utilities from charging cogenerators or other small power-producers "discriminatory" rates and requires utilities to buy back excess power at avoided energy and/or avoided capacity costs.[22] However, some utilities argue that a customer's reliance on cogeneration puts him in a different class from noncogenerating customers. Moreover, utilities may often be able to keep the avoided costs (as determined by state PUCs) relatively low.[23]

Small Wind-Energy Conversion Systems

The kinetic energy contained in winds can be extracted as mechanical energy or converted to electricity. Wind-electric systems use turbines or rotors mounted on a tower to extract kinetic energy from the wind. The mechanical movement of the rotors is transferred to a gearbox that enhances rotational speed enough to run a generator (to produce direct current) or an alternator (to produce alternating current). If direct current is supplied but alternating current is needed, the system must also include a synchronous inverter to convert dc to ac and make the wind system's electricity complement that of utility-generated power. If the system is not grid-connected, electricity storage is needed.

The size of a wind-energy conversion system is indicated by its "rated output," the number of kilowatts (kw) it can generate at a given wind speed. Small wind systems are usually defined as those with rated outputs of less than 100 kw, though wind systems with rated outputs of 2 kw or less are available.

The various turbine and rotor designs available can be divided into two general categories. Horizontal axis systems use turbines resembling airplane propellers. These systems come in upwind and downwind varieties. Vertical axis systems have blades configured to resemble an egg-beater. These systems can use winds from any direction without reorienting the blades. They also cost less per unit of output, and they are simpler, lighter weight, and easier to maintain. But they are less efficient than horizontal wind machines, and most require some external energy source to begin rotating at low wind speeds.[25]

Both performance and safety dictate that wind systems be carefully matched with sites. Wind generators cannot produce power until wind speeds exceed 7 to 10 miles per hour. The minimum wind speed at which the system will work varies with the design, as does the "rated" wind speed or the velocity at which maximum power output (in kw) is achieved.

Both the average wind speed and the wind speed distribution at a given site help determine a wind system's energy output. Since the amount of energy contained in a "given wind" is proportional to the cube of its speed, doubling wind speed increases the energy available in the wind by a factor of eight. Wind speed distribution is also critically important to system efficiency. If the average annual wind speed is composed of steady moderate winds, a system with a low-rated output, a low minimum speed, and little storage is desirable. But if winds are highly variable, the minimum speed is less important and a comparatively high rated output and relatively greater storage are needed.[26]

The efficiency of a typical wind-energy conversion system can be measured in terms of the amount of energy it extracts from a given wind as compared to the amount available. Most experts agree that wind systems have a maximum theoretical efficiency of 59.3 percent. In practice, however, friction in hub bearings and other factors keep high-speed horizontal axis systems from extracting much more than about 40 percent of the wind's energy at the rotor, while additional systemic and conversion losses keep present system efficiencies at about 20 to 25 percent.[27] Using more flexible and inexpensive mounting towers and improved methods of storage should also enhance cost-effectiveness, though on-site storage is ill-suited to cities. Significant breakthroughs are also likely in weight reduction. Experimentation now under way with both wood and fiberglass blades could help reduce noise, weight, TV interference problems, and costs.

Economics.   Of all on-site renewable electricity technologies, small wind-energy conversion systems (SWECs) are probably the most economical now. Economy does tend to be greater with larger systems rated over 40 kw, but smaller systems can now or soon will compete with the electricity grid on a lifecycle (25–30 years) cost basis where wind resources are abundant.

Both the U.S. Department of Energy and wind-system manufacturers are working to reduce the capital costs of turbines by 25 percent and the costs of towers by 10 to 15 percent.[28] (See Table 11.9.) Installation costs are also expected to fall significantly as production volumes increase.

But the decline in federal support for solar energy research and development could delay cost breakthroughs for all solar technologies.

The economics of SWECs is also determined by electric output, current and future electricity prices in the service area, and the sell-back rate available from the local utility.

Few analyses of the lifecycle (or long-term) economics of small wind systems have been conducted. (See Table 11.10.) But one Minnesota study does indicate that even without guaranteed sell-back and a federal renewable-energy tax credit, wind systems can deliver electricity at a cost that is closely competitive with conventional electricity prices over the systems' 25-year useful life.

Table 11.9. Capital Costs of Selected Wind Machines.

MANUFACTURER	RATED OUTPUT (KWP AT MPH)	COST OF MACHINE ($)
Aero Power	1.5 @ 22	3,000
		3,200
Aerowatt S.A. (Fr.)	1.125 @ 16	19,689
Altos Corp.	2.5 @ n/a	4,340
	1.5 @ n/a	
American Wind	25 @ 25	2,975
Astrel	10 @ 23	7,900
Bertola Station	4 @ 18	5,000
Dakota Wind	4 @ 27	7,500
	4 @ 27	7,200
Dunlight Electric	2 @ 25	4,800
Enertech Corp	1.5 @ 21	3,475
Hinton Res.	3 @ 28	1,995
Independent Energy	4 @ 23	5,395
	4 @ 23	5,995
	2 @ 23	3,995
Kedeo, Inc.	5 @ 27	4,975
	3 @ 26	4,495
	2 @ 22	4,195
Millville	10 @ 25	6,600
North Wind	2 @ 20	7,200
Pinson Energy	5 @ 30	9,600
Power Group	4 @ 24	4,990
Product Development	4 @ 20	7,995
Tumac Industries	7.5 @ 28	13,000
Whirlwind Power	4 @ 22	3,900
	2 @ 25	1,995

Source: Wind Products Supplement, *Solar Age,* February, 1980.

With tax credits, current electricity prices can be matched, pending the availability of favorable financing.

Of course, all lifecycle cost analyses are limited because few purchasers make decisions on lifecycle cost bases. To get a clearer idea of the potential economic appeal of small wind systems, cash-flow and payback analyses of actual or prototypical installations are needed. The positive results from the analyses summarized here suggest that such efforts would be worthwhile.

City Applications. Small wind systems with less than five kw capacity can probably be put to best use in relatively sparsely populated areas where most people live in single-family homes. Larger systems (up to 100 kw) could also be used to provide service electricity to larger buildings in relatively open areas or to clusters of single-family residences in newer developments.

Table 11.10. Economics of Selected Wind Systems.

LOCATION	RATED OUTPUT	ANNUAL OUTPUT (KWH)	INSTALLED CAPITAL COSTS ($)	SYSTEM LIFE (YEARS)	LEVELIZED COSTS w/o TAX CREDIT (¢/KWH)	LEVELIZED COSTS WITH TAX CREDIT (¢/KWH)
Minnesota[1]	2 kw at 14 mph	4,805	6,110	25	11.0	7.0
	4 kw at 14 mph	8,995	9,800	25	9.0	6.0
	10 kw at 14 mph	22,502	19,500	25	8.0	6.0
	40 kw at 14 mph	94,346	60,000	25	5.0	4.0
Boston[2]	10 kw at 25 mph	13,200	16,714	30		6.0
Dodge City[2]	10 kw at 25 mph	24,700	16,494	30		3.0

Sources: 1. Minnesota Energy Agency (1981).
2. SERI, *Cost of Energy From Some Renewable And Conventional Technologies* (1981).

These systems will be hard to use in dense inner city areas. Row-houses on small lots with slanted roofs do not have enough space on which to mount the towers. In row-houses with flat roofs, towers could be roof-mounted only if buildings were modified to accommodate reinforced bases and anchor points for guy lines.[29] Even with adequate roof space and strength, wind systems need assured access to turbulence-free winds—a difficult proposition in most inner city neighborhoods. Small wind systems hold somewhat more promise in relatively low-story apartment or office buildings, though roof strength and wind access are still problems.

Would-be wind-users must also consider electrical demand. For example, many lighting systems can be operated on either ac or dc current, so a SWECS could provide dc directly (or through a storage system) to lights when needed. A simple switching device could allow changeover to utility-supplied ac when wind speeds or storage capacity is inadequate.

Institutional Constraints and Socio-economic Considerations.   Wind systems mounted on-site must meet code and zoning requirements. Codes may call for the reinforcement of any rooftops that will support wind towers, or zoning provisions may forbid the use of wind towers over a certain height. Land-use requirements for small wind towers may be minimal, but assuring

Urban wind machine on tenement building in New York's Lower East Side (Con Edison power plant in background).

adequate, turbulence-free wind may require land-use controls far in excess of those needed to protect solar access.

Utility responses to small wind systems will be determined by the utility's financial health and by the state's method of implementing PURPA. Since states have substantial leeway to implement PURPA, cities should carefully follow PURPA implementation procedures in their states and discourage utilities from requesting special backup rates and lower sell-back rates for small wind facilities. If wind systems' outputs reduce the load factors of summer-peaking utilities without cutting peak loads, almost any utility will oppose them, but higher backup rates won't be approved unless the utility can prove that owners of wind systems constitute a different class of customers than similar residential or commercial customers without windmills.

Solar Thermal-Electric Systems

Solar thermal-electric systems use high-temperature active solar collectors to provide heat to turbine boilers or heat engines connected to electric generators. Most such systems require several thousand feet of collector area and will be used primarily in utility-owned facilities to produce electricity for the grid. But some solar-electric systems are envisaged for use "on-site" in industrial facilities to supply both heat and power.

Theoretically, solar thermal-electric systems operate much the same as active solar heating systems. However, collection, storage, and transport temperatures are much higher, and collected heat goes to space- or process-heating only after being expended from a heat engine or boiler. Typical solar thermal-electric systems collect solar energy at high temperatures in a receiver, then transport it either to storage or to an energy conversion system, where electricity is generated.[30] (See Table 11.11.)

Receiver systems can be of two basic types: (1) a distributed receiver arrangement concentrates sunlight in many solar collectors; (2) a central receiver system reflects sunlight onto a single location by means of a field of distributed mirrors called heliostats. With distributed receiver systems, the goal is to collect high-temperature heat at the lowest cost. The parabolic trough collector, often referred to as a line-focussing model, has a transfer fluid that flows through a tube mounted above and across the trough's surface so that light striking the collector focusses on the line. (See Chapters 8 and 12.) The parabolic dish collector focusses sunlight on a point where a transfer fluid is heated. Generally, the effectiveness of these collectors can be maximized by adding one- or two-axis tracking, which allows the collector to follow the sun's path both seasonally and daily.

Table 11.11. Potential Combinations of Collectors and Conversion Systems for 0.1–10 Mw (e) Solar Thermal-Electric Systems.

CONCENTRATOR	CONVERSION	TRANSPORT	STORAGE
1. Point Focus, Central Receiver	Central Rankine	Salt	Salt
2. Point Focus, Central Receiver	Central Brayton	Electrical	Electrical
3. Point Focus, Central Receiver	Central Stirling	Electrical	Electrical
4. Line Focus, Central Receiver	Central Rankine	Salt	Salt
5. Point Focus, Distributed Receiver	Central Rankine	Salt	Salt
6. Point Focus, Distributed Receiver	Distributed Rankine	Electrical	Electrical
7. Point Focus, Distributed Receiver	Distributed Brayton	Electrical	Electrical
8. Point Focus, Distributed Receiver	Distributed Stirling	Electrical	Electrical
9. Fixed Mirror, Distributed Focus	Central Rankine	Salt	Salt
10. Line Focus, Distributed Receiver Tracking Collector	Central Rankine	Oil	Oil
11. Line Focus, Distributed Receiver Tracking Collector	Central Rankine	Oil	Oil
12. Low Concentration, Non-Tracking	Central Rankine	Water	Water

Source: John Thornton, et al. *Final Report—A Comparative Ranking of 0.1–10.0 Mw (e) New Solar Thermal Electric Power Systems* (1980)

Most thermal-electric system designs use a molten salt solution or an organic oil for heat transfer. For energy conversion, they make use of a steam boiler or turbine connected to a generator or an advanced heat-engine electric generator configuration. Many such systems use excess heat from power production for process, space, or water heating or for absorption cooling—in effect, creating a "solar total energy system."[31]

While most distributed receiver systems transport collected or stored heat to a central conversion point, some feature heat engines and generators at each collector. This arrangement costs more because additional equipment is needed, but the heat engines get the highest possible inlet temperatures since no heat is lost in transport or storage.

Predictions of savings and technical problems with these prototypes are conjectural. Experiences with higher-temperature collectors in industrial process applications do, however, give rise to concern about durability and vulnerability to heat loss.[32] (See Chapter 10.) Also, because the interactions between a high-temperature solar energy system, the high-temperature backup fuel source, a boiler or heat engine, and a generator are complex, system design and operation must be painstakingly careful.

Table 11.12. Commercial Readiness Characteristics of Selected Solar Thermal-Electric Technologies.

SYSTEM NUMBER	RELATIVE STATE OF DEVELOPMENT	RELATIVE TECHNICAL/ COST RISK	RELATIVE R&D REQUIRED
1	High	Low	Low
2	Medium	Medium	Medium
3	Low	Medium	Medium
4	Low	High	High
5	Low	Medium	Medium
6	Low	Medium	High
7	Low	Medium	High
8	Low	Medium	High
9	Medium	High	High
10	High	Low	Low
11	Low	High	Medium
12	Low	Medium	Medium

Source: Thornton (1980).

Economics.   The overall economic appeal of these prototypical systems will depend on improvements in reliability, cost reduction, the location of the plant, the nature of its thermal/electric load, the costs of competing energy sources, and sell-back rates.[33]

One way to reduce system costs is to lower the costs of collecting heat. Another is to find the most economical configuration of solar and energy conversion systems—whether by lowering effective system costs or using more expensive but more efficient equipment. Relative cost estimates should be evaluated along with such factors as reliability and commercial readiness. (See Table 11.12.) As a rule, facilities in sunny areas with high year-round needs for heating or cooling and for electricity can probably make the most economical use of these systems.

Institutional Constraints and Socio-economic Considerations.   Land-use questions will surround any efforts to use solar thermal-electric systems inside city limits. Even where open space is adequate, only extensive land-use controls can ensure solar access. Very low-density industrial areas—especially industrial parks—are the ideal settings.

Since most solar thermal-electric systems will be accompanied by on-site, fossil-fueled electric generating equipment, utilities are likely to view these applications much as they do cogeneration systems run partly on solar energy. However, since solar thermal-electric systems will probably be able to provide most peak electricity demands, utilities might find the justification of higher backup and lower sell-back rates for such systems difficult.

Photovoltaics

Photovoltaics (or solar cells) produce electricity directly as sunlight strikes them, providing "instant electricity." Currently, most solar cells are made of single-crystal silicon, which is a good absorber of light and a good semiconductor.[34] Two layers of silicon, each of which contains "impurities" that form an electrical barrier between the layers, make up a manufactured cell. When light strikes the cell, one of the layers accumulates freed electrons that can be channeled to the other layer only by a wire. This flow is an electric current that can be tapped to run appliances or fed back to the grid.

A photovoltaic system consists of an array of panels containing cells wired together. The size of the array depends upon climatic conditions, cell efficiencies, the size of the building's electrical load, power sell-back arrangements with the utility, and other factors. Besides cells, each system also has an inverter and controls to regulate voltage and monitor usage. If the system is not grid-connected or if the owner wishes to take advantage of the highest sell-back rates, electricity storage is also required.

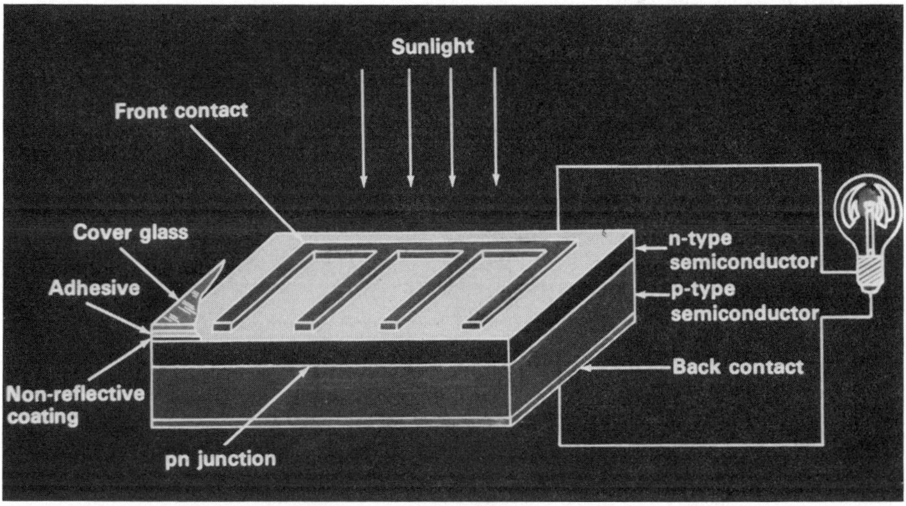

Basic photovoltaic (solar) cell: p = positive and n = negative.

Most operational photovoltaic systems employ round, single-crystal silicon cells mounted on flat-plate panels. Cell efficiency (the amount of electrical power produced at the cell divided by the amount of potential power in the available sunlight) ranges from 8 to 15 percent. While power conditioning equipment and battery storage will reduce system efficiency further, cell efficiencies have by far the greatest impact on system efficiencies. Cells are also the single most expensive component of any photovoltaic system. For these reasons, most strategies for reducing the overall costs of systems focus on increasing cell output or reducing cell costs while maintaining constant levels of electrical output.

One strategy for cost reduction involves changing the shape of solar cells from round to square so that an array can feature more cells per square foot.[35] Still more fundamental improvements are aimed at boosting the electrical conversion efficiency of the cell itself or increasing the intensity of light on the cell's surface. Higher efficiency cells do cost more, but since the theoretical limit of single-crystal silicon cell efficiency is about 24 percent, boosting cell efficiency beyond the 15 percent upper limit now available may be as economical as other cost-reduction strategies. (See Chapter 8.)

The other major strategy for array cost reduction involves using less expensive materials in the cells themselves. For example, silicon that is less pure than single-crystal silicon can be grown at less cost. So-called "thin-film" cells made of such materials as cadmium sulfide or copper sulfide are also under development. But while these cells cost far less than single-crystal silicon cells, their output is lower and they may be less reliable.

Photovoltaic systems have been used effectively and reliably for years to provide power to satellites. In terrestrial applications, most of the systems that have been operating for very long are in remote locations, unconnected to any utility grid. However, a few demonstration or pioneering systems—most built with government support—are in use on homes or office buildings in dense areas.

Economics.   The single biggest barrier to the use of photovoltaic systems is their cost. Even assuming continued increases in the costs of conventional electricity, photovoltaic system costs must come down by a factor of ten to compete with existing utility service. Yet, costs have dropped and the Department of Energy is working to meet ambitious cost-reduction goals by encouraging mass production and developing cheaper manufacturing techniques.

Photovoltaic systems costs comprise array costs and balance of system (BOS) costs (for power conditioning, storage, and so on). Conventionally, both are given in dollars per peak watt ($/wp) of output. Thus, a system with an output of five kilowatts that cost $10/wp for the array and $10/wp for BOS would cost $20 times 5,000 watts—$100,000.

Obviously, at these cost levels photovoltaic electricity cannot compete with utility-produced electricity. However, attaining the U.S. Department of Energy's cost goals of $0.70/wp for cells and $0.90 to $1.90/wp for BOS could change this picture dramatically.

While detailed economic analyses using current costs are pointless, lifecycle cost analyses of prototypical systems indicate that attractive costs are achievable. (See Table 11.13.) While these figures reflect the assumption that no major cost-lowering technical breakthroughs will occur and while they do not take tax credits into account, one system now for sale still sells for under $3.00 per peak watt. Employing aluminum-cone concentrating and cogenerating arrays, this system has reportedly been installed on single-family homes and small commercial buildings for between $9,000 and $15,000, with the manufacturer estimating paybacks at 9 to 12 years for some systems.[36]

City Applications. Photovoltaic systems are much more flexible than either wind or solar thermal-electric technologies. Modular, photovoltaic panels can be used to provide electricity to diverse dc or ac loads in applications as small as cathodic corrosion inhibitors on highway bridges or as large as airports.

The on-site requirements for photovoltaic systems are simple: adequate space, proper building or lot orientation, and solar access. (See Chapters 1 and 8.) As for land, the minimum photovoltaic array area required is a function of the electrical demand to be met by the system and the availability of sunlight. Demand calculations should begin with an estimate of the total annual (and seasonal) electrical demands of the building, which can be obtained by calculating the wattage requirement for each system or appliance and the number of hours each will be used per year.

The costs of photovoltaic arrays depend on the number of cells used. Although photovoltaic systems are capable of providing their highest output when demand is highest, it is still unwise to size the system to meet peak cooling loads if the resulting system would be oversized for the rest of the year unless favorable sell-back rates are available from the utility. A better tack is to base the load design on a monthly average of kilowatt-hour consumption. Regardless of the size of the electrical load, the size of the array needed to meet it will vary with the availability of sunshine. With equal cell efficiencies, it takes more space (and costs more) to meet a given electrical load in New York than in Phoenix. (See Table 11.14.)

A related consideration is potential competition for space between active solar space- or water-heating systems and photovoltaics. Since few cities obtain much space-heating energy from electricity, most urbanites who want to obtain both heating and electrical energy from sunshine may have to use two solar energy systems. In many denser cities, it may be necessary to employ concentrating systems, perhaps with simultaneous heat collection.

Table 11.13. Economic Analysis of Residential Photovoltaic Systems Using DOE Cost Goals.

SYSTEM	BOSTON[1] 1985	BOSTON[1] 1990	BOSTON[1] 2000	ALBUQUERQUE[1] 1985	ALBUQUERQUE[1] 1990	ALBUQUERQUE[1] 2000	MINNESOTA[2]
Capacity	4 kw	4 kw	4 kw	4 kw	4 kw	4 kw	10 kw
System Cost ($)	10,000	6,400	5,600	10,000	64,000	56,000	19,000
$/wp	2.50	1.60	1.40	2.50	1.60	1.40	1.90
Financing	12%	12%	12%	12%	12%	12%	n/a
System efficiency (%)	8.0	8.1	8.2	7.9	8.0	8.1	
Annual Output (kwh)	5535	5590	5672	9683	9779	9875	17,000
Levelized Cost— twenty years (¢/kwh)	13.0	7.9	5.8	9.4	5.6	4.1	7.9

Source:
1. SERI, *Cost of Energy From Some Renewable and Conventional Technologies* (1981).
2. Minnesota Energy Agency (1981).

Table 11.14. Space Needed for Photovoltaic Arrays Under Different Design Conditions.

FT² OF COLLECTOR NEEDED

CITY	INSOLATION (BTU/FT²/DAY)	LOAD ONE: 15 KWH/DAY SYSTEM EFFICIENCY			LOAD TWO: 20 KWH/DAY SYSTEM EFFICIENCY			LOAD THREE: 25 KWH/DAY SYSTEM EFFICIENCY		
		(6.5%)	(8.1%)	(12%)	(6.5%)	(8.1%)	(12%)	(6.5%)	(8.1%)	(12.0%)
Boston	1105	800	643	421	1065	957	562	1331	1072	702
New York	1099	803	647	423	1071	863	565	1339	1078	706
Washington, D.C.	1208	731	589	385	974	785	514	1218	981	642
Atlanta	1345	656	529	346	875	705	461	1094	881	577
Miami	1473	599	483	316	799	644	421	999	805	527
Chicago	1215	726	585	383	969	780	1511	1211	975	638
Minneapolis	1170	754	608	398	1006	810	530	1257	1013	663
Denver	1568	563	453	299	751	605	396	938	756	495
Houston	1351	653	506	344	871	702	345	1089	877	574
Phoenix	1869	472	380	249	630	507	332	787	634	415
Los Angeles	1594	554	446	292	738	595	389	923	743	487
Seattle	1053	838	675	442	1118	900	589	1397	1125	737

Source: Derived via a formula used in Maycock, *Photovoltaics,* (1981).

Shading can have a much more detrimental effect on photovoltaic than on active heating systems. Most photovoltaic systems are wired in two types of connections: series and parallel. Series cell connections allow voltage (or quantity) increases with constant current (or flow), while parallel cell connections allow current increases at constant voltage. Thus, shading even one cell in a series connection can reduce current and power output substantially.[37]

The need for space and solar access suggests that the most promising applications for on-site photovoltaic systems may be in new developments or in low-density residential or commercial areas.[38] Unless concentrating or cogenerating systems are used, photovoltaic applications in denser areas might be confined to community-scale systems. New buildings can be oriented for maximum solar exposure, and some companies now produce shingles combined with photovoltaic cells so that the additional costs of the photovoltaic systems can be partially subsumed in roofing costs in new construction. Nevertheless, photovoltaic retrofits are physically much easier to undertake than active space- or water-heating retrofits because the photovoltaic panels weigh much less, and no storage tank is necessary.

Institutional Constraints and Socio-economic Considerations. Land-use issues surrounding on-site photovoltaic applications will be similar to those involved in the use of active space-heating or water-heating systems. (See Chapter 3.)

Owners of photovoltaic systems will be able to sell power back to local utilities. Yet, photovoltaics cannot meet brief on-site peaks very well because the power output of the system is limited by the availability of sunlight.[39] But nearly all photovoltaic systems will produce most electricity in summer, when most utilities experience peak demands—an advantage for utilities as well as consumers. In sunny areas, photovoltaic systems may be more reliable than thermal power plants.

COMMUNITY-SCALE (CENTRALIZED) APPLICATIONS

Centralized energy-conservation and renewable-energy technologies can provide relatively cheap, nonpolluting, and reliable power to large numbers of customers. These systems could be developed and owned by public utilities. Alternatively, city government, local industries, or public-private partnerships can become profit-making electricity producers.

Many community-scale systems are economically competitive with larger thermal power plants right now. More will become so as the construction, financing, and safety-related costs of large coal- or uranium-fueled facilities continue to escalate. Current estimates put the costs of a 1,000 Mw nuclear plant at between $2 billion and $3 billion.[40] The total costs of Nine-Mile, Unit

Two, a nuclear plant currently under construction in New York State, will exceed $5.5 billion (or $5,500 per kilowatt of capacity).[41] Many cogeneration or renewable-energy electrical systems can be brought on-line for half that price now.

Smaller-scale power facilities are also more reliable than larger plants. A single unexpected outage can severely disrupt service in a large plant, but shutdowns of one or a few smaller facilities in a system will affect overall levels of service less drastically. Indeed, an electric grid supplied by many small generating stations and backed up by a reasonable reserve margin could be virtually outage-proof.

A third advantage of a network of smaller-scale electric generating facilities is flexibility in response to demand fluctuations. When rates were cheap, electricity demand rose steadily and utility planners began assuming that demand would continue to grow. Now, however, many utility industry analysts are uttering jeremiads about electrical supply shortfalls in the next 20 years.[42] Still others contend that large-scale growth projections have already led to the construction of unneeded plants and excess capacity.[43] Certain parts of the South and West are experiencing far greater demand increases than developed regions, and where electric rates are still relatively low, immediate market pressure to stabilize demand patterns is low. Even in high-growth areas, the long-term effects of future price increases cannot be easily predicted, especially since advantages in appliance efficiency may reduce a utility's capacity needs even if end-use demands increase. These uncertainties about demand can devastate a utility since it takes an estimated 13 years to bring a 1,000 Mw nuclear plant on line. While regulatory changes could cut this time by a third, nearly a decade would still be needed. In contrast, smaller-scale power production facilities both take less time to build and make "tracking" demand easier.

Smaller-scale power production also increases opportunities for local control over a community's electricity supply. Smaller-scale generation is within the financial capability of city governments, particularly if favorable sell-back rates are available. Moreover, small power production facilities that make use of renewable energy pose fewer environmental threats than large-scale thermal power plants. The hazards and the costs of controlling air pollution and water pollution and disposing of nuclear waste are already well established, and the nuclear industry's ability to predict, prevent, and deal with a catastrophic accident is by no means assured.[44]

Technical Approach I: Modular Integrated Utility Systems

A modular integrated utility system (MIUS) is a centralized block- or neighborhood-scale utility unit capable of providing heat, power, potable water, sewage treatment, and wastewater treatment.[45] Theoretically, a MIUS can reduce capital and delivered service costs by integrating the operation of several utility services. According to their proponents, MIUSs also produce fewer emissions than the various systems they replace.

Too new to evaluate in detail, MIUSs are major capital projects that can probably be justified only when new facilities are needed or when existing facilities are due for replacement anyway. Their economic benefits are affected by many of the same factors—including the costs of the utility services they replace—that influence the economics of cogeneration strategies. Where even one relatively low-cost conventional service delivery system is available, it should be considered first. And where a MIUS is the preferred option, private users will need to be able to reduce their use of conventional utility services without penalty.

Technical Approach II: Advanced Cogeneration Techniques

A *combined-cycle cogeneration unit* employs both a topping and a bottoming cycle to obtain substantially greater electrical outputs than topping or bottoming cycles alone, with relatively small increases in fuel inputs. First, a combustion turbine produces both electricity and steam. The steam is fed to a waste-heat recovery boiler where temperatures are raised high enough to run an extraction/condensing or backpressure turbine that cogenerates heat and power. (See Chapter 10.) Some utilities calculate that using a combined cycle could increase the amount of economical cogeneration potential in their service areas several-fold. But while combined cycles can be much more economical and efficient than topping or bottoming cycles alone, combined-cycle combustion turbines must be run on oil or gas—a drawback since these fuels are becoming more expensive and scarce.

In *fuel cells,* the hydrogen gas in a fuel is combined with air in a chamber. In the chemical reaction between the air and hydrogen, both dc electricity and heat are produced. If converted to ac, the electricity can be used for low-temperature (330°F or less) industrial processes, water and space heating, absorption cooling, or other purposes. Fuel cells represent a cleaner and more efficient way than fossil-fuel combustion to produce electric energy since combustion is unnecessary.

By one estimate, fuel cells can generate electricity at 40 percent efficiency—compared to 25 to 33 percent for conventional thermal power plants—and cogenerate heat and power at an overall efficiency of 90 percent. These cells are also more efficient supporting partial loads than conventional generators are, and they can respond to dramatic increases in demand faster.

Fuel cells are expected to be available in a range of sizes, and their operation gives rise to relatively little pollution. Indeed, a study of a projected 4.8-Mw fuel cell cogenerating unit for a paper mill in Santa Clara indicated that replacing on-site boilers with fuel cells would reduce nitrous and sulfur oxide and particulate emissions by over 100 lb per day and pollution caused by electric generation by over 1,000 lb per day.[46]

LARGE-SCALE (SEMI-CENTRALIZED) RENEWABLE ENERGY SYSTEMS

Technologies

Wind-energy conversion systems, solar thermal-electric systems, and photovoltaics can all be used in community-scale applications. Numerous other renewable electricity technologies can be used only in a partially centralized fashion. These include small hydroelectric and biomass-fueled systems in cities and both geothermal resources and solar ponds in exurban areas.

Large Wind-Energy Conversion Systems. Wind-energy conversion systems with between 70 and 100 kw of rated output can be used to provide electricity to larger buildings or to neighborhoods, while systems with over 100 kw of rated output can be used expressly to produce electricity for the grid. Most of these larger units cannot be located within city limits because larger-scale wind systems feature strikingly wider blades and higher towers than smaller ones, require more land than small systems, and are more vulnerable to turbulence caused by surrounding obstructions.

Larger wind systems also pose more environmental and safety-related problems than smaller ones. Some larger-scale systems have caused continuous, low-level vibration in surrounding areas, so blade rotation speeds have had to be cut or steel turbine blades replaced with fiberglass blades. In addition, larger wind systems require ice sensors since ice flying off large blades could injure people and buildings.

Few wind systems of over 100 kw are in use, though the Bonneville Power Administration (BPA) recently finished building three 2.5 Mw units in Goodnoe Hills, Washington, that will be closely monitored to establish the effects of turbulence in groupings of larger machines, the reliability of clusters of larger machines, the "match" between electricity production and demand patterns, and environmental impacts.[47]

Solar Thermal-Electric. Solar thermal-electric systems can be used to replace outmoded or expensive power plants. Most current plans for "repowering" projects are for areas that are sunny year-round.[48] Most such repowering projects could make use of distributed receiver systems, though central receiver systems are favored.

Currently, one central receiver plant is under construction in Barstow, California. Others are planned, but experience is limited.[49] Theoretically, with central receiver systems cities would not have to worry about collector durability or the heat loss in transport that may be expected in distributed receiver systems. But some heat loss always occurs if storage is integrated into the system, and the long-term durability of heliostats is still largely conjectural. Finally, interconnecting central receiver systems with the auxiliary heating system and the heat engine can be complex.

Community-scale solar thermal-electric systems require far too much land—five to eight acres of land per megawatt of capacity—to be used within most cities.[50] Curtailing development in surrounding areas to protect solar systems is another access problem. Most likely, community-scale solar thermal-electric systems will be sited in open areas and contribute "imported" electricity to cities.

Photovoltaics.   Photovoltaic arrays can be configured in large modules to provide electric power to community grids. Aside from costs, spatial requirements may impede the widespread urban use of these systems. Although the panels are so light that they could be installed atop flat buildings or the flat roofs of parking areas to save space, solar access for these systems would have to be protected.

Small-Scale Hydroelectricity.   Hydroelectric systems use falling water to power turbines whose mechanical energy is, in turn, used to run electric generators. Currently, hydroelectricity accounts for about 11 percent (59,000 Mw) of total U.S. generating capacity or about 6 percent of total U.S. energy consumption.[51] Most hydroelectric power generation takes place in relatively large (15 to 2,000 Mw) facilities, but substantial untapped capacity exists at smaller (under 15 Mw) sites in or near small- to medium-sized cities, especially in the Northeast.

The chief determinant of the size of a hydroelectric project is the amount of water available at the dam site. Water resources are assessed in terms of (1) the flow-rate, or the gallons of water that pass a fixed point on the bank each minute, (2) the consistency of stream flow or the daily or seasonal variations in flow rates, and (3) the head, or the distance of vertical drop available at a given dam site. Larger hydro projects are located in areas with relatively great flow rates, while smaller projects are on streams with relatively smaller ones. All large-scale hydro projects are "high head," though not all small hydro projects are "low head."

The amount of "head" at a given hydroelectric site is a function of terrain and dam structure. Most storage dams are located in deep valleys so that they can form large, deep reservoirs with a highly dependable capacity. Most good sites for large storage dams in this country are already developed, and few unexploited sites are near large cities. Run-of-the-river dams are much more common where waterways are relatively flat. The Army Corps of Engineers has estimated that over 57,000 Mw of power potential is available at 2,100 sites, 32 percent of which are located in the Northeast.[52]

The difference between low-head and high-head applications is critical to turbine selection. In high-head applications, water tends to reach the turbine at a very high velocity. ("Impulse turbines" utilize such a stream by letting it strike each turbine blade serially.) Water passes through a low-head system at low speed and great pressure. Reaction, propeller, and crossflow turbines operate optimally in response to water pressure rather than velocity.[53]

Overall, hydroelectric plants are extremely efficient: about 86 percent of the kinetic energy in water can be converted to electricity. Thermal plants achieve efficiencies only a third to a half this high.

To investigate hydropower potential, one needs to find out how much power can be produced with existing resources, how much structural rehabilitation is needed on existing structures, and how much energy the adjacent industries and businesses need. Only a qualified engineering firm can assess a location's hydropower potential and average annual output.

Where power generation appears attractive, the existence of a dam already at the site is the most important economic factor. In most cases, building a new dam on a small-scale hydro site is uneconomical. But if the dams, spillways, penstocks, and power houses from old hydroelectric facilities are already in place, only renovation and (perhaps) dredging are needed. (See Table 11.15.)

Marketing and transmitting power pose additional issues. Only in isolated cases can hydropower be used directly without being transmitted through the public utility grid. At least three considerations influence the ease with which power produced at a retrofitted dam can be integrated into a public utility network: (1) the accessibility of transmission, (2) the cooperation of the local utility, and (3) sell-back rates. Reliability isn't a problem in most low-head hydro facilities, but the amounts of power available will vary by season, and most flow rates do not peak in summer, when utilities' demands do. Where maximum flow rates are compatible with demands for service lighting in nearby business or industrial sites, hydropower developers may find it profitable to sell their power directly to users.

The time and money needed to obtain a license for a hydroelectric plant will vary with stream and project size and will involve many federal and state agencies. However, as a result of recent efforts to reduce the federal role in licensing small projects, dams under 1.5 Mw are now required to go through only a "short form" licensing process, and some projects under 0.5 Mw may be exempt from federal licensing entirely.

Table 11.15. Safety Checklist for Retrofitting Existing Dams for Hydropower Production.

DAM TYPE	POTENTIAL PROBLEMS
Earth and Rockfill	• cracking
	• slope and foundation stability
	• settlement at crest
	• excessive seepage at abutment, toe
	• excessive hydrostatic pressure
	• changes in alignment or geometry
	• bulging at toe or beyond
	• loss of freeboard
	• surface erosion
Concrete	• cracking
	• displacement or offsets
	• tilting
	• sliding
	• excessive seepage
	• excessive hydrostatic pressures
	• deteriorated concrete
	• concrete erosion
	• downstream erosion, undermining stability of slopes
Abutments, channels, tunnels	• instability of tunnel wall or lining
	• seepage
	• erosion
	• excessive hydrostatic pressure

Source: Leslie Pruce, Power from Water, *Power,* April 1980.

Biomass Resources.   Steam boilers powering turbines connected to electric generators can burn wood, crop wastes, municipal solid wastes, and other biomass fuels. All can be used for cogeneration or in small "power only" facilities, and methane gas recovered from landfills can also run turbines in electric generating facilities. (See Chapter 13.)

Various equipment capable of burning biofuels is now available. *Spreader-stoker boilers,* for instance, can burn wood with a moisture content up to 65 percent, in some cases producing 100,000 to 300,000 lb of steam per hour on site.[54]

Another versatile boiler is the *fluidized-bed combustion unit,* which burns wood or other biomass fuels in a mass of mineral particles kept in motion via forced air. Because it has a much higher heat transfer rate than a stoker-fired boiler, a smaller boiler of this type can produce relatively more steam—up to 10,000 lb of steam per hour in some wood-burning units now on the market.

Many California canneries now use fluidized-bed boilers to produce process heat from pits, shells, and other wastes.

One of the simplest technologies for wood-burning is the *suspension or cyclone burner,* which can be incorporated into a new wood-fired boiler or retrofitted onto an existing boiler that produces less than 50,000 lbs of steam per hour. These burners can burn only very fine, dry fuel: planer shavings, sawdust, rice hull fines, sunflower seed fines, and the like. In 1978, more than 200 of these units were being used in industry.

Burning municipal solid wastes usually requires larger boilers than burning crop wastes does. Numerous technologies—most of which must be field-erected—are available for burning municipal wastes. In water-wall combustion, a furnace jacketed with waterfilled tubes burns unsorted, unshredded wastes or shredded wastes from which glass and metals have been removed. (See Chapters 10 and 12.)

In facilities that now use gas or oil, boiler retrofitting to burn wood or biomass may be difficult, costly, and less economical than producing a medium-Btu gas on-site with a small "package" *wood gasifier* coupled to the existing boiler kiln. Electric generation facilities converting to wood may require additional handling, storage, and pollution-control equipment as well as new combustion equipment. New unloading equipment, processing equipment, conveyors or loaders, and storage devices will be needed in most cases, while processing- and storage-equipment needs vary with the type of combustion equipment chosen.

Wood-gasification systems are also well-proven. Commercially available since the turn of the century, these systems lost popularity in the 1950s and 60s but may be ripe for a comeback in the 80s. (Retrofitting boilers to use medium-Btu gas produced onsite by gasifiers requires replacing burners and expanding ductwork and fan capacity.)

Technologies for gasifying other renewable fuels are not yet commercially proven. A pilot gasifier installed at the Diamond Walnut Growers plant in 1976 with funding from the California Energy Commission produced a clean low-Btu gas from walnut shells when it was operating, but it suffered chronic mechanical problems, primarily due to clogging. In 1977, management decided to shut the gasifier down and pursue direct combustion instead.

Overall, five considerations govern the viability of using biofuels. First, they are more flexible to demand than other renewable sources of electricity because they can easily be stored. Thus, small power producers can "time" operations so as to obtain the most favorable sell-back rates available. Second, facilities that use biofuels must be near the fuel sources so as to minimize transportation costs: a good rule of thumb is that transporting biomass fuels much more than 50 miles to the point of use is uneconomic.[55] Third, a biomass electrical generating facility must secure reliable commitments for the delivery of raw materials. Supply interruptions or inadequacies can ruin the economics of these systems. Fourth, using biomass resources in retrofitted power plants that previously burned oil or coal nearly always involves derating the plant's capacity unless remedial measures are taken. Finally, using biomass in electric generation entails some environmental penalties—high particulates and deforestation in the case of wood, high metallic content in the case of refuse-derived fuel, and foregone recycling opportunities in the case of burning some types of garbage for electricity.

Geothermal Resources.   Geothermal energy can probably be used for electricity production only in the West, where high-temperature resources are available. Yet, such large cities as Los Angeles, Salt Lake City, and Albuquerque could make use of geothermally produced electricity.[56]

Conceptually similar to conventional power production using heat engines or steam turbines, geothermal electric generation differs primarily in that the heat or steam used to run the turbines comes from the earth itself instead of from coal or fossil fuel, so no (or fewer) boilers are needed.

The rated capacity of geothermal plants ranges widely. The Pacific Gas and Electric Company's Geysers facility—the largest in the world—has 906 Mw of output. But it was developed incrementally between 1960 and 1980, and the smallest "unit" supplies only 11 Mw of power compared to 135 Mw for the largest.

Economically attractive, geothermal electricity production does have potential problems. At the Geysers, for example, substantial pollution-control equipment is needed to remove hydrogen sulfide. More generally, the long-term environmental effects stemming from the intensive exploitation of a geothermal store are not well understood.[57]

Solar Ponds.   The least suitable for city use of all of the "renewable electric" technologies discussed here is the solar pond. Most analysts agree that solar ponds capable of producing electricity can be located only in natural bodies of water on vast tracts of open space in the sun belt, since ponds capable of producing electricity must be several times larger than ponds supplying the same amount of energy as low-temperature heat.[58] Good examples of potential solar pond sites are the Salton Sea in California and the Great Salt Lake in Utah.

Economics of Community-Scale Systems

The economics of these electrical generating technologies are best compared in terms of capital costs per kilowatt of rated capacity and in terms of the deliv-

Table 11.16. Costs/kw of Capacity for Smaller-Scale Renewable Energy Electric Power Technologies.

TECHNOLOGY	CAPITAL COSTS ($)/KW
Wood-fired systems	Under $10.00 in some installations (retrofits—coal)
(New—25–60 Mw)[1]	1,000–1,900
Small Hydroelectric Plants	
(retrofit)[2]	820–2,358
(new)[3]	3,200
Photovoltaics	
(current)	20,000
(DOE costs goals–1986[4] intermediate	1,100–2,600
and central station applications)	
Large wind systems	
(100th unit—200 kw)[5]	
(100th unit—2.5 Mw)[6]	1,717
Solar Thermal Electric	
(current)	
Barstow—10 Mw[7]	14,990
Crosbyton—5 Mw[8]	4,400
Solar Ponds	
(Assuming $30/sq m[9] pond costs)	14,800—(Southwest)
	27,300—(Northwest)
Geothermal	
55—135 Mw (1980)[10]	203–315

Sources: 1. Hewett and High (1978); Rocket Research Company (1980); SERI (1980).
   2. U.S. DOE.
   3. *A New Prosperity* (1981).
   4. U.S. DOE.
   5. U.S. DOE.
   6. U.S. DOE.
   7. U.S. DOE, *Project Summaries* (1980).
   8. Monk (1979).
   9. Edessess, (1980).
   10. Dippio (1980).

ered cost of the electricity (cents per kilowatt hour). (See Table 11.16.) Capital cost is probably the best indicator when capital, fuel, and operation/maintenance costs represent roughly the same proportion of total costs for each system—especially if equipment costs, installation costs, and financing charges constitute the major element of total system costs. By the same token, this measure is far less useful when capital, fuel, and operating costs make up very different shares of the systems being compared. For example, oil-fueled power plants cost less to build but far more to fuel than nuclear plants. Nuclear plants have low fuel costs, but very high construction and operation/maintenance costs. In hydropower and photovoltaic facilities, capital costs may be higher

Table 11.17. Relative Contributions of Capital, Operating, and Fuel Costs to Total Costs of Different Thermal Power Plants.

PLANT TYPE	CAPITAL COSTS	OPERATING COSTS	FUEL COSTS
Oil-fired	Low	Low	High
Coal-fired	Medium	High	Medium
Nuclear	High	Medium	Low
Wind	High	Medium	Low
Photovoltaics	High	Low	Low
Hydropower	High	Low	Low
Solar Thermal/Electric	High	Medium/Low	Low
Solar Ponds	High	Medium/Low	Low
Waste to electricity	High	Medium	Low to medium
Geothermal	Medium	Medium	High

than in thermal plants, but fuel and operating costs are far lower. (See Tables 11.16 and 11.17.)

Wood Systems.   The capital costs of wood-fired electric generating systems depend heavily on whether the wood-burning facility is new or retrofit onto a fossil-fuel plant. In general, it takes more capital to construct a new wood-fired system than a coal- or oil-fired plant. At the least, larger boilers and fuel-handling systems are needed to obtain the same capacity rating because (non-pelletized) wood has a lower Btu content than oil or coal. (See Chapter 13.) However, coal-fired plants can often be retrofit to burn wood at relatively low cost.

Retrofitting a coal-fired boiler to burn wood can be inexpensive because the conversion technologies are similar. In some cases, wood can be substituted for coal with few modifications to the plant, although with significant capacity losses. Converting a 10-Mw unit at the Moran generating station in Burlington, Vermont, cost only $25,000 (or $2.50/kw).[59]

Converting an oil- or gas-fired plant to use wood is much more complicated and expensive than converting a coal-fired plant. Boilers must be revamped and pollution-control equipment added to remove particulates from the flue gases. Thus, it might cost less to entirely replace the existing boiler with a wood boiler (which must be larger in order to provide the same rated capacity output) or to install an on-site wood-gasification system that could provide a low-Btu gas to the existing boiler (thus eliminating the need to replace or overhaul the existing equipment).

Municipal Solid-Waste Systems.   Assessing the capital costs/kw capacity of waste plants that produce steam for sale to power producers is difficult. Differences between two systems currently under development illustrate how much costs/kw can vary. A plant in Columbus, Ohio, scheduled for completion in 1983 is slated to generate 90 Mw of power from an 80-percent trash, 20-percent coal mixture at a capital cost of $152 million (or, less than $1,900/kw).[60] An "all biomass" plant being developed for Plymouth, Massachusetts, is tentatively rated at 25 Mw with a capital cost of $124 million ($4,100/kw) by the completion date in 1985.[61]

Hydropower.   Almost all small-scale hydro development has consisted of the conversion and renovation of existing dams, which cost only about half as much as building new facilities.[62] Since capital costs can represent 90 percent of the total costs of a hydropower system, converted dams can generally provide electricity at lower rates than new projects can.

The costs of adding power-producing facilities to existing dams vary with the amount of head at the site, the planned rated capacity, and the condition of the existing facility. For a given capacity, costs will increase as height decreases. Below 65.5 ft of head, costs per kw begin to climb dramatically for a given capacity; and turbines used in low-head projects cost more per kw capacity than those used in higher-head applications because they have to be bigger to extract the same amount of power.[63] Overall, renovating existing facilities can represent from 20 percent to over 60 percent of project costs.

Geothermal.   Since 1960, Pacific Gas and Electric has invested over $159 million in 906 Mw of geothermal resources at its Geysers Plant in Sonoma County, California. The cumulative average capital cost for 15 units of varying capacity ratings (11 to 135 Mw) is $175/kw. Predictably, the oldest units cost less and the newer ones more because of inflation and pollution control requirements.

Nonetheless, in 1976 PG & E reported that its geothermal plants were the least expensive to construct—26 percent cheaper than oil-fired plants, 50 percent less than coal-fired facilities, and 62 percent less than nuclear plants.[64] Costs are less because no boilers and accompanying control equipment are needed with a dry steam resource.

Other Technologies.   The capital costs of other renewable-energy technologies are much more uncertain because operating experience is so limited. But the most dramatic improvements are expected to occur in photovoltaics development: costs per kw of capacity should drop from about $20,000 to under $2,000 with improved production techniques. As for wind-energy conversion and hydropower, costs will drop as equipment goes into mass-production, though increases in the costs of materials, land, and labor could offset this advantage.

DELIVERED COSTS PER Kwh

Predicting the busbar (generated) costs of electricity for most of these technologies involves even more guesswork than projecting capital costs. Busbar costs are a function not only of capital and financing charges but also of fuel costs, operating and maintenance expenses, and the electric generating system's capacity factor over its useful life. For renewable-energy electric generating technologies, the availability of resources is important too, even when most other factors remain constant. (For example, feasibility studies of two potential small hydropower retrofits on the Scioto River near Columbus, Ohio, indicated that delivered costs of electricity would be 3.1 cents/kwh for one but 5.6 cents/kwh for the other, despite the similarity of development costs, because the higher head at one dam allowed the generation of about twice the electricity.[65])

Another factor is that power that affects the need for peaking facilities is clearly more valuable than power that reduces the base load. Finally, with the higher sell-back rates established under PURPA 210, small power producers can generate electricity at a lower cost per kwh because investments are "repaid" faster than the same capital investment would be where sell-back rates are lower.

These complications notwithstanding, SERI found that all six technologies surveyed in group one in Table 11.18 would cost less than conventional alternatives if construction began in 1990. The California Energy Commission study suggests that all five technologies listed in group two in the table would generate electricity at lower lifecycle costs than oil-fired generation systems. Two systems (wind and wood) would produce cheaper electricity than generation technologies using natural gas, and all the rest would generate electricity for about what natural gas systems cost.

Biomass Systems.   For wood-fired systems, capital expenses, operating costs, and maintenance costs can all be significantly higher than those for an oil-fired system. Numerous firms currently using wood boilers conservatively estimate that annual operation and maintenance costs are approximately 15 percent of the capital costs.[66] Operating costs are also approximately 20 percent higher for wood energy systems than for oil-fired boilers, mainly because of the increased labor costs and operating energy costs connected with the storage and fuel-handling systems. Maintenance costs for wood energy systems can be as much as 200 percent of those for a comparable oil-fired system. However, although wood costs vary by region and type, wood costs per Btu are generally significantly lower than fossil-fuel costs—in many cases, lower enough to more than offset the higher capital and maintenance costs of a system. (See Table 11.19.)

Table 11.18. Costs per kwh of Various Renewable Electric Technologies (Projected).

TECHNOLOGY	LEVELIZED BUSBAR COSTS PER KWH (¢/KWH) (1980)	ASSUMPTIONS
Group One[1]		
A. Large wind-energy conversion system	10.7 (1980–2010) 7.9 (1985–2015) 4.7 (1990–2020)	• 25 wind machines rated at 2.5 Mw at 27.7 mph. • Total energy output at capacity factor of 38%; equal to 208 million kwh per year. • Wind-system life of 30 years. • Capital costs of $1,900/kw (1980 $) declining to $829/kw (1980 $) in 1990. • Annual operation and maintenance costs = 1% of capital costs. • 50% financed from debt at 14% interest. • General inflation rate of 7.5% per year. • 10.5% nominal discount rate.
B. Solar thermal-electric system—central receiver	20.0 (1980) 16.0 (1985) 12.0 (1990)	• 100 Mw rated capacity. • Total energy output at 35%; capacity factor equal to 303.5 million kwh per year. • System life of 30 years. • Capital costs of $2,811/kw (1980 $) in 1980 declining to $1,650/kw (1980 $) in 1990. • Annual operation and maintenance costs = 2% of capital costs.
C. Centralized photovoltaics	67.0 (1980) 32.0 (1985) 8.4 (1990)	• 10 Mw rated capacity. • Total energy outut at 24%; capacity factor equal to 21 million kwh per year. • System life of 25 years. • Capital costs of $10,851/kw (1980 $) in 1980 declining to $1,800/kw (1980 $) in 1990. • Annual operation and maintenance costs = 1% of capital costs. • 50% financed from debt at 14% interest. • General inflation rate of 7.5% per year. • 10.5% nominal discount rate.

Table 11.18. Costs per kwh of Various Renewable Electric Technologies (Projected). (*Continued*)

TECHNOLOGY	LEVELIZED BUSBAR COSTS PER KWH (¢/KWH) (1980	ASSUMPTIONS
D. New wood-fired electrical generation	4.9 (1980) 5.0 (1985) 5.0 (1990)	• 61 mw rated capacity. • Total energy output of 374 million kwh per year at 70% capacity factor. • System life of 20 years. • Capital costs of $1,208/kw (1980 $) in 1980 increasing to $1,447 (1980 $) in 1990. • Annual operation and maintenance costs = 4.5% of capital costs. • 100% of costs financed by debt at 10.5% interest. • 7.5% inflation rate. • 10.5% nominal discount rate.
E. Retrofit low-head hydroelectric facility	6.0 (1980) 5.0 (1985) 4.0 (1990)	• 3.4 Mw rated capacity at head of 20.4 ft and average mean annual flow of 2370 CFS. • Total energy output at 60%; capacity factor equal to 18 million kwh per year. • System life of 40 years. • Capital costs of $1,320/kw (1980 $) in 1980 declining to $1,003/kw (1980 $) in 1990. • Annual operation and maintenance costs = .15% of capital costs.
F. Geothermal		• 50 Mw capacity from hydrothermal resource supplying 92,500 lb/hr 464°F. • Total energy output at capacity factor of 90% equal to 16.4 million kwh per year. • Life of wells, 10 yrs.; distribution system, 20 yrs.; electrical conversion equipment, 30 yrs. • Capital costs of $1,664/kw (1980 $) in 1980; increasing to $1,704/kw (1980 $) in 1990.

Table 11.18.   Costs per kwh of Various Renewable Electric Technologies (Projected). (*Continued*)

TECHNOLOGY	LEVELIZED BUSBAR COSTS PER KWH (¢/KWH) (1980	ASSUMPTIONS
		• Annual operation and maintenance costs = 3.2% of capital costs. • 60% of well costs, 50% of plant costs financed from debt at 11.5%. • 7.5% inflation rate. • 10.5% nominal discount rate.
Group Two[2]		
A. Utility wind systems	7.5 (1990–2020)	• System operates for 30 years starting in 1990, at 35% capacity factor. • Capital costs of $880/kw (1980 $) in 1990. • Annual operation and maintenance costs account for 7% of levelized costs. • Levelized costs include yearly inflation rate of 6.5% between 1990 and 2020. • Nominal discount rate of 12%.
B. Photovoltaics—central station	31.7 (1990–2020)	• System operates for 30 years starting in 1990 at 22% capacity factor. • Capital costs of $1,894/kw (1980 $) in 1990. • Annual operation and maintenance costs account for 5% of levelized costs. • Levelized costs include yearly inflation rate of 6.5% between 1990–2020. • Nominal discount rate of 12%.
C. Wood-fired power plants	14.4	• System operates for 30 years starting in 1990 at a 80% capacity factor. • Capital costs of $1,033/kw (1980 $) in 1990. • Annual operation and maintenance costs account for 1.4% of levelized costs.

Table 11.18. Costs per kwh of Various Renewable Electric Technologies (Projected). (*Continued*)

TECHNOLOGY	LEVELIZED BUSBAR COSTS PER KWH (¢/KWH) (1980	ASSUMPTIONS
D. Conventional hydroelectric	32.5 (1990–2020)	• Levelized costs include 6.5% yearly inflation 1990–2020. • Nominal discount rate of 12%. • System operates for 30 years (1990–2020) at capacity factor of 20%. • Capital costs of $1,300/kw (1980 $) in 1990. • Annual operation and maintenance costs account for 3.1% of levelized costs. • Levelized costs include 6.5% yearly inflation (1990–2020).
E. Geothermal (flash steam)	30.1 (1990–2020)	• Nominal discount rate of 12%. • System operates for 30 years at 70% capacity factor. • Capital costs of $1,130/kw (1980 $) in 1990. • Annual operation and maintenance costs account for 3.7% of levelized costs. • Levelized costs include 6.5% yearly inflation (1990–2020).

Source: Group 1: SERI, *Cost of Energy from Renewable and Conventional Technologies* (1981).
Group 2: California Energy Commission (1981).

Plant ownership also affects the delivered costs of electricity from wood-fired plants. All other things being equal, publicly owned utilities can probably produce the lowest-cost electricity for a given fuel cost because they can use tax-free bonds for 100 percent debt financing; they are exempt from federal income tax payments, and they do not need to return profits to investors.

Municipal Solid Waste Systems. Since very few municipal solid-waste systems generate electricity on-site, precise assessments of the costs per kwh from such systems are premature. The prices of electricity produced from steam or RDF sold to utilities will depend in part on the prices charged for the input steam or RDF, but also on such features as the plant's capacity factor, the

Table 11.19. Simplified Comparison of Oil- and Wood-Fired Boilers (16.7 MMBtu/hr. Capacity).

CAPITAL INVESTMENT ($)	WOOD-FIRED (GREEN WOOD FUEL)		#2 OIL-FIRED
Boiler, installed with fuel handling and limited storage	300,000		60,000
Incremental capital investment for wood		−240,000	
Annual fuel cost (at 40 percent utilization of full capacity)			
Oil at $5/MMBtu (in steam): wood at $1/MMBtu (in steam)	59,572		297,662
Cross annual savings using wood		+238,090	
Incremental operating costs for wood fuel		− 60,000	
Net annual savings on operations		178,900	

capital costs of any adjustments needed to allow the use of RDF, and whether the plant is a base-, intermediate- or peak-load facility.

The costs of steam and electricity production from these plants can also be offset by the tipping fees charged for dumping refuse. According to the Office of Technology Assessment, some landfills currently charge tipping fees as high as $15 to $18 per ton as available land decreases.

Hydropower. Although capital costs account for most of the costs of hydro-electricity, low operation and maintenance costs and free fuel more than compensate. Still, the importance of capacity factors cannot be overemphasized. For example, SERI had demonstrated that a small hydro facility operating at a 60 percent capacity factor can produce electricity far more cheaply than many conventional alternatives. However, the California Energy Commission finds that the levelized costs of a conventional hydropower facility operating at only 20 percent of capacity are far higher than those for most thermal power plant alternatives except where oil is the primary fuel.[67] Generally, highest consideration should be given to retrofitting to produce hydroelectricity.

Geothermal. Fuel costs—the prices that the power producer must pay for the steam—represent the largest fraction of busbar electricity prices from geothermal plants. Unlike capital and operating and maintenance costs, steam costs are highly variable. The formula that PG & E uses to buy steam for the Geysers units is determined in part by the amount and cost of electricity produced from fossil and nuclear facilities during the previous year and the devel-

Table 11.20. Comparative Costs of Electricity from Five
Energy Technologies.

TECHNOLOGY	LEVELIZED COSTS (¢/KWH) (1990–2020) (IOU)	LEVELIZED COSTS (¢/KWH) (1990–2020) (MUNICIPAL OWNERSHIP)
Utility wind system	7.5	5.0
Photovoltaics—central station	31.7	20.9
Wood-fired power plants	14.4	11.6
Conventional hydroelectric	32.5	16.6
Geothermal (flash steam)	30.1	16.1

Source: California Energy Commission (1981).   IOU = Investor-owned utility

opment costs of the field. (A surcharge to cover the cost of reinjecting spent geofluid is also added.) Consequently, the cost of steam will rise and fall with the prices of fossil-fuels and of electricity from conventional sources.

At the Geysers, PG & E has found maintenance requirements to be somewhat greater than those in conventional fossil-fuel-fired power plants. Turbine blades and electrical equipment fail more frequently at geothermal stations because of the corrosive nature of the rock- and dust-laden steam, and maintenance shutdowns have resulted in higher forced outage rates than at first expected.

Other Technologies.   Computing the costs of delivered power from technologies not in widespread use is highly conjectural, but a few general points can be safely made. First, estimates of delivered energy costs depend heavily on future cost reductions in equipment. Second, fuel costs for renewable technologies will amount to little or nothing since sunlight and wind are free, though developers may need to purchase solar access or wind access easements. Third, there is no reason to assume that operating and maintenance costs for these systems will be inordinately high. Fourth, ownership patterns will affect the delivered costs of electricity from these systems. (See Table 11.20.)

DEVELOPMENT OPTIONS

Given the fears about unproven technologies, the reluctance to enter into "third-party" partnerships, and the worries about subsidy, tax, and regulatory barriers, investor-owned utilities are less likely to develop community-scale renewable electric systems than are private developers or city governments working through municipal utilities or with private partners.

Of the private developers, the ones most likely to develop community electrical systems are companies with some fuel resources already at hand—such as industries with large amounts of waste heat that could be used to run generators or with large volumes of waste products that could be used to fire steam boilers. The investment criteria applied by most other private firms clearly constrain development: indeed, one study of potential cogeneration projects found that private developers' costs per kwh produced were somewhat higher than the costs for investor-owned utilities and much higher than those for municipal utilities.[68]

Another alternative is for private developers to enter into partnerships with utilities or city governments: the private firm could provide the development expertise or the resource, and the utility or city government could provide access to less costly capital or a reliable market for the power produced. For example, a steam-producing resource-recovery plant in Saugus, Massachusetts, was developed by two private companies—a construction company that owned a landfill site slated for closing and an engineering firm brought on as a technical consultant. The new refuse-to-energy company that they formed secured commitments for refuse supply from surrounding municipalities, while Saugus posted revenue bonds when it learned that the facility would provide substantial tax revenues for the town.[69]

Renewable-energy-based electric generating facilities could also be developed by cities themselves, especially those with municipally owned utilities and readily accessible resources. Idaho Falls is currently developing 24.6 Mw of hydroelectric capacity on the Snake River.[70] The utility at Burlington, Vermont, has retrofit one 10-Mw power station to burn wood and plans to build another with 50 Mw of rated capacity.[71] Columbus, Ohio's municipal utility plans to stop purchasing power from an investor-owned utility and to start generating most of its power from a city-owned resource recovery plant.[72]

Cities without municipal utilities could also develop their own generating facilities to produce electricity for use within their own borders or for sale to the local utility, particularly if there is ready access to "free" resources, either in the form of municipally owned waterways or municipally owned land in areas with good sunlight or wind access.[73] Paterson, New Jersey, is investigating the possibility of developing 7.5 Mw of hydroelectric capacity on the Passaic River.[74] Well-planned resource recovery plants using nonrecyclable urban wastes represent another possibility.

Of course, the large capital investments required for such projects may prevent even those cities with readily available resources from developing them on their own. In such cases, partnerships with private developers or investor-owned utilities may work.

12. Industrial Process Heat

OVERALL POTENTIAL

Since industry uses roughly 40 percent of the U.S. energy budget, it has enormous potential for improved energy efficiency and increased renewable-energy use. Some 43 percent of industrial energy goes for process heating, the focus of this chapter.[1] The rest is used for machine-driven electric power and such feedstocks as construction asphalt and metallurgical coal.

At present, many urban industries rely heavily on scarce or expensive fuels. The types and amount of fuels used by local industry depend largely on fuel availability and the industrial mix. (See Tables 12.1, 12.2, and 12.3.) Urban industries that rely heavily on expensive, scarce, or imported fuels may be particularly concerned with decreasing this dependency. Uncertainties about price and availability of oil may encourage industry managers to work with city officials to increase energy efficiency and the use of renewable fuels, as St. Paul began doing in 1979.

Relying heavily on electricity, which is getting ever more expensive, also poses problems for industries. Portland's aluminum industry uses 25 percent of the city's industrial energy in processing, and 89 percent of that energy is electric. While cheap hydroelectric power initially attracted the aluminum industry, the increasing electrical demands that accompanied regional growth now have the aluminum industry, the city, and regional utilities concerned about the high monetary and environmental costs of constructing new coal and nuclear generating facilities.

Energy and Employment

Energy price increases or shortages can exacerbate industrial employment problems, too. In Northeastern and Midwestern cities, where manufacturing employment has declined 0.1 percent annually for the past decade and where industry relies heavily on imported oil, city officials may be particularly concerned about how energy price hikes and shortages affect local employment.[2] For example, in Bridgeport, Connecticut, energy consumption in the industrial sector declined 28.6 percent between 1971 and 1975.[3] Consuming more energy would not have created more jobs for cities like Bridgeport, but a carefully planned program of conservation and the increased use of renewable energy could help local industry avoid slowdowns, lay-offs, production cuts, and plant closings next time oil prices escalate rapidly. Well-planned investments could also help create new jobs in manufacturing, installation, and maintenance.

While manufacturing employment in the South and West is in much better shape (increasing 2.4 percent annually from 1971 through 1980 and expected to keep growing during the next 10 years), industrial conservation and renewable-energy improvements will create jobs in these regions as well.[4] For example, the Los Angeles Energy Management Advisory Board expects that city's historical overall job growth to continue. Still, it is trying to make sure that job losses in such sectors as construction, agriculture, mining, transportation, and trade can be made up with new jobs created by the "Energy/LA" Action Plan—a comprehensive, city-wide program calling for $1.2 billion in conservation and renewable-energy investments. Fully implementing the proposed industrial conservation improvements would alone create 326 new construction jobs over eight years and 840 new permanent jobs in operation and maintenance. At the same time, it would create secondary jobs by keeping money in the local economy.[5]

A growing body of literature identifies great potential for improving industrial energy efficiency.[6] After ranking its 18 member countries, the International Energy Agency found that the U.S. is the most energy-efficient manufacturer of aluminum, but ranks 13th in steel, 13th in cement, 9th in pulp and paper, and 9th in petroleum production.[7]

Table 12.1. Percentage of Local Energy Use Consumed by Industry.

CITY	PERCENT TOTAL ENERGY USED BY INDUSTRY
Boulder	8.9
Dayton	49.4
Greenville, NC	17.0
Dearborn	47.2
Lawrence, Mass.	37.1
Knoxville	15.3
Bridgeport, CN	21.3
South Florida*	14.8
Los Angeles	12.4
Miami	6.3
Portland (1971)	39.0
Philadelphia	20.0

*Miami, Broward, Dade, and Monroe Counties.
Source: Comprehensive Community Energy Management Plans Developed by Cities 1978–1980.

Table 12.2. Industrial Energy Consumption by Major Industry Group and Fuel Type, 1978.

INDUSTRY (SK CODE)	TOTAL ENERGY CONSUMED (10^{12} BTU)	ELECTRICITY (%)	FUEL OIL (%)	COAL AND COKE (%)	NATURAL GAS (%)	LP GAS (%)	OTHER FUELS (%)
Food and kindred products (20)	980.5	14.1	20.7	11.0	43.6	0.0	10.1
Tobacco products (21)	20.8	20.8	34.6	—	10.8	0.0	33.8
Textile mill products (22)	326.6	28.1	28.8	9.0	23.0	1.0	10.2
Apparel and other textile products (23)	69.1	33.3	12.5	2.1	22.3	0.0	29.3
Lumber and wood products (24)	231.1	24.6	20.5	—	26.4	1.0	27.5
Furniture and fixtures (25)	55.3	26.3	14.0	5.0	33.7	1.0	19.9
Paper and allied products (26)	1301.1	12.0	39.2	16.7	27.0	0.0	5.0
Printing and publishing (27)	91.5	38.6	9.1	—	30.8	0.0	21.0
Chemicals and allied products (28)	2905.3	17.1	13.8	12.0	50.8	0.0	6.0

Industry						
Petroleum and coal products (29)	1122.7	9.2	6.4	—	78.7	0.0	5.4
Rubber and misc. plastics (30)	261.0	30.0	21.5	7.5	28.7	0.0	11.7
Leather and leather products (31)	21.6	22.0	34.1	6.6	26.1	0.0	10.9
Stone, clay, and glass products (32)	299.8	8.6	12.3	29.6	42.4	0.0	10.9
Primary metal industries (33)	2710.9	21.0	12.9	24.7	37.5	0.0	3.6
Fabricated metal products (34)	400.2	22.5	11.8	2.8	49.9	1.0	11.0
Machinery, except electrical (35)	350.7	29.3	11.1	6.6	39.9	1.0	11.9
Electric and electronic equipment (36)	254.9	34.9	12.6	5.4	26.1	1.0	10.9
Transportation equipment (37)	397.7	27.3	12.6	13.6	38.0	1.0	7.5
Instruments and related products (38)	79.0	24.5	16.7	—	23.3	0.0	35.2
Miscellaneous manufacturing industries	49.0	27.8	18.2	—	31.0	2.0	21.6

Table 12.3. Principal Industries and Industrial Fuel Mix by City.

CITY	PRINCIPAL INDUSTRIES	TOTAL INDUSTRIAL ENERGY USE (10^9 BTU)	Fuel Mix (Percent Industrial Energy Use)						
			ELECTRICITY	FUEL OIL	NATURAL GAS	PROPANE	COKE AND COAL	STEAM	OTHER
Boulder	Computers; paving and gravel; aircraft	1,838	23		77				
Dayton	Transportation equipment; food machinery	16,246	24.7	8.7	45		16.9	4.7	
Greenville	Light manufacturing and assembly; food processing; agribusiness; pharmaceuticals	815	41	42	18				0.2
Wayne County, MI	Motor vehicles; steel machinery; fabricated metals	133,270	19.9	10.3	40.8		23.5		5.5
Knoxville	Apparel; textile mills; food processing; machinery	7,893	28.7	11.1	46.3	1.4	11.2		1.3

Bridgeport	Fabricated metals; transportation equipment; electrical equipment	11,980	20.9	54.9	24.5	
South Florida	Food processing; stone, clay and glass	13,682	66	.03	30	
Los Angeles	Petroleum refining; food processing; stone, clay and glass; transportation equipment	44,700	29.5	24.6	45.6	
Portland	Paper and allied products; primary metals (aluminum); construction; lumber and wood; food processing	98,821	28.4	17.4	43.9	10.3

Source: Comprehensive Community Energy Management Plans developed by Cities 1978–1980.

Renewable energy's potential in industry is also underutilized. Direct plant-site combustion of such readily available renewable fuels as scrap wood at lumber mills is increasing, and technologies to convert renewable fuels to easily used gases and liquids are also becoming increasingly widely accepted. (See Chapter 13.) The use of active solar collectors for industrial process heating is more expensive and less immediately feasible. But 27 percent of industrial process heating requirements nationally are for temperatures less than 550°F, which solar collectors can supply.[8] Using solar heat for preheating and using improved collectors to supply higher-temperature heat should increase solar energy's potential for meeting industry's process heating needs.[9]

Such "theoretical" potential is verified by real-world improvements. Nationally, industry's energy efficiency can be measured by its energy consumption per "gross product." Also called the "energy-value added" ratio, this measure of industry's productivity relative to its energy consumption rose sharply from 1967 and dropped equally sharply from 1970 to 1979.

Industry's interest in energy conservation is manifested in its voluntary participation in the Department of Energy's monitoring program. In 1977, manufacturing corporations participating in the program accounted for 53 percent of total U.S. industrial energy consumption. Five of the industry groups had already exceeded their 1980 targets by 1979, and overall these 10 industry groups reduced their energy consumption per unit of output by an average of 15.4 percent between 1972 and 1979.[10] Since most of the industry groups increased output during that period, overall energy efficiency in the 10 industries actually increased an average of 17.6 percent, primarily because of low- and no-cost conservation measures.[11]

Some cities have used the information from this monitoring program to develop industrial conservation programs and efficiency goals. For example, the Los Angeles Energy Management Advisory Board urges the city to work closely with utilities and industry to reach by 1985 the state goal of 15 percent improvement over 1979 consumption levels.[12] The staff of the Los Angeles energy office also searched through issues of *Energy User News,* the files of a local gas utility, and information provided by the American Institute of Plant Engineers for case studies of successful industry conservation projects. They subsequently used these to evaluate the technical and economic feasibility of industrial conservation improvements and disseminated that information to local industries.

The Department of Energy estimates that the average industrial plant can reduce energy consumption 3 to 7 percent through administrative or informational programs and an additional 5 to 10 percent through small incremental investments in improved equipment. However, the potential savings of low-cost conservation programs may be greater still. Indeed, the average 15.4 percent savings achieved by major industries in the DOE voluntary monitoring program was attributed largely to such low-cost measures as energy-management training, insulation, and better fuel-use controls.[13]

Barriers to Increased Efficiency

The barriers to industrial energy efficiency are formidable, success stories aside. First, few firms have any economic incentive to make major investments in energy efficiency because in even the most energy-intensive industries, energy comprises only 2.8 percent of total operating costs.[14] Second, obtaining the necessary capital can be difficult. The average return on sales in the steel, aluminum, and cement industries was only 3.9 percent in 1979, so financing energy conservation investments internally poses severe problems. Even when investment capital is available, most industrial managers invest first in facilities that will expand production, while next in line are required investments in pollution control and in industrial health and safety. Conservation improvements thus compete with other uses of the remaining "discretionary capital."[15] Moreover, federal tax laws allow conventional energy costs to be deducted from taxes as an annual operating expense.

A second barrier to conservation and renewable energy investments is the lack of reliable information on the technical performance and cost savings associated with the new energy systems. Because of declining profitability, industrial investment in research and development has been decreasing since 1964. The technical and economic information available is most easily used only by large companies with in-house engineering expertise. For example, a city-run survey of industries in Seattle found that the 10 largest firms were tracking energy consumption and starting to make conservation improvements, while only 3 of 18 mid-sized firms surveyed were monitoring their energy use.[16]

Local government programs can greatly help industry overcome these barriers to investments. For example, Los Angeles predicted that its energy-management efforts could save 50 percent more industrial energy than market forces alone would prompt industry to save. Seattle estimated that an energy-conservation package combining information, regulations, and incentives would reduce industrial energy use by 10 to 30 percent in each facility.[17]

Local government programs directed toward industries can take several forms. In high-growth areas, development policies should attract industries that use energy efficiently.[18] Cities could, for example, award business licenses and review development plans on a priority basis for industries able to show that they would use energy frugally. Alternatively, zoning provisions could be structured so that industries expending large quantities of waste heat of potential use in cogeneration or district heating have an incentive to locate near large residential and business complexes. (See Chapter 10.) For established industries, cities could develop information, incentive, or regulatory programs to encourage or require conservation efforts. (For example, Los Angeles' city-owned Department of Water and Power provides on-site audits for large industrial and institutional electricity customers, identifying relatively inexpensive conservation improvements. On average, 60 percent of the identified conservation measures are being implemented.) Where the problem is not attracting new industries but retaining old ones, cities could combine regulatory requirements with financing programs.

Technical Approaches and Issues

Numerous inexpensive approaches can be used to reduce energy consumption at industrial plants. One of the least expensive is to train employees to turn off lights and equipment when not in use. Typically, such training programs are incorporated into a larger conservation program. For instance, a chemical plant in Lynwood, California, owned by Cargill, Inc., trained employees to shut off gas and electric equipment when not in use as part of a larger effort that included changing steam lines and adding a new heat exchanger.[19]

Another approach to improving energy efficiency in industry is to improve fuel combustion in boilers, kilns, and other heaters. By measuring the oxygen content of gases expelled from industry burners, energy engineers can determine whether the fuel-air combustion mix contains the right amount of air. Simple boiler or furnace adjustments can optimize the ratio of air to fuel to improve combustion. Keeping burners soot free also maximizes efficiency. In the petroleum industry, 72 percent of the corporations reporting energy savings between 1972 and 1979 under DOE's voluntary monitoring program saved significant amounts of energy by making combustion improvements in their process heaters.

Many inexpensive industrial maintenance improvements are aimed at making process steam-use more efficient, since 36 percent of all industrial fuel is used to produce steam. Improvements can range from plugging leaks to improving steam distribution to recovering steam condensate. The Union Carbide plant has reportedly saved 78 million Btu of energy annually by training personnel to inspect, maintain, and replace defective steam traps.

To improve industrial energy efficiency by optimizing production scheduling, companies must carefully analyze product output. For companies that consume approximately the same amount of energy at lower and higher rates of output, operating 24 hours a day in a "batch" mode for several days and then shutting down entirely for several days may be the most energy-efficient path. Other companies may be better off operating fewer pieces of energy-consuming equipment at full capacity, rather than more at a lower capacity. (DuPont reports saving the equivalent of 10,000 barrels of oil per year by shutting down some of its chemical-refining columns operating at partial capacity and operating others at full load.[20])

A final "housekeeping" approach is to insulate boilers, pipes, and other heat-transfer equipment. Indeed, a 1975 study by the Federal Energy Administration estimated that if all U.S. industrial plants made maximum use of insulation, the equivalent of 3.5 billion barrels of oil could be saved between 1976 and 1990.[21] Among the improved industrial insulating materials now becoming available is ceramic fiber, which is easier to store and install and has a higher R-value than the firebrick formerly used on industrial furnaces. In Pasadena, a plumbing-manufacturing plant reduced overall annual energy use by 25 percent by lining its furnaces with $225,000 worth of lightweight fiber insulation, which paid for itself in one year.

Insulation can have especially great energy-saving impacts when added to steam-distribution systems. One study reports that an uninsulated 8-inch diameter steam pipe carrying steam at 350 psi loses approximately 3 billion Btu per 100 ft.[22] Larger pipes operating under higher pressure lose even greater amounts of heat. Along with other low-cost improvements at a manufacturing plant in Southgate, California, insulating steam lines reduced annual gas bills by 67 percent.[23]

Urethane foam is also increasingly popular for industrial applications, particularly the easy-to-install spray-on type. Like other organic materials, however, urethane burns at high temperatures, releasing toxic combustion gases. It may also crack and shrink at high temperatures below the flash point. Consequently, urethane is used to keep the cold in or out of an industrial process.

The owners of the Hunts Point Cooperative Market in New York City added urethane foam and PVS outer jackets to 1.5 miles of brine-carrying steel pipes that connect chillers in a utility building to the warehouse and market buildings, making possible the use of an innovative heat-recovery system that can save 220,000 gallons of fuel oil annually.[24]

Economics

Low-cost administrative and housekeeping changes generally have very rapid paybacks. (See Table 12.4.) Not all simple conservation improvements are inexpensive, but even very expensive programs may have rapid paybacks. For example, a medium-sized Union Carbide Plant that initiated a program to inspect, maintain, and replace defective steam traps at a cost of $15 million per year saved $175 million in fuel costs during the first year.[25]

Table 12.4. Economics of Industrial "Housekeeping" Measures.

FIRM	LOCATION	DESCRIPTION OF MEASURE	Cost		Annual Savings		SIMPLE PAYBACK
			MATERIALS ($)	LABOR ($)	THERMS	$ (@30¢/THERM)	
Hughes Aircraft Cargill	Tuscon Lynwood, CA	Employee Energy Watch			300,000	100,000	
		Insulate steam pipes	3,000	0			
		Install heat exchanger	7,000	0			
		Fine tune gas burners	0	0			
		Employee energy-awareness training program	0				
		Reduce time required for process cycles					
		Total	10,000		234,000	78,000	.13 year
Forging Co.	Fontana, CA.	Install ceramic fiber insulation on a furnace	4,000	1,000	55,400	16,600	.30 year
Ceramic Company	San Fernando Valley, CA	Install ceramic fiber insulation on three kilns	2,550	650	25,271	7,500	.50 year
Plumbingware manufacturer	Pasadena, CA	Line furnaces with lightweight fiber insulation	225,000		745,000	223,500	1 year
Juice Company	Orange County, CA	Install insulation on pasteurizer	1,300	425	2,026	608	3 years

Source: Case Histories of Energy Savings in the Industrial Sector, Pat Lang, 1981.

WASTE-HEAT RECOVERY

The potential for reducing industrial energy consumption by recovering waste heat equals up to 25 percent of industrial process energy consumption.[26] This heat can be recovered from various sources at various temperatures. Direct firing of furnaces produces by-product heat at 1,200° to 3,000°F. Exhaust temperatures from boilers, gas turbines, reciprocating engines, and heat-treating furnaces range from 450° to 1,200°F. In addition, many low-temperature sources such as steam condensate, cooling water from furnace doors and engines, and air conditioning and refrigeration condensers produce heat and hot water at 90° to 450°F.

Various techniques can be used to recover waste heat.[27] If the hot air or exhaust gas is relatively clean, it can be used directly for another process. A more common approach is recuperation, in which a heat exchanger is used to transfer the heat content of waste gases, air, or hot water to another medium, rather than using it directly. In regeneration, another option, heat from exhaust gases is conducted to a heat exchanger and stored before being used to preheat combustion air. Alternatively, waste-heat boilers can use a recuperation system in which waste heat is used to help produce process steam, heat water, or generate electricity.

In nonindustrial applications, heat exchangers operate at an average of 80 percent efficiency. However, in industrial applications scale build-up and corrosion caused by hot gases reduce the ability of heat exchangers to transfer heat. Other limitations include the relatively high costs and restrictions on the type of waste heat streams with which heat exchangers can be used.[28] Metal and ceramic recuperators (or "economizers") used to recapture heat from exhaust for use in preheating combustion air currently operate at only 30 to 40 percent efficiency. Heat wheels can be used to exchange both heat and the latent heat contained in water vapor between exhaust and entry gas streams. (See Chapter 8.) A newer, more efficient heat-exchange technology uses heat pipes to transfer exhaust-gas heat to combustion air, much as a simplified air conditioner does. Some types of heat-exchange equipment (finned tube generators, for example) can transfer heat from a gas to a liquid or vice versa.

Despite their limitations, these systems are popular. The chemical industry has made major capital investments in waste-heat recovery and industrial cogeneration. Two thirds of the corporations participating in DOE's industrial energy monitoring program claim to have made significant investments in heat-exchange and improved waste-heat recovery equipment.[29] In the cement industry, where 80 percent of industrial process heat is used for firing kilns, conservation efforts are focussing on recovery of heat from kiln exhaust gases.[30] The glass and paper industries are also investing more in heat recovery.

Examples of individual plants that have successfully installed heat-recovery equipment abound.[31] Cargill, Incorporated, installed a $7,000 heat exchanger in its plant in Lynwood, California. The heat exchanger uses waste heat from plant processes to preheat combustion air for the incinerator and to maintain a "hot room" for chemical processing. In Baldwin Park, California, a foundry installed a boiler blowdown heat-recovery system on five boilers that saved almost 316,000 therms of natural gas in the first year, showing a payback of only a few months. A smelting operation in Los Angeles installed a recuperator to preheat combustion air with exhaust gas from the furnace, saving nearly 90,000 therms of natural gas per year.

A final case study illustrates both the large theoretical potential for reducing energy consumption and some of the technical difficulties of recovering waste heat in industrial processes. A Solar Energy Research Institute team studying the energy inputs and exhaust energy outputs of an aluminum can manufacturing company found that about 48 percent of the total energy input leaves in the waste gases. Investigating the feasibility of recovering this lost energy, the team found that the most important factor affecting gas recycling at this plant is maintaining an adequate air-to-gas ratio for combustion.[32] Since no more than 25 percent of the exhaust gases could be recycled without affecting the current ratio, either the 25 percent limit would have to be observed or some means of generating adequate temperatures at a lower air-to-gas ratio would have to be found. Theoretically, the most important limiting factor on gas recycling under the second option would be maintaining a safe concentration of vaporized organic solvents, which is not a problem unless more than 55 percent of the exhaust gases were to be recycled. At the 55 percent level, it turns out, exhaust gases would supply 27 percent of total process heat requirements, thereby greatly reducing natural gas consumption.

While heat-recovery systems make use of lower-quality waste energy "cascaded" from high-temperature industrial processes, another approach involves using waste heat from cogeneration. (See Chapters 9 and 11.) In electrical power production, all of the many technical approaches for recapturing waste heat produced by on-site power generation involve a so-called "combined cycle," in which a gas turbine generates electricity and hot exhaust. As in the gas turbine topping cycle, turbine exhaust gases are used to heat a working fluid in a boiler. Steam (or organic vapor) thus created drives a second electric generator, which gives off waste heat that is used for process needs.[33]

Economics

The economic appeal of waste-heat recovery depends on the costs of purchased fuels, the amounts and temperatures of available waste heat, the distance that waste heat must be transported, and other factors. In 1977, researchers estimated the cost of a gas-to-gas heat exchanger recovering 4 MMBtu per hour from a 500°F-heat source at $120,000 to $200,000.[34] If the system operated at 80 percent effectiveness for half the year and replaced natural gas costing $1.50 per MMBtu ($0.15/therm), it would pay for itself in 4.6–7.7 years. If the costs of such a system rose 10 percent by 1980, it would have cost $159,720 to $266,200 in 1980, by which time industry was paying $2.12 per MMBtu of natural gas. Hence, payback by 1980 would have fallen to between 4.3 and 7.2 years. If the heat exchanger replaced deregulated natural gas (which cost $6.80/MMBtu at the marginal price in January of 1980), it would have paid for itself in 1.3 to 2.3 years.[35]

In 1977, a gas-to-gas heat exchanger capable of delivering 12 MMBtu per hour from a higher-temperature heat source would have cost $260,000 to $450,000 and had a payback of 3.25 to 5.6 years. Given equipment costs and energy prices as of January, 1980, the system would have cost $346,060 to $598,950 and paid for itself in 3.06 to 5.3 years when displacing regulated natural gas or industrial distillate.

Since 1980, regulated natural gas prices have risen still further to $3.00/MMBtu or higher, and heat recovery equipment has become less expensive. As a result, paybacks at some factories may well be shorter now. (See Table 12.5.)

IMPROVED ENERGY EQUIPMENT AND CONTROLS

Rather than recapturing heat wasted because combustion is inefficient or because equipment and load are mismatched, industry managers can choose to replace or add newer, more efficient equipment and control systems to reduce heat losses. Since almost every industry has many "unit" processes (for example, generation and distribution of steam), they can be improved incrementally. Substantial energy savings can be realized in this way, especially if computerized control systems are installed to optimize fuel use.

Unit changes can be as simple as replacing one machine or as complicated as redesigning a system. For example, turbulators costing only a few thousand dollars can be installed in boilers to slow water flow, thus increasing heat transfer. One fertilizer producer in California realized savings of 12,000 therms of natural gas per year after adding turbulators.[36] Similarly, a can-manufacturing plant in Pasadena replaced its old burner system with a new system that could be controlled more precisely to maintain the correct gas-to-air ratio. This system cost only $12,000 and paid for itself in two years.[37] In a more complex improvement costing $39,140, a forge in Santa Ana, California, replaced its existing furnace with an indirect-fired recirculating air furnace, saving $34,934 in therms of natural gas per year.

Table 12.5. Economics of Waste-Heat Recovery in Industry.

FIRM	LOCATION	DESCRIPTION OF MEASURE	Cost ($)		Annual Savings		SIMPLE PAYBACK
			MATERIALS	LABOR	THERMS	$ (@30¢/THERM)	
Beverage manufacturer	San Fernando Valley	Install economizers on four boilers plus a blowdown heat-recovery unit	190,000	47,500	337,000	101,000	2
Ceramic manufacturer	Commerce, CA	Install a recuperator on a kiln	32,000	8,000	148,700	44,610	.7
Smelter	Los Angeles	Install a recuperator on a furnace	46,000	11,500	89,615	26,880	2
Foundry	Baldwin Park	Install boiler blowdown heat-recovery unit on 5 boilers	10,819	2,700	315,511	94,650	.1
Cardboard container corp.		Install blowdown heat-recovery unit on a boiler	3,100	800	37,800	11,304	.33
Carpet cleaning/dyeing firm		Install counterflow heat exchanger	12,000	2,000	19,543	5,800	3
Foundry	Anaheim, CA	Install recuperator on furnace	3,245	800	4,500	1,350	4

Source: Lang, 1981, Case Histories of Energy Savings in the Industrial Sector.

In some cases, factory managers may find adding new energy-recovery equipment more economical than simply replacing inefficient units. One California synthetic-resin manufacturing company installed condensate return-lines at the same time that it insulated its steam lines. Overall, the plant's total gas consumption fell by 47 percent.[38]

The most sophisticated energy-control systems are computerized. (See Chapter 8.) In industrial applications, computers can monitor boilers, condensate return systems, and waste-heat recovery equipment. New computers now available can be "hard-wired" or they can use software packages to optimize plant ventilation, equipment start-up times, and fuel-use rates and to shed electrical demand during peak loads. In particular, optimizing "enthalpy" (or ventilation) is important since process-heating requirements can increase greatly if outside air at low temperatures is admitted. A Winnebago-manuacturing plant in Forest City, Iowa, recently wired its chassis-preparation room—the most energy-intensive section of the plant—for computer control of fuel use. The system cost $160,000 and saved $166,900 in the first year.[39] In Taft, Louisiana, a Union Carbide plant installed a computer system to monitor plant operations, schedule environmental maintenance, and control the use of steam, by-product gases, and raw materials. The firm now uses 12 percent less energy than in 1972, while productivity has increased 30 percent.[40]

Economics

Improved energy equipment and control systems vary widely in cost. Most such equipment is slightly less economically attractive than waste-heat recovery systems: while some measures have paybacks of only one or two years, such improvements as replacing furnaces and kilns have four- or five-year paybacks. Thus, few firms will make these investments unless the equipment was due for replacement anyway.

As Table 12.6 shows, the cost of computer systems installed at industrial and commercial facilities varies with the degree of sophistication.[41] Installed costs reflect the number of points monitored in the building, labor rates, the plant's design, and the location of equipment. Wiring and communication expenses can account for more than one half of the computer system's installed cost.[42] In fact, the wiring has almost no resale value, so a company that invests in a system will take a large loss if it decides to replace it. Overall, the higher costs and longer paybacks of these sophisticated systems compared to those of simpler "housekeeping" and heat-recovery measures may restrict investments.

The most common source of industrial financing for any type of conservation improvements is internally generated capital. Industrial firms with little internal capital may seek bank loans to finance conservation equipment. But high

Table 12.6. Economics of Computerized Energy-Management Systems.

FIRM	LOCATION	FUNCTIONS	Cost ($)		ANNUAL ENERGY SAVINGS	SIMPLE PAYBACK	FINANCING
			COMPUTER	WIRING			
R. H. Macy Properties	Paramus, NJ	Peak-load reduction at electrically heated shopping center	30,000		About ⅓		internal cash flow
Rouse Corporation	Throughout U.S.	Climate control at 10 shopping centers	12,000–18,000	38,000–57,000	12–15%		lease-purchase
Winnebago	Forest City, Iowa	Controls fuel use in chassis-preparation room (distributed processing system)	160,000		$166,900	1 year	
Schlosser Forge	Cucamonga, CA	Control of high-temperature forging process	400,000		$199,000	2 years	
Goodyear Tire	Albany, NY	Temperature control in 300,000 ft^2 warehouse	7,302		$9,000 + $9,000 savings in maintenance costs	0.4 years	

Sources: Interviews with Rouse Corp., R. H. Macy (3/81); Lang, 1981.

interest rates (17 to 22 percent) and short terms (3 to 5 years) may foreclose this option. (See Table 12.7.)

To spur reluctant industry managers to invest in computers and other sophisticated conservation equipment, a growing number of engineering firms and equipment manufacturers are offering innovative financing plans. These allow industries to install and use the conservation equipment without tying up their capital in what they view as unproven technologies. All such alternatives, which range from leasing to "shared savings" and "guaranteed cash-flow," cost more over 5 years than a purchase financed with a 5-year, 18-percent loan. But many plans would require no capital investment and most include maintenance and engineering services. With small or no capital investments to offset, monthly payments for equipment leasing or rental are immediately compensated for by utility bill savings. Some arrangements even guarantee a certain level of savings.[43]

PROCESS CHANGES

All of the conservation improvements described so far make improved use of energy in existing industrial processes. Another approach to decreasing industrial energy consumption is to redesign the manufacturing process. Depending on the industry, new processes can reduce energy use per unit of output by 20 to 90 percent.

In the past, new industrial processes were developed primarily to produce a needed product more efficiently and cheaply. Today, however, process improvements may be triggered by rising fuel prices. Chances are that most such efforts will take place in the five major manufacturing categories that presently account for 75 percent of total industrial fuel purchases.[44]

Major Types of Industrial Processes

Pulp and Paper.   The pulp and paper industry is the third largest energy user among manufacturing industries. Purchased fuels, which account for half of the energy this industry uses, include oil (39 percent), natural gas (27 percent), coal and coke (18 percent), and electricity (18 percent). The other half of this industry's energy needs are met by such on-site by-products as bark, waste wood, "white liquor," and "black liquor."

Many process improvements being developed in the pulp and paper industry are aimed at reducing unnecessary drying operations. Currently, pulp is dried at the pulp mill and then shipped to a paper mill, where it is moistened before being formed into paper. Integrating the two operations would reduce the total amount of energy required to produce paper by about 10 percent.[45] Alterna-

Table 12.7. Six Methods of Financing Energy Conservation Projects.*

	1. PURCHASE USING 5-YEAR $100,000 BANK LOAN	2. 5-YEAR LEASE-PURCHASE AT 20 PERCENT	3. GUARANTEED CASH FLOW 5-YEAR CONTRACT	4. SHARED SAVINGS WITH 5-YEAR CONTRACT	5. ENERGY SERVICES 5-YEAR CONTRACT	6. 5-YEAR LEASE AT 18 AND 25 PERCENT INTEREST
			(Note: Spokesmen of companies offering this plan said their firms must borrow money at prime rate or higher, so they must charge higher rates than prime.)			(Note: Leasing rates can vary. Some lessors offer prime rate to leasers, other add up to 7 points to the prime rate.)
Costs						
Payments toward equity	$100,000	$100,000	$100,000	None with most plans.	None with most plans.	0
Capital outlay	$100,000	5–10% of Equipment cost.	Estimated $15,000 for feasibility study	0	0	5–10% of equipment cost
Total payments	$152,360	$156,000	$160,143	$125,000	0	$150,000 at 18% $172,500 at 25%
Energy cost savings to user over 5 years	$250,000	$250,000	$250,000	$125,000 (half of $250,000 savings over 5 years)	$100,000 (if energy services firm takes tax advantages; $40,000 if user takes tax advantages)	$250,000
Tax Benefits (Not applicable to nonprofit institutions.)						
Depreciation	$100,000	$100,000	$100,000	0	$100,000 if user elects to take it	0

Interest deduction	$52,360.40	$56,000	$69,143	0	0	User can write off 100% of lease payments, in this case either $150,000 or $172,500, depending on leasing rate
Total deductions	$152,360.40	$156,000	$169,143	0	$100,000 if user elects to take it	$150,000 at 18% $172,000 at 25%
Credits	$10,000 investment tax credit (ITC) (10%); $10,000 energy conservation tax credit (ECTC) ($10%, if applicable)	$10,000 investment tax credit (ITC) (10%); $10,000 energy conservation tax credit (ECTC) ($10%, if applicable)	$10,000 investment tax credit (ITC) (10%) plus $10,000 energy conservation tax credit (ECTC) (10%, if applicable)	0	$10,000 investment tax credit (ITC) (10%) and $10,000 energy conservation tax credit (ECTC, if applicable) if user elects to take them	If available, user can negotiate for $10,000 investment tax credit (ITC) (10%) and $10,000 energy conservation tax credit (ECTC) (10%)
Net tax benefits	50% bracket $86,180.20 plus $10,000 ECTC, if applicable	$38,000 plus $10,000 for ECTC if applicable	$94,571.50 plus $10,000 for ECTC if applicable	0	$60,000 if user elects to take them, add $10,000 for ECTC, if applicable	$75,000 at 10%, $86,250 at 25%, without ITC or ECTC. (if user takes credits, lease payments increase)

Net Benefits over 5 Years
(Savings plus net tax benefits minus costs over 5 years)

	$183,819.80 (Without ECTC) $193,819.80 (With ECTC) PLUS OWNERSHIP	$182,000, plus $10,000 for ECTC if applicable. PLUS OWNERSHIP	$175,428.50, PLUS OWNERSHIP	$125,000 PLUS OWNERSHIP WITH SOME PLANS	$100,000 Whether or not user takes tax benefits	$175,000 at 18% $163,750 at 25% WITHOUT ITC OR ECTC, ADD $10,000 FOR ITC AND $20,000 FOR ITC AND ECTC.

Assumptions:
$100,000 Energy Conservation Project.
$50,000 Per Year Energy Savings.
10 Percent Prime Interest Rate.
50 Percent Tax Bracket.

285

tively, the separate processes could be made more energy efficient. In Oxnard, California, a pulp mill installed a $300,000 "Hi—Pli" press to reduce moisture content of the pulp slurry, thus reducing drying energy needs. Along with other improvements, the new press is credited with reducing natural gas consumption by 14.3 percent and electricity use by 8.2 percent annually since 1979.[46]

Within the pulping process (which accounts for 42 percent of all energy consumption in this industry) are numerous opportunities for making energy use more efficient, primarily by increasing the use of waste by-products.[47] The two commonly used processes for pulping are mechanical pulping and the Kraft (or sulfate) process. Because it makes greater use of by-product liquors, the Kraft process requires much less energy. Yet, 25 percent of the pulp produced in the U.S. in 1978 was still processed mechanically. Completing the transition to the Kraft process could reduce energy consumption in pulping by 6 to 10 percent.[48] Purchased energy use can be reduced even further by increasing use of the "black liquor" by-product, a condensate of pulping vapors. In theory, the pulp and papermaking industry's purchased energy requirements could be reduced to almost nothing by increased use of waste by-products.[49] A final area of energy-efficiency improvement for pulp and paper-making lies in the development of oxygen-bleaching technology, which uses much less energy than current bleaching techniques.

Overall, new processes have the potential to reduce energy consumption at new paper plants by 30 to 40 percent in the near term and by as much as 80 percent over the long term.[50] But in the DOE voluntary conservation program, which was aimed primarily at relatively inexpensive housekeeping, heat-recovery, and control improvements, pulp and paper manufacturers failed to meet their 1980 goal of improving energy efficiency by 20 percent. In 1976, the industry spent only 12 percent of its profits on research and development, compared with a 16.4 percent average among other energy-intensive industries.[51] Still, some firms have begun to increase drying efficiencies, reclaim fibers, and make other process changes.[52]

Primary Metals.   Manufacturing primary metals requires large amounts of high-temperature thermal energy. To meet that energy need, a modern integrated iron- and steel-mill typically uses about 30 percent purchased natural gas, 14 percent purchased oil, and 38 percent coal-derived coke oven gas. Another 18 percent of energy consumption is supplied by reclaimed blast furnace gases, tars, creosote, and other energy-rich by-products. The aluminum industry, in contrast, relies on electricity for 78 percent of its energy.

Steelmaking involves several steps with varying energy requirements. On average, raw materials preparation consumes 6 percent of total energy; coke-making, 18 percent; ironmaking, 35 percent; steelmaking, 13 percent; and rolling and finishing, another 18 percent.[53] New processes in iron and steelmaking already in use in Japan and Europe cut energy use in some of these steps. Dry quenching coke after it leaves the coking ovens, for example, saves 20 percent of the net energy used in the coking process—one million Btu per ton of coke produced. This process, widely used in Europe, is much less polluting than the present waterquenching process, but it is expensive.[54]

Ironmaking in blast furnaces is the most energy-intensive step in steelmaking. Unfortunately, U.S. industry has shown little interest in the one new process—direct reduction of iron ore—that could make iron production more energy-efficient. Advances in steelmaking are more promising. U.S. steel plants are gradually switching from the open hearth furnace to the basic oxygen furnace, which increases the oxygen input to combustion and cuts energy needs by up to 90 percent of energy consumed per ingot ton of steel produced. But we still lag far behind the Japanese, who use the open hearth furnace for only 1 percent of production (compared with 24 percent in the U.S.).

In "rolling and finishing," energy is wasted at several points. For example, in specialty steel production, "reheating furnaces" used 40 to 80 hours per week may be kept at a high temperature 24 hours per day. In these furnaces, steel is oxidized and about 4 percent of total plant output is lost as a result.[55] Lightweight furnaces that could quickly be heated to the required process temperature and that heated and formed steel under pressure or a vacuum would be much more energy efficient.

Energy is also wasted in ingot casting. Most U.S. steel plants cast steel into ingots, allow it to cool, and then reheat it to roll into the final product. As an alternative, the continuous casting process both reduces energy requirements by about 1.4 million Btu per pound of steel output and increases total output by about 10 percent. At present, however, only about 6 percent of the nation's domestically produced steel is continuously cast.[56] A final technique to reduce energy waste allows steel slabs to be inspected and surface defects removed at temperatures over 1,000°F. It is expected to save over one third of the energy now used for reheating.[57]

Like the paper industry, iron and steel manufacturers are slow to change because their processes are capital-intensive. By late 1979, the primary metals group as a whole had not met its 1980 energy conservation target of only 9 percent savings over 1972 consumption levels. On average, U.S. steel plants are 15 to 30 percent more energy-intensive than their foreign competitors, partly because they have a low profit margin, so little capital is available for replacing inefficient production equipment.[58]

Prospects for increased energy efficiency in the relatively well-capitalized aluminum industry are somewhat brighter. Indeed, a high percentage of research and development spending within the industry is dedicated to conservation research on, for example, electrolysis cells that operate with lower current density and smelting processes that require from 30 to 90 percent less energy than the Hall-Heroult electrolytic reduction process now in use.[59]

Cement Manufacturing.   Stone, clay, and glass manufacturers make up the fourth largest energy-consuming industry grouping. Cement production alone accounts for over one third of total energy consumption in this sector and 1.5 percent of total annual industrial energy use in the U.S. The industry relies primarily on natural gas (44.3 percent) and coal (34.7 percent), secondarily on fuel oil.[60] Eighty percent of these fuels are consumed by kilns, where limestone and clay are "calcined" at 2,700°F to form cement clinker. Another 13 percent of fuel use is for grinding.

The greatest potential for increasing energy efficiency lies in switching from the present "wet-process" to dry-process kilns. In the wet process, ground cement is mixed with water to form a slurry that is pumpable but that must be dried out later at a great energy cost. In the dry process, no water is added after grinding: combustion heat is recaptured from the kilns and used to preheat lime entering the kilns, and there is no need to pump wet cement first to holding tanks and then back to the manufacturing plant. Consequently, dry-process plants, on average, are 20 percent more energy efficient than wet-process plants, using roughly one million Btu less energy per ton of cement output than the wet process.[61] The cement industry predicts that by 1985 about one third of its 1976 capacity will consist of dry-process kilns.[62]

In concrete-block manufacturing, another energy-intensive industry, cement and aggregate are mixed with a small amount of water and compacted into "green" bricks that must be cured before shipment. At the turn of the century, the blocks were simply left outdoors under a tarp to cure to "28-day strength." To increase inventory turnover, however, manufacturers soon began to speed the curing process by floating the blocks in high-pressure steam, or by "bubble" curing using large amounts of oil and electricity.

A new technique in concrete-block curing makes use of the heat given off by the block itself as cement, aggregate, and water combine. By putting the green block into a kiln to trap this "exothermic" heat and adding only small amounts of heat and moisture, the block quickly reaches "28-day strength" with much less input energy. In the humid southern states, no additional energy is needed in the kiln, so energy use per pound of cement production is reduced by about 80 Btu.[63] This technique is now in widespread year-round use in Florida. Elsewhere, it is used during the summer months. If kiln-insulation techniques improve, the method could be used throughout the U.S.—even in the winter.

Petroleum Refining.   Petroleum refining, which uses energy both as a feed-stock and for process heat, is the fourth largest industrial consumer of energy. Using a measure of energy-use per dollar value added, the petroleum industry group is the most energy intensive. In this industry, 60 percent of all energy is used for direct process heat, and 25 percent for process steam. (Another 15 percent is consumed in by-product streams.) In 1978, almost four-fifths of this energy was supplied by natural gas, with a small amount of electricity and fuel oil supplying the rest.

Petroleum refining separates crude oil into constituent products through chemical reactions. Many of these reactions require high-temperature heat. The first step in refining, crude oil distillation, requires temperatures of about 650°F. When naphtha, one of the end products of this step, is later treated before it is formed into still other products, a temperature of about 650°F is again needed. Vacuum distillation requires temperatures above 750°F, and the "catalytic-cracking" operation temperature is around 1,200°F.

While present refining techniques make extensive use of waste-heat recovery and heat exchange, further improvements can be made. For example, new distillation techniques can greatly reduce heat losses, and a new fluidized-bed process now installed at 90 percent of all oil refineries uses about 200,000 fewer Btu per barrel of output in the catalytic-cracking step than the old moving-bed process did.[64] (The fluidized-bed cracking technique also improves productivity by a factor of ten.) Alternatively, the Advanced Cracking Reactor (ACR) now being demonstrated at a Union Carbide facility in Seadrift, Texas, could be used to make waste-heat recovery more efficient. Heat exchangers are located so that they will not be fouled by coke and tar, so much of the heat from the 2,000°F product steam is recovered and returned to the process. The ACR can use naphtha, vacuum oil, kerosene, and even some undistilled crude oils as feedstocks.[65]

The petroleum and coal products industry is profitable enough to research, develop, and invest in the energy and process improvements described here. Yet only 7.6 percent of the industry's profits have been spent on research and development, and most industry energy-conservation efforts have been focussed on process-heater modifications, waste-heat recovery, steam-system improvements, and insulation of equipment.[66] But the industry's net energy efficiency did improve by 14.7 percent between 1972 and 1979, and further gains in productivity and energy efficiency can be expected.

Chemicals.   Chemical and allied companies make up the largest energy-consuming group in U.S. industry. These industries rely primarily on natural gas (51 percent) and electricity (17 percent) for their processing needs, with smaller amounts of energy supplied by fuel oil (14 percent) and coal and coke (12 percent). Like the petroleum refining industry, the petrochemical industry relies on oil and gas as feedstocks, as well as fuels. However, as gas liquids become increasingly expensive, the industry is making increased use of crude oil and other less refined oil products.

In distillation, very high-temperature heat is added to the bottom of the distillation column and various products are recovered as vapors or liquids at different points along the column. Because it demands temperatures high enough to cause the "lightest" component of the mixture to vaporize at the top of the column, this approach is very energy inefficient: the heat is used just once, and many chemical plants waste as much as 95 percent of the input energy used in distillation.[67] A more energy-efficient approach is to begin with lower temperature input heat and to use heat pumps to boost the temperature as required. "Multiple reboiling distillation," for instance, uses multiple heat levels, added at several points along the distillation columns, rather than simply at the bottom. Another alternative is chemical (or liquid-to-liquid) extraction, whereby a chemical additive that has a high affinity for one component of the mixture combines with that component and sinks to the bottom of the tank where it can be removed. DuPont claims to be saving the equivalent of 40,000 barrels of oil per year using this process to recover a solvent used in manufacturing synthetic fibers.[68]

Another technique for increasing the energy efficiency of chemical production is vapor recompression—a technique for recapturing the latent heat contained in vaporized chemicals at the top of the column and using it to supply heat elsewhere along the column. In this approach, feed liquids are passed through an evaporator into a vapor/liquid separator, where concentrated liquids are drawn off. Some of the vapors are recycled and compressed so that they give off their latent heat, via a heat exchanger, to the evaporator.

Like the petroleum refining industry, the chemical industry has enough discretionary capital to make conservation investments. In 1976, chemical and allied industries spent 39.7 percent of their profits on research and development (compared to the 16 percent average R and D investment of energy-intensive industries). As one result, the industry had by 1979 exceeded its 1980 energy efficiency goal (a 14-percent improvement) by 8 percent.[69]

Environmental Impacts

Few potential industrial energy-efficiency improvements are stalled for want of vacant land. Adding waste-heat recovery equipment or improving boiler maintenance at urban factories will not affect adjacent land uses adversely and may well improve air and water quality. Indeed, pollution problems have spurred industrial improvements that also conserve energy: Europe embraced dry quenching in coke production partly because wet quenching causes pollution problems. One Midwestern corn wet-milling plant that began recycling process water to meet environmental regulations realized considerable energy savings.[70] In Milwaukee, Wisconsin, a coil-coating plant that was ordered to clean up its emission of toxic paint solvents or shut down installed a thermal oxidation system that burns pollutants as fuel. It now meets emissions regulations, and it has reduced its process gas consumption by 30 percent.[71]

Institutional Constraints and Socio-economic Considerations

Electric utilities' reactions to increased energy efficiency are likely to be minimal. Few energy-intensive industries rely primarily on electricity; and where they do, many electric utilities appear to be supporting conservation efforts. But where utilities have excess generating capacity and industries are large users of electricity, utilities may issue declining block rates or attempt to sell large amounts of electricity at cut rates, thus thwarting conservation.

RENEWABLE SOURCES OF INDUSTRIAL PROCESS HEAT: DIRECT SOLAR ENERGY

The use of direct solar energy for industrial process heat is less technically developed and frequently more expensive than industrial energy conservation improvements. But these two options are technically and economically compatible for reducing local industries' reliance on scarce, expensive fossil fuels, and they should be considered in tandem.

Energy-conserving improvements to a city's industrial plants may actually help make solar process heat more viable. For example, to reduce energy consumption, the food processing industry is converting some of its steam boilers to high-temperature hot-water heat delivery systems, which are more compatible than steam with solar industrial process heat. Industrial solar systems using hot water as the heat delivery mechanism are cheaper, better developed, and more efficient than those designed to produce steam.[72] Industrial efforts to make better use of lower-quality heat also help make the use of solar energy more feasible. In general, solar energy's contribution grows as industries learn to make better use of low-temperature heat (less than about 200°F). Conversely, solar heating contributions can make greater conservation improvements possible.

In some plants, solar energy may be more economically attractive than energy efficiency improvements. According to SERI researchers, a crude oil de-watering plant in Wyoming that used propane very inefficiently because the

Solar industrial process hot water system for Campbell Soup Company, Sacramento, Calif.
(Courtesy of Acurex Solar Corp.)

separator tank was poorly designed would have to be completely redesigned to
reduce the tank's energy consumption. As alternatives, the team identified
three solar-system configurations (all with paybacks of seven years or less) that
could preheat the oil/water mixture.[73]

Several types of collectors that supply industrial heating needs are now commercially available. Which will work best in a particular plant depends on
which type of heat is needed—hot water, process steam, or hot air—and at
what temperature. Flatplate collectors, typically manufactured in modular
panels of about 20 ft², can provide low-temperature heat to a maximum of
about 150°F. (See Chapter 7.) In industrial applications, where higher temperatures are needed, flat mirrors or concentrators designed to capture more
sunlight can be used to boost temperatures further—by 1.45 to 6 times.[74]

Evacuated tubes absorb more energy per unit area than flatplate collectors,
providing output temperatures of 150° to 200°F. (See Chapter 8.) These collectors are popular for industrial applications in the North and East because
their efficiency varies less with temperature and insolation than that of flat-
plate collectors. Similarly, evacuated-tube collectors can be "boosted" by sil-

vering the bottom half of the glass tube or adding reflectors or reflective troughs.

Still another type of industrial collector uses multiple reflectors. The absorber pipes of this collector are covered with a black selective surface to increase absorption. Mirrored reflector blades concentrate the sunlight onto the tubes in a 24-to-1 ratio.

Parabolic trough collectors for industrial use are being manufactured by several companies. The most common type is a single-axis tracking collector, which can either follow the sun from east to west each day or face south and tilt up and down. These collectors produce temperatures between 200° and 600°F.

Still in the development stages are linear Fresnel lens and heliostat systems, 2-axis tracking collectors, parabolic dishes, and other high-temperature collectors. These systems are designed to supply heat at temperatures above 572°F for use in heat engines connected to electric generators or to supply process heat directly. (See Chapter 11.)

Industries installing solar collectors can choose from a number of alternative heat delivery systems: direct hot water, hot-water exchange, direct hot air, hot-air exchange, steam-flash systems, and steam-generation systems. The simplest, direct hot-air or hot-water systems require only pumps or blowers to move the heat. More complex systems may include heat exchangers, expansion tanks, and storage. The collectors themselves may be mounted either on the roof or walls of the factory. They can also be mounted on the ground, though most ground-mounted systems require more extensive piping networks, insulation, and controls than roof-mounted systems, and heat losses tend to be greater.

Since solar industrial-process-heat technologies are so new, evaluating their performance is difficult. SERI found that the solar systems it analyzed in 1980 performed less well than expected. (See Table 12.8.)[75] Both DOE-funded field tests and private installations encountered such problems as excessive thermal losses from piping in large ground-mounted systems, collector degradation, performance losses due to the coating of collector glazings and concentrating mirrors with industrial emissions, and control system malfunctions. In some systems, these problems led to stagnation, thermal shocking, and breakage of evacuated tubes and mirrors. Finally, operating energy requirements in forced-air systems were high, so net system efficiencies were not.

Until incremental improvements in the design, engineering, and installation of industrial systems boost performance, simple roof-mounted hot-water systems hold the most near-term promise. For now, the industry best suited to use solar energy may be the food industry, which uses low-temperature hot water. Solar drying also holds great potential in lumber-drying and food-drying.[76]

Table 12.8. Predicted and Actual Energy Delivery of Solar Industrial-Process-Heat Field Tests.

PROJECT	Annual Energy Delivery			
	Predicted		Actual	
	(MMBTU/YR)	(BTU/YR)/FT²	(MMBTU/YR)	(BTU/YR)/FT²
Campbell Soup	2156	290,000	—	—
Riegel Textile	1400	210,000	370	55,000
York Building Products	1500	160,000	364	39,000
Gold Kist	3700	280,000	788	60,000
LaCour Kiln Services	900	360,000	370	147,000
L and P Foods	2300	110,000	1035	39,000

Source: Kutscher and Davenport, Preliminary Operational Results of the Low-Temperature Solar Industrial Process-Heat Field Tests (1980).

Economics

For numerous reasons, solar industrial process heating in most locations is still only marginally cost-effective. First, system efficiencies have been low—only 8 to 33 percent in DOE field tests, though prospects for the higher operating efficiencies and improved performance needed to help solar systems offset high costs with high output levels appear good. Given present efficiencies, solar industrial process heat will probably be most immediately economical in the Southwestern states, where today's collectors can deliver at least twice as much energy per square foot as they can in New England or the Pacific Northwest.[77] Indeed, of the 61 industrial systems designed or operational in 1981, most were located in California, the rest in Texas and the Southwest.[78]

A second major factor governing the economics of solar industrial process heat is the cost of fuel displaced by the system. While operating efficiencies in the Northeastern states are lower, oil costs are higher and gas supplies less certain, so solar process heating may be economical in some industrial settings. In the Midwest and Mid-Atlantic states today, a handful of solar systems are now operating.[79] As fuel costs rise, more may come to be seen as wise buys.

Since the total installed costs of most solar industrial process heating is $20 or more per square foot higher than expected, future increases in energy prices alone will not make solar industrial process heat economically attractive. Total installed system costs must come down as the cost of displaced fuel increases. According to SERI, installed system costs must fall by half from the current costs of $40 to $50 per square foot before solar process heating finds widespread acceptance.[80]

Solar process-heat economics is also hurt by the strict payback criteria most plant managers apply to investment decisions. As Table 12.9 illustrates, the present paybacks of solar applications for industrial use all exceed 5 years, and some are longer than the 20-year assumed life of the solar system. Moreover, the collector costs assumed in these case studies are relatively low. Of course, innovative financing plans could improve the solar investment picture. Then too, the accelerated depreciation provisions in the Economic Recovery Tax Act of 1981 (P.L. 97-34) allow industries to depreciate purchases of energy-saving equipment over 5 years, reducing their taxes and improving cash flows. Finally, the economics of solar process-heating applications could be improved by conservation improvements at each potential solar site. For example, case studies of a bakery in Denver and of an aluminum container manufacturing plant indicate that solar systems are not cost-effective alone but are when combined with waste-heat recovery.

Table 12.9. Economics of Solar Industrial Process Heat: Case Studies.

TYPE OF INDUSTRY	APPLICATION	COLLECTOR TYPE	TOTAL INSTALLED COLLECTOR COST ($/FT^2)	FUEL COST ($/MMBTU)	PAYBACK (YEARS)	LOCATION
Group 1: Hooker, et al., 1980 (1979$)						
Crude oil de-watering	Pump oil through collector for preheat	Flat-plate	48	5.62	4.5	Wyoming
	Pump oil through external heat exchangers attached to collectors	Parabolic troughs	71	5.62	6.6	
Aluminum container manufacturing	Supply hot water for can washing	Parabolic trough	40	2.00	less than 20	
	Preheat combustion air, with gas recycling	Parabolic trough	35	2.00	over 20	
Wet milling corn	Preheat corn-steep water	Flat-plate	27	3.92	over 20	Midwest
Polymeric resin manufacturing	Supply hot water to heat-storage tanks	Flat-plate	26	2.80	13 over 20	El Paso Chicago
	Supply steam for low-temperature reactors	Parabolic trough	33		over 20	
Milk processing fluid	Provide hot water for clean-up	Flat-plate	28	2.38	less than 20	Southwest

Industry	Application	Collector type		Cost ($)		Location
Bread and bun bakery	Provide heat for pasteurization via heat exchanger	Parabolic trough			less than 20	
	Preheat makeup water, with heat recovery				10	Denver
	Preheat oven air, with heat recovery				less than 20	
Meat processing	Preheat truck-washing water	Flat-plate		1.87	less than 20	
	Preheat boiler feedwater	Flat-plate		2.23	less than 20	
Group 2: Clark and Studstill, 1981; (1980–81 $)						
Processed food	Heat vegetable oil	Flat-plate		3.50	13.2	Georgia
Meat processing	Hot water	Evacuated tube		3.50	16.7	Georgia
Bakery	Preheat hot air for ovens	Evacuated tube	48	3.50	16.1	Georgia
Group 2: Clark and Studstill, 1981; 1980–81 $ (cont.)						
Small carpet manufacturing	Heat dye water, with heat recovery	Evacuated tube		3.50	9.2	Georgia
Broadloom Carpets	Heat dye water, with heat recovery	Flat-plate	24	3.50	8.5	Georgia
Apparel Fabric	Preheat process water	Flat-plate	33	3.50	17.3	Georgia

City Applications

Solar industrial-process heating's potential in cities is severely limited by land-use and other constraints. For example, the Gold Kist soybean plant in Decatur, Alabama, had to construct a massive steel I-beam structure to support 13,104 ft^2 of collectors.[81] At L&P Foods in Fresno, California, 21,000 ft^2 of collectors were ground-mounted, and additional land area was required for rock-bin thermal storage.

With older inner-city factories, an additional limitation on using solar applications is that the factories' plants and processes are likely to be energy inefficient. More than new energy-conserving plants—most of which are in suburbs or rural areas anyway—older plants may require expensive and technically difficult engineering overhauls to become energy efficient or to make use of solar process heating. Another constraint in using solar industrial process heat in large cities is industrial pollution, which can degrade collector and mirrored surfaces and thus impair efficiency.

None of these constraints is insurmountable, however. Solar collectors require space at the factory site, but not necessarily land. In many settings, they can be mounted on the walls or roofs of existing buildings. For example, 4,600 ft^2 of collectors were mounted on the roof of an Anheuser Busch beer-pasteurizing building in Jacksonville, Florida. No structural modifications of roof supports were required in this building, which was based on a prototypical design that many future Anheuser facilities will employ.[82] Owners of a photoprocessing laboratory in Richmond, Virginia, found that mounting flat-plate collectors vertically on the side of the lab was much more cost-effective than ground-mounting them because hardware needs were less. Roof- or wall-mounted systems also have lower "parasitic power" requirements than ground-mounted systems do—that is, their operating energy requirements are lower. As Table 12.10 suggests, the roof-mounted systems with the greatest potential for urban use can also be the most efficient.

As for the greatest near-term potential for using solar process heat, the likeliest candidates are textile manufacturers, food processors, and chemical industries. All use low-temperature heat, and the textile industry is concentrated in the Southeast, where insolation levels are high. The challenge, of course, is designing systems to accommodate the changing technology of the industry. For instance, the DOE-funded solar retrofit at Riegel Textile Corporation's plant in La France, South Carolina, supplies heat to a batch-type dyeing tank that is now being replaced by a continuous-dyeing technology that can dye more fabric more quickly.

Most of the heating needs in food processing are for temperatures less than

Table 12.10. Performance of Solar Industrial-Process-Heat Field Tests.

PROJECT	LOCATION	COLLECTOR AREA (FT^2)	MOUNTING TYPE	NET SYSTEM EFFICIENCY	PARASITIC FRACTION (%)
Campbell Soup	Sacramento, CO	7335	Warehouse roof	—	4.2
Riegel Textile	La France, SC	6680	Ground	8.1	3.4
York Building Products	Harrisburg, PA	9216	Integral part of roof	8.6	3.9
Gold Kist	Decatur, AL	13104	Special structure	19.7	8.7
La Cour Kiln Services	Canton, MS	2520	Storage building roof	32.5	1.0
L&P Foods	Fresno, CA	21000	Ground	14.2	9.5

Source: Kutscher and Davenport, 1980.

350°F (which commercially available collectors can provide), and many food-processing plants are located in urban areas to take advantage of the labor force and markets there. Dairies and beverage plants, which have a high conservation potential and process temperature requirements within the range of flat-plate collectors, appear to be particularly well-suited to urban use.[83] Conservation improvements at these plants would help make solar more economically attractive by reducing required collector areas.

The highly energy-intensive chemical industry is already investing heavily in energy efficiency. Chemical plants tend to locate close to a large labor force and to require process temperatures in the range produced by parabolic trough, evacuated tube, and flat-plate collectors. These factors combine to make solar process heating applications at chemical plants quite promising in Southern cities. Yet, Northern cities that produce industrial chemicals may also be able to use solar process heat, since producing industrial inorganic chemicals requires temperatures in the range produced by evacuated tube collectors, whose performance varies relatively little with climate.

Land Use.   Shortages of collector space in dense cities can be partially overcome by mounting collectors on roofs and walls and at the city's edge. Industries there will have to weigh the value of such fringe land for plant expansion against its potential use for solar energy collection.

In many ways, current-generation industrial design—including flat roofs, one-story construction, and development on flat terrain—favors solar retrofits at existing factories.[84] A recent study of 200 industrial plants found that 87 percent had over 25,000 ft^2 and 55 percent had more than 100,000 ft^2 of flat roof or south-facing wall suitable for collectors.[85] Additional flat roof area is available at warehouses.

In planning solar industrial-process-heat applications, cities should first encourage industries to cluster so that they can share process heat and electricity. Industrial parks also have other economic advantages—including cost savings associated with shared parking and storage facilities and joint site selection—that combine with the appeal of reducing industrial noise and pollution to make energy-efficient "clustering" even more desirable. More specifically, collector areas could be combined, and collector space and thermal storage areas could be shared, which would save factories both money and energy. The large landscaped areas surrounding industrial parks ensure solar access, and may prove fit for ground-mounted collectors or producing biomass energy crops. Uniform building height and setback requirements for new construction can further protect solar access, and new buildings can also be designed with roofs having adequate load-bearing capacity to support collectors.[86]

Environmental Impacts. Using solar energy to supply industrial-process needs reduces air pollution by enhancing the feasibility of energy-conserving improvements that reduce emissions and by reducing the need to burn fossil fuels.[87] To implement Prevention of Significant Deterioration (PSD) regulations, federal and state air quality planners now follow a "bubble concept": they allow firms that reduce emissions below allowable levels at one smokestack to exceed emission standards at another. Accordingly, plant managers who want to expand one part of a process might add solar equipment to another part to trade off emissions.[88]

A similar regulatory approach, the emissions offset policy, is used to regulate new sources of pollution in already polluted areas. Under this approach, state air-quality officials will allow a new plant to open in a "non-attainment area" only if the projected emissions from the proposed new plant offset greater emission levels from existing sources. In effect, new industries in a non-attainment area must shoulder the costs of helping other industries in the area reduce their emissions in exchange for the economic advantage of locating there.

Institutional and Socio-economic Considerations

Electric utilities should put up little opposition to increased use of solar industrial process heat, since most industries now rely on coal, oil, and gas. However, utilities in areas where industry relies heavily on electric process heat might oppose solar installations if those utilities have excess capacity.

Since solar industrial process heat costs more than most industries are willing to invest today, its use will not create many new jobs in the near future. However, if costs come down and the prices of competing fuels increase, new jobs will open up in collector manufacturing, installation, and maintenance. Increased use of solar industrial process heat could also help maintain existing urban jobs by providing an affordable, reliable energy source for urban industries. Despite its present high costs, solar energy equipment appeals to owners of industrial plants as a potential long-term, reliable energy source.

RENEWABLE SOURCES OF INDUSTRIAL PROCESS HEAT: BIOMASS FUELS FROM INDUSTRIAL PROCESSES

Biomass fuels could supply much of the energy that American industry needs. (See Chapter 13.) Currently, nearly half of the energy that the forest-products industry uses is supplied by wood and wood wastes. As fossil-fuel prices increase, industry's wood consumption is expected to grow too. By the year 2000, SERI projects, 70 to 80 percent of industry's needs will be met by wood.[89] Increasingly, the food-processing industry, which also has a ready supply of biomass available, is burning or gasifying peach pits, walnut shells, and other food by-products. By the year 2000, these by-products could supply 10 percent of the energy that the food processing industry uses.[90] Other urban industries can put energy from municipal solid wastes to use.

Industrial product wastes can be used to produce various fuels. Gasifying or directly burning wood and food-processing wastes are obvious options. Industries can also produce liquid fuels, though rarely to provide industrial process heat. (See Chapter 13.)

Most technologies for supplying industrial process heat from renewable biomass resources rely on direct combustion. (See Chapter 11.) In smaller industrial plants, factory-assembled biomass boilers capable of producing 60,000 lb of steam per hour and burning various solid fuels in small sizes (3 in. or less) can be used. Wood-fired package boilers, about 100 of which were operating in North America in 1978, can heat water or produce low- or high-pressure steam.[9]

Economics

The economic attractiveness of converting or retrofitting boilers to use biomass depends on numerous factors. One is the initial capital cost: new wood-fired boiler systems (including handling and storage equipment) cost from $10,000 to $100,000 per MMBtu per hour of output capacity. Retrofits of existing boilers vary even more in cost. A coal-fired boiler may need little or no retrofit, while retrofitting an oil- or gas-fired boiler would be much more extensive and costly. Installing a package gasifier at a cost of between $40,000 and $90,000 and with the capacity to produce 2 to 10 MMBtu of heat per hour is an intermediate-priced alternative for plants now using oil or gas.

The capital costs of wood-fired boilers must, of course, be compared with those of an existing or competing boiler. Typically, wood-fired systems cost 250 to 750 percent more than oil- or gas-fired combustion systems, largely because larger boilers and more complicated handling and storage systems are required. For the same reason, wood systems are generally more expensive than coal-combustion equipment. However, if the competing boiler burns high-sulfur eastern coal, the wood system may be more competitive because the coal-burning system would require expensive pollution-control equipment while the wood-based system would not. (See Chapter 11.)

Operating and maintenance costs, like capital costs, are higher for biomass combustion than for fossil-fired process heating, especially if the biomass must be pulverized, dried, or treated.

The big advantage to biomass combustion, of course, is low fuel costs. In most of the country, wood costs less than fossil fuels; and at many plants, wood or food wastes are essentially free. Transporting these fuels long distances is uneconomical, but opportunities for using such wastes near where they are generated are great.

A final factor affecting the economic attractiveness of biomass combustion or gasification is system performance. Fossil-fired boilers retrofitted to burn or gasify wood can lose overall capacity since wood has a lower heat content than coal. Boilers thus derated by 5 to 25 percent may perform quite well, but will not pay for themselves as quickly as improvements that do not derate the plant.

City Applications

A shortage of space for storing and handling bulky biomass fuels could restrict retrofits at factories in crowded urban industrial areas, as could a lack of space for the larger biomass boiler in smaller factories. A more readily available renewable-fuel source for use in urban areas is municipal solid waste. The resource-recovery plant in Saugus, Massachusetts, is guaranteed a fuel supply through 30-year contracts with 13 participating cities. In Chicago, the city's northwest incinerator produces process steam for sale to a nearby candy factory.

Land Use. As noted, plants converting to make use of biomass fuels will need additional space at or around the plant site. For example, a spreader-stoker boiler system requires six to seven times as much space as the equivalent oil-fired system and four times as much area as a coal system with same energy output. Small urban factories might not have this additional space available, either within or outside the factory. Increased use of biomass, like increased use of solar energy, may thus hold more potential in new industrial parks on the urban fringe. The parks could use nearby greenbelt areas for biomass production or purchase the rights to nearby wood or agricultural wastes. Then too, land could also be set aside in the industrial parks to accommodate centralized fuel storage and handling equipment.

Environment and Safety. Mishandling biofuels for industrial process heat can pose health and safety hazards. Such bulk sources as loose sawdust or fruit pits and nut shells must be carefully stored to prevent spontaneous combustion. Another potential hazard is ignition of biomass on conveyor belts. Electrical equipment must be carefully maintained to keep sparks from flying and flammable biomass from starting a fire.

Storing and transporting finely-ground sawdust or nutshells can endanger the health of workers inhaling the dust, and incomplete combustion of biomass in boilers can produce a highly explosive mixture. In addition, the medium-Btu gas produced by wood gasifiers may contain toxic pollutants. Finally, the tars and oils produced as gasification by-products may also be carcinogenic and require careful handling. None of these problems presents an insurmountable barrier to the use of biomass fuels, but none can be ignored, either.

Air pollution is seldom a problem if biomass combustion and gasification systems are equipped with efficient particulate controls.[92] In fact, centralized biomass combustion involves less air pollution than fossil-fuel combustion. Emissions of nitrous oxides and sulfur dioxide are naturally low when shells, pits, wood, organic materials, or municipal refuse are burned. Emissions of carbon monoxide and organics can be higher than in coal combustion, but these pollutants can be controlled by maintaining high combustion efficiency.

To meet air quality laws, wood-burning boilers producing more than 10,000 lb of steam per hour will have to undergo a Federal Prevention of Significant Deterioration regulatory review. Generally, boilers with outputs smaller than 20,000 lb of steam per hour can meet such requirements by adding simple cyclones, while large facilities may need electrostatic precipitators or bag-houses to control particulate emissions. In the Southeast, many industries with large coal-fired boilers use wet venturi scrubbers, which collect particulates in water. In highly polluted areas ("non-attainment areas"), industries that want to add new biomass-fired boilers may have to obtain emissions offsets from other nearby plants.[93]

Liquid effluents created in biomass combustion or conversion can also cause problems if controls are not adequate. Effluents produced by wet scrubbers used for particulate control and by clean-up of medium-Btu gas produced by wood gasifiers are high in oxygenated hydrocarbons and must be treated before disposal. These effluents may also contain heavy metals, sulfides, and thiocyanates.[94]

Institutional Constraints

Electric utilities are unlikely to oppose increased use of biomass as a source of industrial process heat since the forest-products industry and the other industries most likely to increase their use of biomass already use large amounts of biomass and relatively little electricity. Yet, owners of wood-fired industrial cogenerators may encounter problems obtaining an adequate sell-back rate if they wish to sell electricity to the local utility. (See Chapter 11.)

Another institution that could affect whether the use of wood as a fuel source increases is the wood-contracting and delivery business. If better and more reliable wood-supply systems become available, various industries in forested areas might use wood as a fuel source. In the meantime, wood's use as a fuel will be largely confined to the forest-products industry.

13. Fuels from Renewable Resources

Most renewable-energy technologies can be used to substitute for conventional liquid, solid, or gaseous fuels, but some "renewables" can also be used as feedstocks to produce these fuels. Cities can produce and facilitate the production of these fuels, as well as consume them. The options are plentiful: as consumers, cities can use gasohol in city-owned vehicle fleets or purchase pelletized wood for use in municipal power plants; as production facilitators, they can encourage industries to get into renewable-fuel production; as producers, they can extract methane from landfills.

Of course, cities need guaranteed access to local feedstocks to pursue these options. All cities have waste steam, but the availability of such feedstocks as wood, crops, or other biomass is confined to specific locations, and transportation costs can make or break consumption or production ventures.

In general, fuels produced from renewable energy will probably play a smaller role in large cities than in smaller ones, and rural and exurban areas hold the most development potential. Yet, even sprawling Los Angeles extracts biogas from one of its sewage-treatment plants to fuel on-site processes and sells the "leftovers" to a neighboring municipally owned utility.[2]

LIQUID FUELS

Cities use fuels primarily for heating and transportation. While heating oil can be replaced by gas, wood, coal, electricity, or solar energy, the only alternative to liquid fuels in transportation is electricity, whose use in vehicles is promising but constrained by costs and energy-storage problems. Accordingly, cities will probably use liquid renewables—ethanol and methanol—mainly for transportation.

All but a negligible fraction of the 105 million gallons of fuel ethanol produced in the U.S. in 1980 was consumed as gasohol—a one-part ethanol, nine-part gasoline blend.[2] Another 95 million gallons of domestic ethanol (not counting some 22 million gallons imported) was used in the manufacture of pharmaceuticals and other chemicals.[3] About 1.1 billion gallons of methanol was also produced for use in chemical industries. Another million gallons of

Table 13.1. Energy Usage for Transportation
(Selected Cities).

CITY	YEARLY TRANSPORTATION ENERGY USE (MMBTU)	PERCENT TOTAL ENERGY USE
Bridgeport	15,100	26.9
Boulder	3,657	28.2
Dearborn	5,820	22.0
Knoxville	24,922	31.5*
Los Angeles	173,400	47.9
Seattle	34,122	33.0

Sources: CCEMP Audits.

ethanol was produced and used on the farm.[4] Cities made no significant use of alcohol fuels except in chemical manufacturing and vehicles.

Theoretically, alcohol could be used in boilers for space heating, for process steam, and in electrical power generation where cheaper coal-fired generators cause unacceptable amounts of pollution. Methanol can also be used in fuel cells and in the gas turbine engines that utilities use for peak-load generation and that industry uses to produce electricity. As of 1981, however, alcohol fuels could not compete economically with conventional fuels in any of these applications.[5]

In transportation, the story could be quite different, particularly since fixed demands for transportation fuels account for roughly one fourth of many cities' total energy budgets.[6] (See Table 13.1) However, before substituting alcohol fuels for conventional liquid fuels, cities must grapple with numerous technical, social, and economic issues. In particular, the economic feasibility of both the short-term and long-term production or use of alcohol fuels will depend upon federal and state production incentives and regulations over which the city has no control. Yet, whether and how to use alcohol fuels are decisions cities can now make with some reliable evidence in hand.

Resources and Potential

Gasohol is now marketed in over 15,000 gas stations. It is used heavily where feedstocks are in greatest supply, primarily in the Midwest. (See Table 13.2.) According to the Renewable Fuels Association, alcohol production will reach 500 million gallons by 1985 if gasoline prices continue to rise and the federal production incentives set forth in the Energy Security Act are not withdrawn.[7]

Theoretically, ethanol's potential equals that of total U.S. gasoline use—110 billion gallons per year. But meeting even the 500-million gallon goal set by

Table 13.2. Regional Production of Potential Feedstocks.

	CORN[1]	MILO[1]	WHEAT[1]	OATS[1]	BARLEY[1]	RICE[2]	Potatoes[2]				SWEET POTATOES[3]	SUGAR BEETS[3]	SUGARCANE[3]
							WINTER	SPRING	SUMMER	FALL			
New England	0	0	2	2	0	0	0	0	0	31	0	0	0
Midlantic	208	0	22	35	12	0	0	0	3	18	1	0	0
Southeast	373	22	73	18	9	56	1	8	5	0	9	0	33
Cornbelt	3,462	68	259	142	2	1	0	0	2	4	0	1	0
Lake states	1,077	0	166	257	56	0	0	0	3	40	0	7	0
Great Plains	1,090	662	1,044	270	137	23	1	1	2	23	1	5	2
Western states	136	36	462	20	199	18	1	14	6	187	1	13	0
Hawaii	0	0	0	0	0	0	0	0	0	0	0	0	20
Alaska	0	0	0	0	0	0	0	0	0	0	0	0	0
United States	6,346	788	2,026	755	415	98	2	23	21	303	12	26	56

1. Units: 10^6 bushels.
2. Units: 10^6 hundredweight.
3. Units: 10^6 tons. *Source:* U.S. Department of Agriculture, Agricultural Statistics, U.S. Government Printing Office, Washington, 1978.

President Carter requires bringing ethanol plants on-line by 1985—no mean feat.[8] Lesser concerns include modifying car engines slightly so that they can burn methanol or fuel mixtures that contain more than 10 percent ethanol, finding ways to maximize pollution controls, and making sure that ethanol use at this level does not drive up agricultural commodities prices.

Commercial-scale methanol production for transportation use will become a reality when the plant under construction in Winchester, New Hampshire, begins operating.[9] But while the theoretical potential of methanol equals that of ethanol, and methanol feedstocks are far more plentiful, the technology for converting biomass to methanol still requires improvements. In short, while the use of methanol in cities is not ruled out, it will take time for adequate supplies to become available and for consumers to make the switch.

Technology

Ethanol production takes place in three stages. First, the feedstock (usually sugary or starchy plants) is cooked. During cooling, carbohydrates are converted into fermentation sugars (usually glucose) with the aid of liquefying and saccharifying enzymes. Second, these simple sugars are converted through fermentation into ethanol and carbon dioxide. Finally, the dilute ethanol is concentrated through distillation and separated from the water so it will burn. Nonfermentable materials such as distillers grains can be converted either before or—as is more common—after the fermentation step.

While no facilities for converting wood and other forms of cellulose to methanol are operating in the U.S., the process is well understood. First, the wood or other feedstock is gasified, then the gas and water effluents are separated and cleaned, and finally the synthesis gas is converted by high pressure and condensation processes to methanol. In methanol production, unlike ethanol production, most of the process energy is obtained from the heat generated in production.

Which fermentation technology is optimal depends in part upon which feedstock is available at affordable prices. At the Archer-Daniels-Midland plant in Decatur, Illinois, some 5 million gallons of high- (192- to 200-) proof alcohol is produced from corn syrup, damaged corn, and wheat starch. In 1979, the plant tripled its output by abandoning the beverage-grade standard and producing fuel-grade alcohol only. It also tied its operation closely to a company-owned corn-processing facility so that it can extract and market nutrients from the corn feedstock. This facility efficiently produces 150,000 gallons of ethanol per day. Scale affords the operation many economies, and the plant makes use of waste heat from the adjacent corn-processing facility to meet some of its energy needs.[10]

A second model is the Milbrew plant of Juneau, Wisconsin. Built in 1978 and expanded in 1979, this plant reportedly produces over 5 million gallons of 193-proof ethanol annually from cheese whey. Fired by a combination of natural gas, propane, and fuel oil, the facility produces high-protein yeasts and condensed fermentation solubles as by-products.[11]

A third example of ethanol-production technology is that employed in a Georgia-Pacific plant in Bellingham, Washington, to convert the waste liquor from a sulfite pulp mill into alcohol. This 6-million-gallon-per-year stainless steel plant has the capacity to ferment continuously and to reuse the yeast. It also uses diethyl ether (instead of the more commonly used benzenes) to dehydrate ethanol, so simpler equipment and less steam and water can be used, and it siphons off salable methanol from the incoming waste liquor. Drawing its operating energy from a combination of wood waste, fuel oil, and natural gas, and selling spent fermentation liquor to manufacturers of such products as artificial vanilla and herbicides, this plant represents an integrated approach to alcohol fuels production.[12]

These three prototypes are based on ethanol technologies that are no longer patentable. Cities may also soon have at their disposal several new technologies. Besides relatively advanced membrane and vacuum technologies and improved hydrolysis technology, which are expected to be viable in ethanol production by the late 1980s, there is a pressure-cascading technique (called the Katzen process) that allows process heat to be recovered and reused, a twin-tower distillation system that reuses ether vapor at considerable energy savings, and a thermally integrated production complex that combines the equipment for fermentation, distillation, and distillate concentration. Two other noteworthy new production breakthroughs are an ultra-efficient industrial heat exchanger and a continuous fermentation process (the "Chemapec" process) based on energy recovery and the use of relatively low-temperature process heat.[13]

Other technology-related issues include the performance characteristics of alcohol fuels, their efficiency ratings, the operating problems associated with the use of methanol and ethanol, and the net energy gained in alcohol fuels production. Performance characteristics are important because cities' investments and commitments will be influenced by the utility and versatility of the fuel and because public information campaigns may be needed to speed acceptance of these unfamiliar fuels.

The performance of ethanol in vehicles is considered superior to that of gasoline (though blends do not test as well as pure ethanol).[14] In general, alcohol fuels give off fewer carbon monoxide and hydrocarbon emissions than gasoline but more photochemically reactive and evaporative emissions. Alcohol combustion is smokeless, odorless, and octane-rich. However, alcohol fuels are more corrosive than gasoline to engine parts. They are also less dependable in temperature extremes, though temperature-related problems (such as vapor-lock and phase-separation) can be reduced greatly if additives are used and the alcohol-burning vehicle's carburetor is modified. As for mileage, alcohol fuels tend to get more miles per Btu, but since they contain fewer Btu per gallon, a tank of alcohol will not go as far as a tank of gasoline.

Some critics of alcohol fuels contend that more energy is required to produce a gallon of ethanol than the ethanol contains—an issue too complicated to more than outline here.[15] Some processes do allow net energy gains, and the liquid fuels produced have more applications than the energy contained in the feedstock or used to fuel the process. One critical factor here is monitoring all production variables closely to make sure that each step and input is energy-efficient. Another is using renewable resources in the production process to tip the energy balance.

A final technical issue is choosing between ethanol and methanol production. The choice here may be between a more modest near-term pay-off with well-established technologies and a potentially larger pay-off over the long term (as methanol-conversion technologies are perfected).

Economics

Overall, the economics of alcohol fuel production are improving with production technology improvements and increases in the costs of competing fuels. Yet, some of the economic variables are out of cities' control, and experience with large-scale plants is scant and confined primarily to the farm belt. Further, both price and demand will fluctuate with the availability of feedstocks and changes in farm commodities prices.

Uncertainty abounds over the economics of alcohol production. But there is widespread agreement on two points. First, the cost of the feedstock and the commercial value of the by-products of alcohol production are the primary determinants of price.[16] Second, the price of the finished product relative to the competing fuel (in most cases, gasoline at the refinery gate) is the main determinant of salability.[17]

If corn is used as the feedstock and the by-products of the grain-to-ethanol process (corn oil, corn syrup, distillers dried grains, etc.) are sold at market value, ethanol can be produced for as little as 96 cents per gallon, though $1.30 to $1.65 per gallon is more typical.[18] If corn prices remain stable, new ethanol-production processes are commercialized, and prices for distillers dried grains (protein-rich by-products used as a livestock feed supplement) hold, the costs of ethanol derived from corn should fall to about $1.20 per gallon within 5 years.[19]

Other feedstocks besides corn can be used in ethanol production. Wheat, grain sorghum, oats, sweet sorghum, sugar cane, and other food crops make good feedstocks, as do food-processing wastes (such as cheese whey) and both spoiled and substandard grain. For biomass-to-methanol production, the likeliest choices are timber, wood refuse, wood-processing residues, and municipal solid wastes. Each of these costs considerably less than ethanol feedstocks. For example, methanol made from wood costs an estimated 52 to 95 cents per gallon at the plant gate; methanol from a municipal solid-waste feedstock costs even less.[20] However, methanol's relatively lower heat content and performance profile make it a less valuable vehicle fuel than ethanol, even though methanol has a higher octane rating.

Plant construction costs will be determined by the type of feedstock used, the grade of alcohol to be produced, the heat source the plant employs, the costs of labor, the environmental or safety controls in place, and the costs of money. Plant costs will also vary fairly consistently with plant capacity. For example, in a 1981 report by the Solar Energy Research Institute's Strategic Planning Branch, the building costs of a new, energy-efficient 50-million-gallon-per-year, coal-fired distillery using corn as the feedstock were estimated at $89.6 million ($1.79 per annual gallon output.)[21] An independent study conducted a year earlier reached a consistent conclusion, estimating the cost of a 25-million-gallon-per-year coal-fired plant at $41.4 million ($1.66 per annual gallon output).[22] Capital costs could be substantially reduced if a city could use an existing production facility. (According to the National Commission on Fuel Alcohol, the idle capacity of local breweries is about 100 million gallons nationally.)[23]

Besides feedstock costs, other influences on fuel costs are the resale value of by-products, capital costs, and transportation costs. Especially important are federal and state subsidies to alcohol fuel users. Federal subsidies to alcohol production now amount to 4 cents per gallon of gasohol, which equals 40 cents per gallon of alcohol. Twelve states also offer subsidies ranging from 10 to 95 cents per gallon of alcohol. These indirect subsidies directly allow the fuel to compete economically with unleaded gasoline. Of course, the payback on investments in alcohol-fuel plants will be longer if these subsidies are eliminated before petroleum prices rise sufficiently or alcohol-fuel production costs fall.

Ethanol at the distillery. (in late 1979—$1.60 gallon) appeared far from competitive with refinery-gate gas priced at $0.65–$0.75. While methanol at $0.70 per gallon would have been competitive, no methanol from biomass was on the market. In addition, methanol's poor performance characteristics in cars suggests that it cannot be compared directly to gasoline as engine fuel. However, most researchers argue that the cost of ethanol and methanol should not be compared to the average cost of gasoline, but with the "marginal" costs of imported crude oil.[24] According to one study, at $45 per barrel, the refinery gate price of unleaded gasoline would be $1.65 per gallon.[25] With unleaded gasoline at this price, ethanol at $1.60 per gallon would be an attractive alternative.

Numerous other factors may make ethanol more economically competitive than straight price comparisons indicate. First, the newer ethanol processes described here could bring the delivered price of ethanol down to about $1.20 per gallon. And as crude oil prices rise, ethanol and refinery gate gasoline prices should become more competitive. Second, the U.S. Office of Technology Assessment suggests that ethanol's true economic value may lie in its octane-boosting properties. Thus, refiners might be willing to pay more for ethanol because they could mix it with a commensurately less octane-rich (and less expensive) gasoline.[26] In addition, retailers may be willing to pay more for ethanol because they can charge more for ethanol-gasoline blends (gasohol) than for less octane-rich gasoline.

As for the potential for competition between food crops and fuel crops, and the impact of such competition on food and fuel prices, the key issue is the production level of alcohol fuels. Without production controls, commodities prices will certainly rise with demand. At least one agricultural economist contends that "in the absence of any governmental limitations on the conversion of agricultural commodities into fuel, the price of oil could eventually set the price of food."[27] Yet, by using crop wastes and nonagricultural biomass as the alcohol feedstock, recycling the protein that remains once the sugars in the feedstock are converted to fuels, and keeping production low enough to avoid triggering grain-price inflation, competition between food and fuel can be managed. The overriding questions are thus whether federal and state governments will set and enforce such policies and how trustworthy the feedstock market will be if they do not.

Environmental Impacts

Three types of environmental impacts are associated with alcohol fuels. Environmental problems can stem from (1) feedstock production, (2) fuel production, and (3) the use of alcohol fuels in vehicles. By most accounts, the most serious over the long term are the impacts of feedstock production, while vehicle-related pollution is relatively less important.[28] Yet, pollution from the production process can be expensive to control, requiring expenditures at plant-construction time of roughly one eighth the total capital costs (of a 50-million gallon per year facility).[29]

The gravest problem related to feedstock cultivation is cropland degradation or conversion.[30] Either increasing the intensity of agricultural production or the amount of land under cultivation (as opposed to using surplus or spoiled crops) paves the way for soil erosion, which increases the need for fertilizer and pesticides to sustain the soil's productivity. In turn, agricultural run-off rich in in these chemicals can contaminate creeks, rivers, and water supplies, even as the sediment that transports them silts up or eutrophies waterways.

Cultivating fallow acreage or forests and other virign lands can cause other pressures too. Expanding agriculture's demand for water can devastate water tables, which are already low in several Western states, and exacerbate "water wars" between up-river and down-river farmers, as well as between agriculture and fossil-fuel developers.[31] It can also increase salinity.

In the production process, emissions from the heat source and from distillery wastewaters pose the greatest pollution problems. Beyond that, the level, type, and seriousness of the damages depend upon which type of boiler and which feedstock are used. (See Table 13.3.) Coal-fired boilers tend to generate the most harmful emissions and solid wastes (including coal dust, fly ash, and bottom ash), though effective emission-control technologies are available.

The major air emissions arising from ethanol production are grain dust, carbon dioxide, oxygenated compounds, and volatile organic fermentation products. In the recovery of co-products, some commercial distilleries also emit sulfur dioxide, carbon monoxide, and nitrous oxides. While "scrubbers" and other environmental control systems can neutralize some of these contaminants, pollution-control equipment can itself generate secondary wastes, so no perfect solution to the emission problem exists. Thus, alcohol production facilities have little place in urban areas with air pollution problems.[32]

Environmentally hazardous effluents from the fuel-production process are discharged in distillery wastewaters. The alcohol plant's wash waters are high in biological oxygen demand (BOD). Some also contain both dissolved and suspended solids, trade chemicals, and pesticide residues. Stillage from alcohol plants is high in alkali salts and BOD. Since a 50-million-gallon-per-year distillery generates over a million gallons of wastewater a day, commercial alcohol plants must either install water purification systems or—the cheaper alternative in some cases—pipe the wastewater to publicly owned treatment works with spare capacity.

Experience does not permit generalizations about the environmental impacts of using alcohol fuels on a large scale. But it is known that the exhaust and evaporative emissions from pure alcohol fuels contain roughly the same amount of carbon monoxide and hydrocarbons as gasoline emissions, less nitrous oxide, and more aldehyde. With blends, the evidence on hydrocarbons and nitrous oxides is more ambiguous.[33]

An advantage of the use of pure ethanol is the absence of sulfur dioxide and

Table 13.3. Bioconversion of Grain to Ethanol—Emission and Effluent Sources.

OPERATION	EMISSIONS TO ATMOSPHERE	LIQUID EFFLUENTS	SOLID WASTES
Foodstock Storage	Transfer operations (fine dust)		Grain, dirt
Milling and Cooking	Mechanical collectors for milling operations (particulates)	Wash water (dissolved and suspended solids, organics, pesticides, alkali); Flash-cooling condensate (dissolved and suspended solids, organics)	Grain dust (from mechanical collectors)
Hydrolysis and Fermentation	Fermentation vents (CO_2, hydrocarbons)	Wash water (dissolved and suspended solids, organics, alkali)	
Distillation and Dehydration	Condenser vents on columns (volatile organics)	Rectifier bottoms (organics); dehydration bottoms (organics)	
Storage	Storage tanks (hydrocarbons)		
Co-product Recovery	Dryer flue gases (NO_x SO_x, CO, particulates, hydrocarbons); Evaporation condenser vents (hydrocarbons)	Evaporator condensate (dissolved and suspended solids, organics)	Grain dust (from direct-contact dryers)
Steam Production (coal-fired)	Flue gases (NO_x, SO_x, CO, particulates)	Boiler blowdown (inorganics); cooling water blowdown (dissolved and suspended solids, organics)	Coal dust, ash (bottom ash and fly ash)
Environmental Control Systems	Evaporation from biological treatment ponds (organics)	Scrubber blowdown (dissolved and suspended solids, organics)	Sludge (from flue gas treatment); biological sludge (from wastewater treatment)

Source: R. M. Scarberry and M. P. Papai, *Source Test and Evaluation Report: Alcohol Synthesis Facility for Gasohol Production,* Radian Corporation, January, 1980.

Table 13.4. Capital Costs of
Environmental Technologies: 1981
($Million for a 50-Million-Gallon-
Per-Year Ethanol Production
Facility).

Flue-gas scrubbing system	4.2
Wastewater treatment system	2.0
Fire protection system	0.6
Ash collection package	0.17
Vapor-controlled storage tanks*	0.16
Vent condensers	0.13

*Incremental cost beyond that for cone-roof tanks.
Source: The Gasohol Handbook, V. Daniel Hunt,
Industrial Press, Inc., 1981.

combustion particulates. According to the Office of Technology Assessment, it may actually be possible to "engineer an unambiguously beneficial environmental effect by channelling gasohol to certain urban areas with specific pollution problems (for instance, high carbon monoxide concentrations but no smog problems)."[34]

A final environmental concern is compliance with the regulations that govern alcohol fuels production. Numerous federal laws (among them, the Clean Air Act and amendments of 1977, the Clean Water Act of 1972, and the Resource Conservation and Recovery Act of 1976) contain provisions covering alcohol production, and even more stringent environmental regulations can be imposed on fuels manufacturers at the state levels. Moreover, any large projects receiving federal funds or loan guarantees must complete environmental impact statements to receive aid, so shut-downs or fines are risks.

The cost of reducing or eliminating these environmental impacts depends on how big the plant is and whether pollution controls are built in or added on. (Direct-control systems for new ethanol facilities equal an estimated 13 percent of the initial fixed investment.[35]) While the control technologies have been fully developed and have stood the test of time in many chemical and food-processing industries, some are quite expensive. (See Table 13.4.)

SOLID FUEL USE IN CITIES

Most discussions of solid fuels refer to coal, which supplies about 19 percent of U.S. energy.[36] Here too, a discussion of coal is also the starting point because most renewable fuels will be used to replace or supplement coal. Even when wood is used in an oil-burning system, it is as an alternative to coal. Seventy-seven percent of the coal burned in the United States is used to produce 48 percent of U.S. electricity.[37] Most of the remaining coal produces steam and industrial process heat.

Coal is cheaper than oil or natural gas, abundant in the United States, and more reliable than nuclear power. But transporting coal is expensive, and coal contributes to carbon dioxide build-up, acid rain, and other forms of air pollution. Mining can also harm the environment and workers.

The renewable forms of solid fuel that offer an alternative to coal represent the very old and the very new—wood and refuse-derived fuel. Wood, the world's first fuel, is now used extensively in the wood products industry and has gained popularity, especially in New England, for home heating. Burlington, Vermont, is using wood to generate part of its electricity. Refuse-derived fuel (RFD) is a newly developed fuel that is produced by separating energy-rich materials from garbage and chemically converting that waste into a form suitable for transport, storage, and burning. RDF can be prepared in various forms and can be burned alone or in combination with coal or oil.

Wood

Wood currently supplies about 1.4 to 1.7 quads (quadrillion Btu) of energy a year in the United States. Of this, about 1.2 to 1.3 quads is used by the forest-products industry and 0.2 to 0.4 quads is burned in home stoves or fireplaces. The equivalent of about two quads of energy is returned to the soil through decomposition or burned at logging sites. Some of this energy could be used commercially.[38] The U.S. Office of Technology Assessment estimates that with widespread intensive forest management, wood could provide between 5 and 10 quads of energy per year by 2000 without depleting forests.[39] The Solar Energy Research Institute (SERI) sets nine quads as the upper figure.[40] The forest-products industry is likely to be using up to four quads of wood energy annually in 2000, and the remainder of the projected use would be in homes, electricity generation, and other industries.[41]

No generalizations can be drawn about the practical and economic feasibility of using wood as an energy source. The local forest conditions, the quality of the wood, transportation costs, wood prices, and forest management techniques all vary significantly from place to place, and all must be considered in planning wood-energy use.

The technology for using wood fuel is already in operation at forest products facilities throughout the country, and especially in New England, the South, and the Northwest. In most cases, wood waste is burned to produce process steam, but dozens of operations are designed to cogenerate both steam and electricity. The forest products industry currently gets between 45 and 55 percent of its energy from wood.[42] Because the wood does not have to be bought or transported, the only major expense is forgoing the profit that could be made by selling the waste or using it to make other products. Since the industry is finding new ways to use wood waste for making such products as fiberboard, in the future a smaller percentage of the total wood harvest will be considered waste.

Several utilities are using wood-fired boilers to generate electricity. Vermont's Burlington Electric adapted two 10-Mw oil/coal boilers to burn a mixture of 75 percent wood and 25 percent oil. New plans call for a 50-Mw wood-fired generator.[43] The Eugene Water and Electric Board in Oregon operates a 33.8-Mw wood-fired boiler that provides space heat, industrial process heat, and electricity.[44]

The high moisture content of wood residue and wood chips (about 50 percent) makes it impossible to burn them in a boiler designed for coal. Increased moisture leads to slower combustion and a change in the composition of the exhaust gas, necessitating adjustments in the pollution-control system. Wood pellets, in contrast, contain only 10 percent water and can be burned in coal systems without any special adjustments.

Boilers can be built specifically to burn wood, but adapting an existing coal-fired boiler costs less. If a new boiler is required, buying a system that burns both wood and coal is advisable because wood supply can be unreliable.

Wood-fired boilers also require more storage area and a larger combustion chamber than equivalent coal-fired boilers. The largest wood-fired plants would be about 50 Mw (compared to 1,000 Mw for large coal plants) because wood would have to be transported more than 50 miles to supply a larger plant, and the transportation costs would make the increased size uneconomical. As a rule of thumb, each mile that wood must be transported adds 1.25 cents to the costs per million Btu.[45]

Economics. Wood-fired steam or electric-generating plants will be higher in capital and operating costs than oil or coal-fired plants, but much lower in fuel costs. (See Table 13.5.) In many cases, wood costs will be low enough to make wood-burning plants the most economical alternative. (See Chapter 11.) However, variability in the price of wood makes fuel cost projections tentative at best. For example, in Burlington, Vermont, during 1978 and 1979 the price went from $12 per ton to $26 per ton and then back to $16 per ton.[46] If wood becomes increasingly popular as a fuel, competition with other uses of wood could raise prices. Thus, urban energy planners should focus primarily on energy systems in which wood can be used along with other fuels. If wood

Table 13.5. Wood and Fossil Costs in the Southeastern U.S., 1979.

FUEL	PRICE/UNIT	PRICE/MMBTU
coal	$55/ton	$2.63
oil	$.60/gal	5.00
natural gas	$.003/ft³	3.86
wood pellets	$85/ton	2.94
wood chips	$12/ton	1.99
wood residues	$8/ton	1.33

Source: Levi and M. S. Grady, *Decisionmaker's Guide to Wood Fuel for Small Industrial Energy Users* (1980).

prices go up, coal or oil can be used instead. Another advantage is that retrofitting existing systems to burn wood as a supplemental fuel costs less than converting completely to a wood-fired boiler.

Environmental Impacts.   Using wood for energy could degrade U.S. forests and air quality. Without proper forest management, wood is not a renewable resource. If wood use increases and industry's forest-management practices remain shortsighted and wasteful, precious topsoil will erode, rivers will fill with silt, flooding will occur, forests will deteriorate, and U.S. wood reserves will be quickly depleted. Managers must make such decisions as whether selective cutting or clear cutting is most appropriate, whether timber waste should be cleared or left, and if fertilizer is necessary. And they must carefully select species for replanting.[47] These caveats aside, the Office of Technology Assessment estimates that intensive forest management will enable the United States to double or triple the energy yield of its forests without destroying any forest land. Indeed, it may be possible to use less wooded land for commercial production.[48]

City officials will be directly involved in protecting air quality. Compared to coal combustion, wood burning produces less sulfur and nitrogen oxides and therefore contributes less to the creation of acid rain. Emissions from wood burning, however, contain more particulate matter and more polycyclic matter (POM), which causes cancer in animals.[49]

Air-tight wood stoves are particularly high in POM emissions. Controlling emissions from wood combustion will not be difficult at large industrial or utility facilities, but home wood stoves are too numerous to regulate. Wood also leaves more ash than coal does: this ash must be landfilled and the heavy metals can leak from the landfill, affecting water quality adversely.

REFUSE—DERIVED FUEL

The average U.S. city dweller produces 3.5 lb of solid waste a day, of which approximately 75 percent contains energy—the equivalent of 1.5 lb of coal.[50] Few cities use this energy. On the contrary, most devote considerable labor, energy, and land to dispose of the waste in landfills. Yet, several methods of using the energy in municipal waste are already being used. Direct combustion of waste to produce steam is the most common. (See Chapter 10.) It is not, however, necessarily the best. Direct burning requires a large capital investment and a nearby market for the heat or process steam produced. Conversely, converting the waste to solid transportable fuel involves a smaller capital investment and allows greater flexibility in plant siting.

The cost of a refuse-derived fuel (RDF) plant is about 30 percent that of a direct-burning plant.[51] And RDF can be marketed outside the immediate vicinity of the plant. These considerations could make RDF an attractive alternative to a direct-burning operation for many cities. The RDF can be sold to industries or utilities that can burn it in combination with coal or alone in specially designed boilers. Ames, Iowa, manufactures RDF that its municipal utility burns to produce electricity.[52]

Technology.[53]   Several processes exist for separating the combustible portion of municipal waste and converting it to RDF. Some also separate glass, iron, aluminum, and other recyclable materials from the trash mix before it is used for energy purposes. The most common forms of RDF are fluff, wet, densified, and dust. (See Table 13.6.) *Fluff RDF* is produced by shredding the waste into particles about 4 to 8 in. across and passing it through an air classifier that separates the light materials (the 50 to 85 percent that contains the energy) from the inorganic and heavy organic material. The light portion is then passed over a screen or trommel to remove small pieces of glass. After screening, the RDF is shredded again into pieces of ¼ to 2 in. Fluff RDF can be burned in combination with coal in many boilers. Handling fluff RDF is difficult, however. It hangs up in the tops of hoppers, it can bind like papier-mâché if left standing for a few days, and it can become explosive if stored too long.

Wet RDF is produced by mixing the waste with water and passing it through a hydropulper that separates out large, heavy items. The remaining slurry is passed through a liquid cyclone to remove such heavy materials as glass, metals, thick plastics, and wood. The remaining slurry, mostly paper, is then dewatered for burning. Removing the water is expensive and requires energy: to reduce moisture by only 10 percent uses between 4 and 8 percent of the slurry's energy content. Wet RDF cannot be combined with coal in most boilers because it contains so much moisture.

Table 13.6. Comparisons of Various Forms of RDF and Coal.

FUEL	% MOISTURE	% ASH	BTU/LB	% NET ENERGY	ADVANTAGES	PROBLEMS
Fluff RDF	20–30	19	5,000–6,500	49	best-known technology	production: poor resource recovery; use: difficult to handle; can be explosive
Wet RDF	50	20	3,500	48	least cost to produce	prod: poor resource recovery; use: can only be used in special equipment
Dust RDF	2	10	6,900	63	easy to handle: can be mixed with oil	prod: still developmental; expensive; use: can be explosive
Densified RDF	2–30	10–20	5,000–6,500		easy to handle	prod: still developmental; requires more money and energy to make
Coal	3–12	3–11	11,500–14,300			

Source: Tania Lipshutz, *Garbage-to-Energy—The False Panacea* (1979).

Densified RDF is an experimental product that could be produced by pelletizing, briquetting, or extruding fluff RDF. The densified RDF would be easier to handle, store, and transport than fluff RDF.

Dust RDF—which is still being developed—goes through the same initial shredding and air classification as fluff RDF. However, after the first shredding, the material is treated with an embrittling agent and then pulverized. Dust RDF has a higher energy and lower moisture content than other forms of RDF, but it costs more to produce and can easily become explosive in storage. It can be burned with coal or mixed with oil.

One of the advantages usually associated with producing energy from wastes is the recovery of reusable resources such as glass, iron, and aluminum. Resource recovery, however, has not been highly successful, either in RDF or bulk-burning facilities.[54] In many cases, the glass recovered is unusable because the content of ceramics that remains after mechanical separation is so high. Aluminum and iron mixed with other metals in products cannot be separated mechanically, and mixed metals are less valuable than pure metals. Recovered material may also be too sullied to have much commercial value.

More important, garbage-to-energy systems can discourage the more economical and effective practice of source-separation recycling done by individuals. For a garbage-to-energy system to work economically, a steady and reliable supply of waste must be available for processing. Individual recycling efforts could significantly reduce the garbage flow. In particular, paper recylcing would be set back, and the paper that constitutes 40 to 60 percent of a city's garbage accounts for most of its energy content.[55]

RDF can be used alone or to supplement coal in producing steam electricity. Burned with coal, RDF can provide between 5 and 25 percent of total primary energy. RDF is currently being used to supplment coal to produce electricity in Ames, Iowa, and Madison, Wisconsin. Baltimore, Maryland, produces RDF pellets that it sells to a local industry to mix with coal to produce steam.[56]

Several problems exist with using RDF: it is still a developing technology. RDF and the plastic in RDF can leave a residue that clogs machinery. RDF also contains some glass, which can stick to boiler walls or emission systems or block the openings in the combustion grate.

RDF cannot be burned in all boilers. Its lower energy content and higher ash content make it unsuitable for systems with small combustion chambers or limited ash capacity. Because RDF contains less energy than coal, the boiler will operate at a lower capacity when RDF is burned with coal. If fluff is substituted for 25 percent of the coal, for example, the output of the boiler will be reduced by approximately 12.5 percent.

Economics.   Although conclusions about RDF plant economics must be broad and tentative, capital costs are affected by whether the RDF facility pays for

Table 13.7. Economics of Operating RDF Facilities.

PLANT LOCATION	PLANT CAPACITY (TON RDF PRODUCED PER DAY)	CAPITAL COSTS ($ MILLION)	PRICE OF RDF ($/TON)	RDF HEAT CONTENT (BTU/LB)
Baltimore	250	8.4 (1975)	27.00	6500–7000
Ames, IA	110	6.3	N/A	5625
Madison	30	3.8	25.00*	7000–8400
Albany	N/A	11.6 (1975)	N/A	4650
Tacoma	300	2.4 (1977)	8.00–12.00	5000–6000

Sources: Center for Renewable Resources Interviews with RDF Facility Engineers.

the boiler modifications that make burning RDF possible. (See Table 13.7.) For example, the two plants with the highest capital costs per output, Ames and Madison, both include costs of boiler modification in total capital costs. There are also apparent economies of scale. The two smallest plants, again Ames and Madison, cost the most per output ton, even if boiler-modification costs are excluded.

The cost of operating a plant also varies with location, the distance that the RDF must be transported, the composition of the waste, and the tipping fees charged to waste haulers. In relatively developed areas, land will be expensive and extra precautions must be taken to control emissions. If a production plant is located far from the point of use, transportation costs will be high. The energy content of the waste and the difficulty of separating out the energy-containing materials will also affect the ultimate cost of the RDF.

Although the prices charged for RDF can compete with coal prices, few RDF plants are operating economically. With its product selling at $25.00 per ton, the Madison plant does not operate in the black because until markets expand, the plant will have idle capacity.

Is RDF cheaper than direct combustion to waste? Although an RDF plant costs only about 30 percent as much to build as a waste-burning plant, it could be much easier and more profitable to market steam or heat than RDF. A private company would be more likely to buy steam, which it can use directly, than RDF, which cannot be used in most boilers unless modifications are made. Then too, to sell RDF, the city might have to agree to take the loss if users experience technical problems. Overall, steam technology remains more reliable than the still-developing RDF systems.[57]

Does the cost of an RDF facility compare with the current cost of garbage disposal? In fact, the cost of waste disposal and the problems of landfilling vary tremendously. In many cities, land is extremely limited and the cost of new landfill sites very high. If landfills are located near homes or water, extra care must be taken so that dangerous substances do not leak from the landfill. If only distant landfill sites can be found, transportation costs will be high.

Unfortunately, the major government studies of the economics of waste-to-energy systems do not include comparisons of large plants with comprehensive recycling operations.[58] However, recycling can save more energy resources and provide more jobs than waste-to-energy plants.

What other economic effects will result from an RDF plant? No doubt, cities will tolerate some air pollution to keep industry operating and workers employed; the key is making sure that polluting activities provide compensatory economic benefits. The production and burning of RDF creates air pollution and few jobs. Recycling, on the other hand, creates no pollution and could create many jobs.

None of these considerations necessarily means that RDF plants are uneconomical. In any particular instance, producing solid fuel from waste could be the most efficient and economical method of waste management. But to determine the value of an RDF system, local conditions and alternative technologies must be assessed carefully.

Environmental Impacts.   RDF proponents often list reductions in air pollution as an advantage of garbage-to-energy systems. Waste does contain less sulfur than the coal it would replace, but sulfur is not the only pollutant. Compared to coal, RDF has much greater concentrations of heavy metals and produces substantial amounts of nitrogen oxides, carbon monoxide, and hydrochloric acid.[59] In addition, waste can contain polyvinyl chloride from plastics, PCBs from fluorescent lightbulbs, and pesticide residues, all of which could enter the atmosphere as the waste is processed or burned. Thus, an RDF production facility needs a baghouse, an electrostatic precipitator like those used in coal plants to control emissions, and some means of keeping heavy metal concentrates from leaking out of the landfill area.

GASEOUS FUELS FROM RENEWABLE ENERGY

Gas produced from renewable resources (such as wood, wastes, or manure) can be burned to heat space or water or to produce steam. It can also be used as primary fuel in electricity generation or converted to liquid fuel. But economics dictates that most gaseous fuels produced from renewable resources must be used fairly close to where they are produced.

Cities can initiate gaseous fuels production within their boundaries in several ways. They can help local industries that have available resources (such as waste wood or food-processing wastes) or high energy needs to install gasification equipment, or they can become a gas producer by using gasification equipment at facilities where appropriate resources are concentrated (at, for example, waste or sewage treatment plants) or by extracting methane from landfills.

The Office of Technology Assessment estimates that with maximum federal support and private development, biomass gas could provide 9 quads of energy by 2000.[60] In reality, the contribution is likely to be far lower since the optimum combination of market-driving forces will probably not occur.

The different types of renewable gaseous fuels at a city's disposal vary according to the fuel input (feedstock), the gasification process used, and the most feasible end-uses. As a frame of reference, natural gas contains 1,000 Btu per cubic food (CF). Low-Btu gas (100–200 Btu/CF) can be produced from wood or other biomass resources by gasifiers coupled to boilers. Medium-Btu gas (300–500 Btu/CF) is made by gasifying municipal solid waste or biomass resources. This gas can be used in industrial process heating or in electricity generation via gas turbines. It can also be synthesized into liquid fuels. Biogas (600–900 Btu/CF) can be produced by anaerobically digesting manure or sewage sludge, or through the natural decomposition of wastes in landfills. It can be used for process or space heating or electricity generation. In addition, it can be upgraded to pipeline quality by removing impurities or, in some cases, by boosting its Btu content with propane.

Technological Approaches

Thermal Gasification.   Direct combustion of a feedstock results in heat, gas, ash, and other by-products. Applying heat to a feedstock in a highly controlled environment can allow better control over the type and quality of conversion outputs. In gasification systems, the gas output of a thermal-conversion process is maximized.

The most commonly used thermal gasification system is the airblown-gasifier, which blows air through a prepared feedstock at a rate that allows partial combustion. This accelerates the gasification of the remaining feedstock. The product is a low-Btu gas that can be used almost immediately in gas- or oil-fired boilers. Indeed, it must be used quickly because the residual oil, char, and tar in the gas would clog valves and fuel lines if the gas cooled. Consequently, these systems are usually close-coupled to boilers.

Airblown-gasifiers can be highly efficient since both the heat generated during gasification and the output gas can be used in the boiler. The U.S. Office of Technology Assessment estimates that properly working systems could achieve efficiencies of 80 to 90 percent.[61] However, some system development is needed to reduce tar and oil production first.

Most currently installed airblown-gasifiers are relatively small-scale industrial models (2 to 10 MMBtu/hour output). They are used primarily in forest-products industries, which use waste wood as feedstocks. Such units could also be connected to retrofitted industrial boilers that formerly burned oil. Wood gas is an almost perfect substitute for oil, so boiler retrofits are simple. Some firms are also producing larger models (up to 100/MMBtu/hour output) that can be field-erected near where energy feedstocks are harvested. At present, smaller-sized units that could be used in residential or commercial applications are not cost-effective because the lower demand for gas means that recovering capital and fuel costs would take much longer.[62]

In general, thermal gasification systems work best with dry feedstocks. Although some of the larger models use municipal solid waste, wood is the preferred fuel for most airblown-gasifier applications, partly because waste wood is often the most accessible feedstock. Then too, some technical problems attend the use of other biomass feedstocks in airblown-gasifiers. For example, the bulkiness of grasses and crop residues can cause clogging. Experiments using denser agricultural wastes as feedstocks have had more promising, although not problem-free, results. The California Energy Commission has developed a small gasifier based on a Swedish technology that converts any type of dry agricultural or wood waste to a low-Btu gas.

Another technology similar to air gasification is the oxygen gasifier. This system limits biomass combustion by controlling the oxygen content in the chamber, allowing the gas given off during combustion to remain confined. The product is a medium-Btu gas that can be used for industrial process heating or methanol production.

Various feedstocks can be used in oxygen gasification. For example, one demonstration unit in a Union Carbide Plant in South Charlestown, West Virginia, has produced medium-Btu gas from 200 tons per day of municipal solid waste.[63] But despite its potential, the oxygen gasification process has not attracted as much commercial interest as air gasifiers, perhaps because the high organic quality of the gas produced requires extensive clean-up.

Fundamentally different gasification processes use the so-called pyrolysis technique, which involves applying high heat to a feedstock in the absence of oxygen to produce oil, charcoal, and medium-Btu gas.[64] More oil and charcoal are produced when the process is slowed, while a faster pyrolytic reaction at higher heat produces more gas. The process can make use of various feedstocks, including wood, other biomass resources, and municipal solid waste.

Interest in pyrolysis is substantial because a variety of feedstocks can be used and products produced. Moreover, in this process outputs produced at one stage of the process can be used as input fuels in subsequent stages, and some of the environmental problems that plague other fuel-production processes are avoided. Pyrolysis takes care of both waste disposal and energy production, and because the waste is burned in an oxygen-free environment, air pollution is minimal. However, pyrolytic materials do give off pollutants when subsequently burned.[65]

For all these theoretical advantages, however, the technology remains developmental. And some remaining problems may be difficult to solve. A municipal solid-waste treatment plant using pyrolysis in Baltimore encountered so many operational problems that the original developer withdrew, leaving the city to manage the plant.[66] Some analysts claim that the process is too costly, given the amount of energy it delivers and the relatively restricted uses of the oil and gas produced.[67]

Medium-sized airblown-gasifiers with outputs from 2 to 10 MMBtu per hour should cost between $40,000 and $90,000, according to Booz-Allen-Hamilton.[68] This estimate squares with OTA's assessment that such units should cost about $10,000 per MMBtu per hour of capacity.[69] Using these cost levels, Booz-Allen-Hamilton calculates one representative payback for an industrial facility substituting wood gasification for oil: even with $40,000 capital costs and relatively high operating and maintenance costs, the ready availability of wood fuel means a payback of only 1.4 years because wood costs so much less than oil.

The economics of similar systems installed in residential or commercial facilities are far less promising, principally because capital costs are offset by lower yearly fuel savings since the load factor (or utilization rate) is low. Table 13.8 summarizes OTA comparisons of dollars per MMBtu of delivered energy for

Table 13.8. Comparative Costs/MMBtu of Industrial and Commercial Wood Gasifiers.

	TOTAL COST ($/MMBTU)	DELIVERED FEEDSTOCK ($40/DRY TON)	CAPITAL COST ($/MMBTU)	OPERATING COST ($/MMBTU)
1. Medium-sized industrial user:				
250 dry tons greenwood/day; 90% load factor; 85% energy efficiency	3.60	2.95	0.40	0.25
2. Smaller, commercial user:				
30 dry tons greenwood/day; 25% load factor; 70% energy efficiency	5.50	3.60	1.40	0.50

Source: Office of Technology Assessment, *Fuels from Biological Processes* (1981).

industrial and commercial wood gasifiers. With a higher load factor (90 percent compared to 25 percent), industry can offset the capital costs with a higher level of annual energy production.

Compared to the small-scale gasifiers discussed so far, large field-erected gasifiers would cost proportionately much more. Indeed, OTA estimates that energy from field-erected units could cost four to five times more per unit of output than gas from the smaller units.

Anaerobic Digestion. When bacteria consume feedstocks in an airless environment, "biogas," a mixture of carbon dioxide and methane with a heat content of 500–900 Btu/CF, is released. Biogas can be produced from manure, selected biomass resources, municipal wastes, and many other feedstocks using any of several methods. The two principal technologies used are constructed digestors and direct methane-recovery from landfills.

Anaerobic digestion units can be added to feedlots where livestock is confined or to facilities that treat municipal sewage. Smaller units can produce gas that can be used on-site to generate heat, steam, or electricity. Facilities with access to larger resources can also produce gas for delivery to pipelines.

Perhaps the largest impediment to using aerobic digestors with a manure feedstock is matching feedstock output and digestor size with demand levels. Most confined livestock operations are relatively small, and few produce enough manure to make digestor use continuous and (thus) economical.[70] Conversely, the manure available from large-scale feedlots could produce much more gas than would be needed on site, particularly since these feedlots typically use liquid fuels to drive tractors and run other machines. Biogas cannot be stored effectively either, so some other solution to the supply/demand imbalance must be found—perhaps purifying the gas and selling it to a pipeline or using the gas to generate electricity for on-farm uses or sale to the utility.

The economics of biogas production from manure depends on whether the gas is used on site or sold to a utility and whether the gas is used to produce thermal or electrical energy. An additional variable is the value of nonenergy by-products. So far, using biogas at the production site has proved to be more economical than selling it to utilities, even for large production facilities. For example, one company in Guymon, Oklahoma, had been selling purified methane from its 100,000-head beef feedlot to Peoples Gas for transmission to Chicago at $1.97 per MMBtu. However, in 1980, the company decided it would be more economical to use the gas on site to process residual protein products into finished feeds.

The economic feasibility of on-site electrical production depends heavily on utilities' willingness to purchase excess electricity. OTA predicts that most utilities would be willing to purchase electricity generated from biogas only at low wholesale rates, so biomass production combined with electricity generation will be economically attractive only for livestock operations with a high base-load electricity demand. Thus, large poultry farms in the northern U.S., which use large amounts of expensive electricity for heating, lighting, and mechanical operations, may be the first to use electricity from biogas on a large scale. Some large beef feedlots with high electricity use might also find this approach attractive, OTA contends. In piggeries that use electric self-feeding systems, it might also be possible to make economic use of a combined digestor-generator system.[71]

The economics of producing methane from manure also varies with the value of by-products—a fact too few economic analyses take into account. Valuable liquid fertilizers are produced in the digestion process, and solid by-products are being tested as cattle feed.

Municipal Sewage. While manure-based biogas operations will for the fore-seeable future produce fuels for on-farm use only, municipal sewage-treatment plants hold considerable promise for urban use. Even small-scale applications could help reduce urban energy consumption by making sewage-treatment plants more energy self-sufficient. In 1973, some 70 percent of all municipal sewage-treatment plants in the U.S. contained an anaerobic digestion unit in the sewage-treatment chain.[72] Still, most biogas produced by these digestors is burned off. According to the Solar Energy Research Institute, using all this gas would meet all or most of the country's energy needs for sewage treatment.[73]

The attractiveness of adding on-site gas-recovery systems is illustrated by the experiences of several communities. Cranston, Rhode Island, has been cap-turing biogas and converting it to pure methane since 1941. About 75,000 CF of pure methane is produced per day and used both to fuel the digestion process and to heat a city administration buidling.[74] In Los Angeles, the 350-million-gallon-per-day Hyperion Plant produces 4 million CF of (60-percent methane) gas daily. One-quarter of this gas is sold to an adjacent electric generating plant owned by the city. The rest fuels an on-site sewage treatment process, principally via electricity generation.[75] A newer approach to combined sewage treatment and gas production may soon be tested in Hercules, California. In early 1980, the city opened a sewage treatment facility that grows water hyacinths and duckweed as part of the wastewater treatment process. Now, the city plans to add a digestor and electric generator to the facility, since these plants are excellent feedstocks for anaerobic digestion.[76]

Plants currently producing gas from sewage enjoy favorable economics. The Hyperion Plant in Los Angeles produces 4 million CF of biogas daily for only $2.42 per MMBtu.[77] The Archie Elledge plant in Winston-Salem, North Carolina, produces 350,000 CF per day for only $1.60 per MMBtu. These costs do not include costs for upgrading gas to pipeline quality, which could add another $1.04 to $1.79 per MMBtu.[78] But even including these costs, gas produced from such facilities would still compete with natural gas at current prices.[79]

Direct Recovery from Landfills. As municipal waste is compacted in sanitary landfills, natural anaerobic digestion occurs. The biogas thus produced can migrate from the landfill, causing fires and killing vegetation. But it could be recaptured on-site, solving these environmental problems and providing energy.

Landfills can be "tapped" for methane gas by digging collection wells, installing piping systems, and using processing equipment to remove contaminants or to bring the gas up to pipeline quality. Such wells consist of perforated pipes vertically sunk at calculated intervals either in the landfill itself or nearby. Generally, the pipes are imbedded in gravel or cracked stone and spaced to maintain maximum gas-flow rate, a process aided by a pressurized pumping system that maintains a slightly negative pressure within the landfill. Both the wells and the pumping system are linked by a manifold piping system that delivers the gas to the processing facilities, where the water, solids, carbon dioxide, and other contaminants are extracted. Finally, if pipeline-quality gas is required, the heating value of the landfill gas can be increased by adding small amounts of propane.[80]

The growing interest in this technology is indicated by the number of cities that have started landfill-tapping projects. In March of 1980, a National Center for Resource Recovery list showed seven demonstrations of commercial-scale landfill gas-recovery systems operating or under construction. By September of 1981, the number had grown to 19.[81] (See Table 13.9.) Some cities are using the recovered gas in their own buildings or power stations, while others are developing projects in cooperation with local utilities. In Los Angeles, for example, gas obtained from the Sheldon-Arleta Landfill in Los Angeles is used in a nearby city-owned generating plant. Under a royalty agreement with the city of Mountainview, California, Pacific Gas & Electric is recovering 1.2 million CF of landfill gas (45- to 50-percent methane) per day, purifying it for use in PG & E's pipeline.[82]

Experiences with currently operating landfills are providing some valuable lessons for future projects. For example, the Sheldon-Arleta Landfill operating in Los Angeles had to close temporarily for retooling when it was discovered that the contaminant-control system was not large enough to handle the gas supply and that corrosive impurities left in the gas caused cooling coils in the compression system to corrode.[83]

Table 13.9. Projects to Recover Methane from Landfills.

LOCATION AND MAJOR PARTICIPANTS	OUTPUT OR GAS PRODUCED MILLION FT^3/DAY	CAPITAL COSTS ($ MILLIONS)	STATUS
Azusa Azusa Land Reclamation Co. (wholly owned subsidiary of the SW Portland Cement Co.	Low-Btu gas	n/a	Operational
Carson Watson Biogas Systems; SCS Engineers	Medium-Btu gas to power generators, producing electricity for sale to utility	n/a	Collection system complete; operations expected in Feb. 1982
City of Industry	Medium-Btu gas; 0.5 (Approx.)	0.45	Operational
Duarte Watson Biogas Systems Lockman and Assoc.	Medium-Btu gas to power generators, producing electricity for sale to utility	n/a	Collection system under construction; operations expected in Feb. 1982
Los Angeles (Bradley East Landfill)	Medium-Btu gas; 2.4	n/a	Operational
Martinez Getty Synthetic Fuels, Inc.	Medium-Btu gas	n/a	Operations expected in late 1981
Monterey Park Getty Synthetic Fuels, Inc. Operating Industries, Inc. Southern Calif. Gas Co.	High-Btu gas; 4.0	n/a	Operational
Mountain View City of Mountain View; EPA; Pacific Gas & Electric Co.; Dept. of Energy	High-Btu gas; 0.3	0.85	Demonstration plant; operating and producing 0.3 MMSCF of treated gas with a HHV of 850–950 Btu/SCF; expansion under investigation
Palos Verdes Getty Synthetic Fuels, Inc. Los Angeles Co. Sanitation District; So. Calif. Gas Co.	High-Btu Gas; 0.75	n/a	Operational
San Fernando Getty Synthetic Fuels, Inc.	Medium-Btu gas	n/a	Operations expected in late 1981
San Leandro Getty Synthetic Fuels, Inc. Oakland Scavenger Co.	Medium-Btu gas	n/a	Operational
Sun Valley (Sheldon-Arieta Landfill Gas Recovery Project) City of LA Departments of Public Works and Water and Power	Low-Btu gas; 2.8	2.5	Compressor station undergoing equipment modifications
Wilmington Watson Biogas Systems, Inc. SCS Engineers	1.5	n/a	Operational

Table 13.9. Projects to Recover Methane from Landfills. (*Continued*)

LOCATION AND MAJOR PARTICIPANTS	OUTPUT OR GAS PRODUCED MILLION FT³/DAY	CAPITAL COSTS ($ MILLIONS)	STATUS
ILLINOIS *Calumet City* Getty Synthetic Fuels, Inc. Waste Management, Inc. Natural Gas Pipeline Co.	High-Btu gas to power generators, producing electricity for sale to utility	n/a	Operational
MICHIGAN *Riverview* Watson Biogas Systems; SCS Engineers	Medium-BTU gas to power generators, producing electricity for sale to ultility	n/a	Contracts with city signed; applications for construction submitted; utility negotiations proceeding
NEW JERSEY *Cinnaminson* Sanitary Landfill, Inc.; Public Services Electric and Gas Co.; Hoeganaes Corp.	Medium-Btu gas (570 Btu/SCF); 1.0 (Used in-plant by Hoeganaes Corp.)	n/a	Operational; modifications planned to improve service reliability and add other customers.
NEW YORK *Staten Island* (Fresh Kills Landfill) Brooklyn Union Gas Co.; New York City Office of Resource Recovery and Waste Disposal; N.Y. State Energy Research and Development Authority; U.S. Dept. of Energy; Leonard S. Wegman, Inc.	0.05	.33	Test began in Feb. 1981 using raw landfill gas in an internal combustion engine generator to produce 100 kw of electricity; 2000 hrs logged on unit as of July 11, 1981.
Staten Island (Fresh Kills Landfill) Getty Synthetic Fuels, Inc.; City of NY; Brooklyn Union Gas Co.	High-Btu gas	n/a	Operations expected to begin in 1982
NORTH CAROLINA *Winston-Salem* City	Medium-Btu gas (burned to generate electricity to power sewage treatment plant)	Less than $25,000 for wells and pipeline	Operational

The extraction of methane gas from landfills has led to some experimentation with integrated systems—combined landfill-recovery operations and sewage-treatment plants. These facilities mix municipal solid waste with sewage sludge in a controlled, airless environment. While gas-production in landfills may take 20 to 40 years, the same process could occur in 5 to 30 days in an integrated plant. A DOE-operated pilot plant in Pompano Beach, Florida, is running at about half capacity and processing about 46 tons of refuse per day—producing some 10,000 CF of 50-percent methane gas per ton.[84]

Based on its experience with methane recovery in Mountain View, California, Pacific Gas & Electric has developed a detailed economic analysis of landfill gas collection, treatment, and compression. Cost data from the Mountain View system (completed in May 1978 at a cost of $840,000) indicates that a larger plant producing 96,100 MMBtu of raw gas annually and purifying it to pipeline-quality methane would cost $3.3 million to construct. If financed 50 percent with debt at 16-percent interest and 50 percent with equity at 20-percent return on equity, the system could produce gas at $12.02 per MMBtu (in 1981 dollars). These costs are competitive with the marginal cost of oil ($10.90/MMBtu) and, according to PG & E, with the current cost of liquified and synthetic natural gas.[85]

PG & E's projected costs for gas produced from a large new facility are higher than present costs of gas from the Mountain View facility. In July, 1980, PG & E charged $4.43 for pipeline-quality gas from the system. However, both capital and operating costs are expected to increase over the lifetime of the facility, partly because more and more wells will have to be dug to maintain desired output as gas production falls with continued extraction. Thus, the costs of gas for landfills should be differentiated in terms of short-term costs and the much-higher lifetime costs.

Environmental Impacts

Both the production and use of gaseous fuels derived from renewable resources are expected to have relatively benign environmental impacts compared to those of fossil fuels. Of course, where new gasification and end-use conversion systems do not replace existing fossil-fueled facilities, environmental problems will increase.

Little solid evidence exists about any pollution-creating potential of gasifiers. But OTA has made some general observations. First, air pollution emissions vary greatly with feedstock composition, conversion technology, and the degree of pollution control exercised. Most gasification technologies will emit fewer sulfur or nitrogen compounds and less carbon dioxide or carbon monoxide than oil-fired boilers. But oil-burning facilities will emit fewer particulates and hydrocarbons.[86] In most cases, a wood-gasification unit retrofit onto an oil boiler will emit far less carbon monoxide, particulate matter, or hydrocarbon than a direct wood-combustion unit.

Second, environmental assessments cannot ignore the impacts of using or disposing of gasification by-products—pyrolytic oil or char. Burning char can emit nitrous and sulfur oxides, while high levels of particulates can be emitted when pyrolytic oils are burned in the gasification process. Some studies have also suggested that the waste effluent from pyrolysis contains a very high organic content, potentially damaging to water quality.[87] However, where waste is the feedstock, the pollution-related costs will have to be balanced against the environmental costs and disturbances associated with other methods of waste disposal or energy production.

Most anaerobic digestion or landfill recovery operations can have positive environmental impacts. For example, tapping accumulations of methane in landfills can help prevent dangerous gas build-up that could lead to explosions or to land- or water-pollution. Similarly, confining animal wastes for gas production could reduce environmentally damaging runoff from cattle feedlots.

The potential environmental problems of anaerobic digestion are mostly those of waste disposal. Digestion reduces the amounts of toxic wastes requiring disposal, but it does not eliminate them.

Notes

CHAPTER ONE

1. U.S. DOE, *Local Government Energy Activities* (Washington, D.C.: U.S. DOE, 1979), vol. I, p. II-2.
2. *Nation's Cities Weekly,* June 22, 1981, p. 3.
3. Testimony of Nicholas Carbone, *Energy and the City,* Record of Hearings held in September, 1977, Subcommittee on the City, pp. 189–90.
4. Of course, all regions of the country spend money on products imported from other areas. But evidence suggests that the "exchange rate" between energy and other regional goods is increasingly favoring the energy-producing regions. See Hans Landsberg and Joseph Dukert, *High Energy Costs: Uneven, Unfair, Unavoidable?* (Washington, D.C.: Resources for the Future, 1981), pp. 40–43.
5. David Morris, et al., *Planning for Local Self-Reliance,* A Case Study of the District of Columbia (Washington, D.C.: Institute for Local Self-Reliance, 1979), pp. 32–33.
6. City Planning Commission of New Orleans, Study Design for the New Orleans Energy Management Program, November, 1978, p. 3.
7. I. M. Levitt and A. Di Tommaso, Philadelphia Energy Management Program Fact Sheet, June, 1979, p. 3.
8. Eugene Kramer and Linda Berger, The High Cost of Heat: A Threat to Urban Neighborhoods, in Joel T. Werth, ed., *Energy in the Cities Symposium,* PAS Report no. 349 (Chicago: American Planning Association, 1980), pp. 12–13.
9. See Landsberg and Dukert, p. 30. For earlier estimates, see City of Hartford, The Management of the Heating Oil Crisis: Local Government Energy/Housing Response (Draft), Oct. 1979, p. 4.
10. Testimony of Steven Ferrey, *Renewable Energy and the City,* Record of Hearings held October, 1979, Washington, D.C.: GPO, 1979, p. 153; Eunice Grier, Colder . . . Darker, (Washington, D.C.: Community Services Administration, 1977).
11. City of Portland, *Energy Conservation Choices for the City of Portland,* II vols. (Washington, D.C.: HUD, 1977), vol. 3, p. 15.
12. Cf. James Ridgeway, *Energy Efficient Community Planning* (Emmaus, PA: J.G. Press, 1979), ch. 2.
13. St. Paul Energy Committee of 100, Large Energy Users Sub-Committee Report, January, 1980.
14. City of Portland, op. cit., p. 17.
15. City of Los Angeles, *The Energy/LA Action Plan,* July, 1981, p. 17.
16. Study Design for New Orleans, op. cit., p. 3.
17. For a description of how density-bonus language might be included in local subdivision regulations, see: Martin Jaffee and Duncan Erley, *Protecting Solar Access for Residential Development: A Guidebook for Planning Officials* (Washington, D.C.: HUD, 1979), pp. 76–78. Among the states that have authorized local property tax statements for solar energy

systems are Connecticut, Georgia, New Hampshire, Vermont, and Virginia. State-wide property tax statements are in place in 25 other states.

18. John B. Alschuler, Jr., *Community Energy Strategies: A Preliminary Review* (Draft), May, 1981, pp. 1–2.

19. For a listing of communities participating in the CCEMP program, see John Moore, et al., *Community Energy Auditing Experience with the Comprehensive Community Energy Management Program* (Chicago: Argonne National Laboratory, September, 1980).

20. City of St. Paul, Committee of 100, Setting Directions for an Energy Conscious City, August, 1979.

21. Jack Gleason, District Heating Saves Energy (In Europe), *Public Power,* September–October, 1979, pp. 46–48.

22. City of Portland, op. cit., vol. 3, p. 8.

23. *Washington Post,* Despite Megabucks Campaign, Nukes Lose One, November 6, 1981, p. A-10.

24. Jack Gleason, *Efficient Fossil and Solar District Heating Systems* (Golden, Colorado: Solar Energy Research Institute, 1981).

25. For example, a 1000-Mw nuclear-powered facility (Nine-Mile, Unit 2) in New York State was originally scheduled to cost $3.7 million ($3,700/kw). New projections set total costs by completion in 1986 at $15.5 million ($5,500/kw).

26. Hittman Associates, Inc., *Comprehensive Community Energy Planning,* 2 vols., Washington: DOE, 1978.

27. Cited by Mark Braley in Presentation to the Comprehensive Community Energy Management Program, Feb. 14, 1980, pp. 6–7.

28. Ann Kline, Allegland Development, City Energy Program, Richmond, Indiana.

CHAPTER TWO

1. See, for example, the reports of energy audits conducted under the U.S. Department of Energy's Comprehensive Community Energy Management Planning (CCEMP) Program. Published reports include documents from Bridgeport, Boulder, Greenville, Los Angeles, and Seattle. (See Appendix.)

2. Energy Task Force, *Windmill Power for City People* (Washington, D.C.: Community Services Administration, 1977), p. 5.

3. New England River Basin Commission, Potential for Hydropower Development at Existing Dams in New England, January, 1980; New York State Energy Research and Development Authority, An Action Program for the Development of Small Hydropower in New York State, (undated), pp. 2–7; General Accounting Office, *New England Can Reduce Its Oil Dependence Through Conservation and Renewable Resource Development* (Washington, D.C.: U.S. General Accounting Office, 1981) vol. 1, pp. 7, 8–9; vol. 2, pp. 26–7; Resource Policy Center, *The New England Energy Atlas* (Hanover, N.H.: 1980), pp. 20–28.

4. Much depends on how deep the water resource is. Ground water temperatures close to the surface can vary substantially with ambient temperatures. Fluctuations are less likely with deeper reserves. Cf. Southern Tier Central Regional Planning Board, *Renewable Energy Resource and Technology Assessment* (Washington, D.C., U.S. Department of Energy, 1978), pp. 25–26.

5. See, for example, City of St. Paul, Energy Committee of 100, Large Energy Users Subcommittee Report, January, 1980, p. 10.

6. Cf. Hampshire College, Energy Self-Sufficiency in Northampton, Massachusetts (Washington, D.C.: U.S. Department of Energy, 1979), pp. 126–130; Donald L. Kirkpatrick, et. al., A Unique Low-Energy Air Conditioning System Using Naturally-Frozen Ice, presented at the Solar Rising Conference of the International Solar Energy Society, Philadelphia, PA, May 26–30, 1981.

7. Michael Edesess, Solar Ponds' Resource Potential, Non-Convecting Solar Pond Workshop Summary (Boulder City, Nevada: Desert Research Institute, 1980), pp. 2–8.

8. City of Burlington, Report for the Electric Department for the Year Ending June 30, 1978; Timothy S. Cronin, New Wood-Fired Generating Unit Brings Savings, *Public Power,* January–February, 1978, pp. 32–33.

9. Office of Technology Assessment, *Energy from Biological Proceses* (Washington, D.C.: Office of Technology Assessment, 1981), p. 4; City of Portland, Energy Conservation Choices for the City of Portland, Oregon, Volume 3D; Industrial Conservation Choices (Washington, D.C.: U.S. Department of Housing and Urban Development, 1977), p. 14; Michael P. Levi and Michael J. O'Grady, *Decisionmaker's Guide to Wood Fuel for Small Industrial Energy Uses* (Golden, CO: Solar Energy Research Institute, 1980); John Lohnes, Potential Industrial Wood Boiler Users in New England (Hanover, NH: Resource Policy Center, Dartmouth College, 1979).

10. Charles Hewett and Colin High, *Construction and Operation of Small Dispersed Wood-Fired Power Plants* (Hanover, NH: Resource Policy Center, Dartmouth College, 1978), p. 21.

11. OTA, *Energy from Biological Processes,* op. cit., pp 78–79.

12. Cf. Colin J. High and S. E. Knight, Environmental Impact of Harvesting Non-Commercial Wood for Energy. (Hanover, N.H.: Resource Policy Center, Dartmouth College, 1977); C. G. Wells and J. R. Jorgenson, Effect of Intensive Harvesting on Nutrient Supply and Sustained Productivity, Proceedings: Impact of Intensive Harvesting on Forest Nutrient Cycling (Syracuse: State of New York, 1979); Resource Policy Center, op. cit., p. 17.

13. N. Bhagat, H. Davitian, and R. Pouder, *Crop Residues As A Fuel for Power Generation* (Upton NY: Brookhaven National Laboratories, 1979), pp. 1–2.

14. Solar Energy Research Institute, *Fuel From Farms—A Guide to Small-Scale Ethanol Production* (Golden, CO: 1980), pp. A.2–A.3.

15. Lester R. Brown, *Food or Fuel: New Competition for the World's Cropland* (Washington, D.C.: Worldwatch Institute, 1980).

16. Mueller Associates, *Biomass Availability for Highway Vehicle Fuels: A Resource and Availability Survey* (Washington, D.C.: U.S. Department of Energy, 1979), pp. 21.

17. Ibid.: Bhagat et. al., op. cit., p. 2; See, however, OTA, op. cit., pp. 119–20.

18. Mueller Associates op. cit., p. 21.

19. Eric Leber, Golden Garbage, *Public Power,* March/April, 1980, pp. 14–17; U.S. Conference of Mayors, Economic Development and Resource Recovery (Washington, D.C.: U.S. Conference of Mayors. October 1979) (Draft). In addition, see the series of case studies prepared by the Conference on cities that have implemented resource recovery plants.

20. Even where waste is burned for heat or steam, landfills may still be needed, however, to dispose of ash from garbage incineration and items in the waste stream that cannot be recovered. The extent to which landfills are still needed will offset the economics of garbage-to-energy plants significantly. See Tania Lipshutz, *Garbage to Energy—The False Panacea* (Santa Rosa, California: Sonoma County Recycling Center, 1979).

21. Environmental Protection Agency, *Resource Recovery Plant Implementation: Guides for Municipal Officials,* Vol. 3: Markets (Washington, D.C.: U.S. Envriomental Protection Agency, 1979).

22. U.S. Geological Survey, Assessment of Geological Resources in the United States—1978, Geological Survey Circular 790, (Washington, D.C.: U.S. Department of the Interior, 1979); Personal Communication, William J. Toth, Supervisor, Technology Transfer, EG&E, Idaho, Inc., December 11, 1981.
23. DOE Moves Ahead with Geothermal, *EPRI Journal,* March 1980, pp. 28–32.
24. For an excellent technical discussion of geothermal resources, see *Direct Utilitzation of Geothermal Energy: A Technical Handbook* (Davis, California: Geothermal Resource Council, 1979).
25. *EPRI Journal,* op. cit.
26. U.S. Department of Energy, Geothermal Energy: Program Summary Document (Washington, D.C.: U.S. DOE, January, 1980) pp. 32–33; Stahrl W. Edmunds, The Government's Role in Developing Geothermal Energy Policy in Imperial County, California, Joel T. Werth, ed. Energy in the Cities Symposium, PAS Report no. 349, (Chicago: American Planning Association, 1980) pp. 37–42.
27. The standard method used to calculate wind speed distributions on the basis of average wind speed measurements is the so-called Rayleigh distribution described by Park, op. cit., p. 54.
28. *EPRI Journal,* June, 1981.
29. Cf. U.S. Army Corps of Engineers, *Feasibility Studies for Small-Scale Hydropower Additions: A Guide Manual* (Washington, D.C.: U.S. Department of Energy, 1979); Tippetts-Abbett-McCarthy-Stratton, *Evaluation of Small Hydroelectric Potential* (New York: 1979) available from T-A-M-S office, 655 Third Avenue, New York, New York 10017.
30. For a more detailed discussion see, *Resource Recovery Plant Implementation,* op. cit., vol 1, Planning and Overview.

CHAPTER THREE

1. Compact Cities: Energy Saving strategies for the Eighties, Report together with Dissenting Views by the Subcommittee on the City of the Committee on Banking, Finance, and Urban Affairs, House of Representatives (Washington, D.C.: U.S. Government Printing Office, 1980), p. 13–14.
2. City of Los Angeles, Solar Envelope Zoning: Application to the City Planning Process (Golden, Colorado: Solar Energy Research Institute, 1980), p. A-35.
3. City of Boulder, *Community Energy Management Plan,* vol. 1, pp. 45–47. Some growth management programs, however, are potentially discriminatory and may be struck down judicially because of it. See Compact Cities, p. 60.
4. John Alschuler found that political impediments to these changes were perhaps the strongest barrier. See Community Energy Strategies: A Preliminary Review (Draft) May 1980, pp. 37–38. Cf. The Energy/LA Action Plan, op. cit.
5. For example, the Philadelphia Solar Planning Project calculated that almost 80 percent of the housing units in the city were in row-homes. See Philadelphia Solar Planning Project, Working Papers: Applications, May, 1980, p. 7.
6. Cf. George E. Peterson, et al., Urban Development Patterns (Washington, D.C.: The Urban Institute, 1980).
7. City of Boulder, op. cit., p. 35.
8. City of Portland, op. cit. (Chapter 1, no. 11), p. 32; Compact Cities op. cit., p. 57; James

Ridgeway, Energy Efficient Community Planning (Emmaus, PA: J.G. Press, 1979), pp. 57–61.

9. Testimony of Marion Hemphill, Energy and the City, Hearings before the Subcommittee on the City of the Committee on Banking, Housing, and Urban Affairs, House of Representatives, (Washington, D.C.: U.S. Government Printing Office, 1977), p. 94–95.

10. City of Los Angeles, *The Energy/LA Action Plan,* July 1981, p. 48.

11. Office of Energy Conservation, State of Colorado, Energy Conscious Planning, 1979.

12. Energy Audit—Greater Bridgeport Region, op. cit., p. 24.

13. City of Portland, op. cit., Volume 3D, Industrial Conservation Choices, pp. 66–67.

14. For an excellent discussion of the technical and land-use issues involved in assuring solar access, see Martin Jaffee and Duncan Erley, *Protecting Solar Access for Residential Development: A Guidebook for Planning Officials* (Washington, D.C.: U.S. Department of Housing and Urban Development, 1979), esp. pp. 82–98.

15. Ibid, p. 83; Winslow Fuller, Collector Location: No Taboos on East or West, *Solar Age,* December, 1980, pp. 26–27.

16. Ibid, p. D-48.

17. Jaffee and Erley, op. cit. pp. 10–21.

18. Ibid, p. 99. Cf. Gerald Mara and David Engel, Institutional Barriers to Solar Energy: Early HUD Demonstration Experiences, *Solar Law Reporter,* March/April, 1980, pp. 1095–1117.

19. Robert Twiss, et al., *Land Use and Environmental Impacts of Decentralized Energy Use* (Washington, D.C.: U.S. Department of Energy, 1980), p. D-3.

20. For a partial list of communities proposing or adopting solar shade control ordinances, see Jaffee and Erley, Appendix.

21. Philadelphia Solar Planning Project, Working Papers—Neighborhoods, Part II, May, 1980, pp. 27–48; Michael Shapiro, Boston Solar Retrofits, Discussion Paper E-80-11, Energy and Environmental Policy Center, John F. Kennedy School of Government, Harvard University, December, 1980, p. 21.

22. Twiss, op. cit., p. D-30.

23. Ibid, p. D-76.

24. The best source on solar access protection in existing development is Gail Boyer Hayes' *Solar Access Law* (Cambridge, MA: Ballinger, 1979).

25. Seattle Energy Office, Energy Ltd., (Draft), October, 1980, pp. 113–115; Jaffee and Erley, op. cit., pp. 46–50.

26. Jaffee and Erley, op. cit., pp. 88–90.

27. Greater Roxbury Development Corporation, Solar Typologies Project: Neighborhood-Scaled Energy Development Strategies for Roxbury, MA., November 1, 1980, p. 8; Cf. Compact Cities, op. cit., p. 10; Solar Envelope Zoning, op. cit., p. 54.

28. Cf. Energy/LA, op. cit., p. 48.

29. Hayes, op. cit., p. 15–18.

30. For an extensive discussion of the solar envelope concept, see Ralph Knowles and Richard Berg, *Solar Envelope Concepts* (Golden, CO: Solar Energy Research Institute, 1980).

31. This is truer for some areas than others. For instance, Los Angeles found that applying the solar envelope concept in high density areas would reduce density levels substantially. Still, a solar envelope could be applied in most low- or medium-density areas without adverse impacts. See Solar Envelope Zoning, op. cit., p. 7.

32. This approach is described and critiqued by Hayes, op. cit., pp. 181–192.

33. Cf. Solar Envelope Zoning, op. cit., pp. 35; 43.

34. Philadelphia Solar Planning Project, Neighborhoods, op. cit., Shapiro, op. cit., p. 11.

CHAPTER FOUR

1. At last count, 46 states plus the District of Columbia had adopted some form of energy conservation standard. See National Conference of States on Building Codes and Standards, Draft Report on State Energy Codes for New Building (Washington, D.C.: October, 1980).
2. See American Institute of Architects Research Corporation, *Solar Dwelling Design Concepts* (Washington, D.C.: U.S. Department of Housing and Urban Development, 1976), pp. 53–56.
3. Recent practice for new homes features R-19 insulation in ceilings and R-11 in walls, considerable improvements over energy efficiency levels achieved in the early 70s. Still, there is considerable room for improvement. Analyses of energy conservation features used in homes participating in HUD's Residential Solar Demonstration Program indicated that builders regularly used R-30 insulation in the ceilings and R-19 in walls.
4. Michael Andreassi, Lorene Yap, and Olson Lee, *The Impact of Residential Energy Efficiency Standards on Households* (Washington, D.C.: The Urban Institute, June, 1980). pp. 17–19.
5. David R. Kaminsky, Shelter and Neighborhoods: Indicators of Physical Deterioration in Cities, Robert P. Boynton, ed., *Occaional Papers in Housing and Community Affairs,* vol. 4 (Washington, D.C.: U.S. Department of Housing and Urban Development, 1979), pp. 126–147 at 138.
6. Douglas Sliger, Modular Retrofit Experiment (Princeton Project), June 17, 1980.
7. Kaminsky, op. cit., p. 137.
8. Steven Ternoey, et al., Energy Efficient Commercial Buildings: The Effect of Environmental Systems on Architectural Form, Proceedings: The 5th National Passive Solar Conference (Newark, Delaware: American Section of the International Solar Energy Society, 1980), pp. 35–43.
9. Residential Building Standards Development Project, *High-rise Buildings,* Building and Appliance Standards Office, California Energy Commission, May, 1980, pp. 3–8.
10. Office of Technology Assessment, Residential Energy Conservation (Washington, D.C.: Office of Technology Assessment, 1979), p. 35.
11. Greater Washington Board of Trade, *Energy Conservation Manual for Multi-family Dwellings* (Washington, D.C.: 1981), p. 65; National Electrical Contractors Association and National Electrical Manufacturers Association, *Total Energy Management: A Practical Handbook on Energy Conservation and Management* (Washington, D.C.: 1979), p. 27.
12. Peter Hollander, et al., *Installation Guidelines for Solar DHW in One- and Two-Family Dwellings* (Washington, D.C.: U.S. Department of Housing and Urban Development, 1979), p. 18.
13. Robert Twiss et al., *Land Use and Environmental Impacts of Decentralized Energy Use* (Washington, D.C.: U.S. Department of Energy, 1980), p. D-15.
14. Cf. Vivian Loftness and Volker Hartkopf, Passive Solar Retrofit of All-Glass Office Buildings, Proceedings: The 5th National Passive Solar Conference (Newark, Delaware: American Section of the International Solar Energy Society, 1980), pp. 869–873.
15. Alex Wilson, Trombe Wall Retrofit, *Solar Age,* September, 1978, pp. 28–30.
16. Philadelphia Solar Planning Project, *Working Papers—Neighborhoods,* Part II, May, 1980, p. 28.
17. State of Florida, *Model Energy Efficiency Building Code,* November, 1978, pp. 5.2–5.8.
18. Andreassi, et al., op. cit., p. 16.
19. City of Portland, *Energy Conservation Choices for the City of Portland, Oregon,* vol. 1, Preliminary Data and Analysis (Washington, D.C.: U.S. Department of Housing and Urban

Development, 1977), pp. 37–8; St. Paul Energy Committee of 100, Existing Housing and Zoning Subcommittee Report, January, 1980, p. 2.

20. Andreassi, op. cit., p. 22.

21. Philadelphia Solar Planning Project, *Working Papers—the Philadelphia Solar Audit,* February, 1980, p. 1.

22. Compact Cities: *Energy Saving Strategies For the Eighties,* Report together with Dissenting Views by the Subcommittee on the City of the Committee on Banking, Finance and Urban Affiars, House of Representatives (Washington, D.C.: U.S. Government Printing Office, 1980), p. 30. Note that much of the power of Executive Order 12185 has been diluted by the change in administrations and the lower levels of funding for HUD housing programs.

CHAPTER FIVE

1. This would not necessarily be an attempt to make an uneconomical technology look more promising than it is. If accelerated natural gas deregulation occurs, cities heavily reliant on natural gas supplies could be devastated by precipitous fuel price hikes. Encouraging the substitution of solar energy for natural gas could be seen as an effort to cushion the blow.

2. South Florida Regional Planning Council, South Florida Regional Energy Audit (Draft), January 13, 1981, pp. 66ff.

3. George Peterson, et al., *Urban Development Patterns* (Washington, D.C.: The Urban Institute, 1980).

4. Douglas W. Hooker, E. Kenneth May, and Ronald E. West, *Industrial Process Heat Case Studies* (Golden, Colorado: Solar Energy Research Institute, May, 1980).

5. City of Portland, *Industrial Conservation Choices for the City of Portland, Oregon,* vol. 1. Preliminary Energy Data and Analysis (Washington, D.C., U.S. Department of Housing and Urban Development, 1977), p. 82.

6. Grant Thompson and David Strom, *Comments on the Proposed 10 CFR Part 435 Energy Performance Standards for New Buildings* (Washington, D.C.: The Conservation Foundation, April 30, 1980), pp. 11–12.

7. This is not to deny that shell improvements can have some effect on cooling loads. However, most research suggests that those effects will be minimal as compared with effects on energy consumption for space heating. See National Association of Home Builders Research Foundation, *Insulation Manual* (Rockville, MD: NAHBRF, 1979), pp. 12–13.

8. City of Portland, op. cit., vol. 1, pp. 53–54; City of Boulder, *Community Energy Management Plan,* vol. 1, Energy Use Study, May, 1980, p. 57; Statement of the National Association of Home Builders on Proposed Building Energy Performance Standards (BEPS), March 24, 1980, p. 5.

9. Cf. Hans Landsberg and Joseph Dukert, *High Energy Costs: Uneven, Unfair, Unavoidable?* (Washington, D.C.: Resources for the Future, 1981), pp. 52.

10. Cf. Hooker, et al., op. cit., p. 80.

11. Office of Technology Assessment, *The Energy Efficiency of Buildings in Cities* (Washington, D.C.: Office of Technology Assessment, 1982).

12. Peter Margen, et al. *District Heating/Cogeneration Applications Studies for the Minneapolis/St. Paul Area,* Oak Ridge National Laboratory, 1979, p. 46.

13. Todd Onuskiewicz and Roy Meador, *District Heating Backgrounder* (Detroit: Michigan Energy and Resource Research Association, 1980).

14. New England Innovation Group, *The Lawrence Economic Development/Energy Program* (undated), pp. 122–26.

15. See Part Two, Chapter Four.
16. U.S. Conference of Mayors, *Nashville: A Case Study of Economic Development and Resource Recovery*, (Washington, D.C.: U.S. Conference of Mayors, 1980), p. 5.
17. James O. Kolb, St. Paul District Heating Project: Economic Feasibility and Implementation Strategy, Presentation to Integrated Energy Systems Task Group, August 11, 1981.
18. Conversation with Richard Urbania, Street Lighting Project, City of Oakland, January, 1981.
19. Cf. U.S. Court of Appeals for the District of Columbia Circuit No. 80-1789, *American Electric Power Service Corporation, et. al.,* vs. *FERC,* January 22, 1982.
20. For an excellent, detailed discussion of how cities can become qualified small power producers under the provision of PURPA 210, see David Morris, David Bardaglio, et al., Municipalities and PURPA 210: Opportunities for Local Revenue Generation, (Draft), (Washington, D.C., Institute for Local Self-Reliance, 1981).
21. Cf. Despite Megabucks Campaign, Nukes Lose One, *Washington Post,* November 6, 1981, p. A. 10. In addition, provisions of the Pacific Northwest Electric Power Planning and Conservation Act now require utilities in the marketing area of the Bonneville Power Administration (BPA) to explore conservation and renewable-energy options first and to build new thermal power plants only as a last resort. See the summary of the act prepared by the Bonneville Power Administration (BPA), 1981, p. 5–11.
22. For a good recent discussion of the dilemmas confronting utilities considering building new nuclear power plants, see Graham Allison, et. al., The Governance of Nuclear Power, A Report Submitted to the President's Nuclear Safety Oversight Committee (Cambridge, MA: John F. Kennedy School of Government, Harvard University, 1981).
23. Because of the dominance of high fuel costs in oil-based, peak-load units.
24. Cf. City of Philadelphia, Pre-application for a U.S. Department of Housing and Urban Development Innovative Grant for Community Energy Conservation, June 1980. For technical support, see C. Lee et al., *Philadelphia's Gallery II Shopping Center Complex: An Analysis for the Applications of an Integrated Community Energy System,* two vols. (Chicago: Argonne National Laboratory, 1979).
25. South Florida Regional Planning Council, South Florida Regional Energy Audit (Draft), January 13, 1981, pp. 57–58.
26. Cf. City of Los Angeles, *The Energy/LA Action Plan,* July, 1981, p. 64; See also John H. Alschuler, Community Energy Strategies: A Preliminary Review.
27. City of Portland, op. cit., vol. 1, p. 22.
28. For a critique of the overall effectiveness of lifeline rates, see Landsberg and Dukert, op. cit., pp. 61–63.

CHAPTER SIX

1. James W. Fossett and Richard P. Nathan, The Prospects for Urban Revival, Undated, but submitted for publication in Roy Bahl, ed. *Urban Government Financed in the 1980s,* p. 5.
2. Michael Andreassi, Lorene Yap, and Olson Lee, *The Impact of Residential Conservation Standards on Households* (Washington, D.C.: The Urban Institute, 1980); Alice Levine and Jonathan Raab, *Solar Energy Conservation and Rental Housing* (Golden, CO: Solar Energy Research Institute, 1981).
3. *Annual Housing Survey*—Part A—General Housing Characteristics for the United States and Regions (Washington, D.C.: GPO, 1979).
4. Testimony of Steven Ferrey, Renewable Energy and the City, Joint Hearings before the

Subcommittee on the City, Committee on Banking, Finance, and Urban Affairs and Subcommittee on Oversight and Investigations, Committee on Interstate and Foreign Commerce, House of Representatives (Washington, D.C.: Government Printing Office, 1979).

5. Margaret Hilton, Sharing the Costs and Benefits of Energy Conservation in Rental Housing, Report Prepared for the U.S. Department of the Interior, Heritage Conservation and Recreation Service (Washington, D.C.: Center for Renewable Resources, 1981).

6. Cf. Henry Dearborn, *Escalating Energy Costs of Rent Stabilized Apartment Buildings: The Abandonment Crisis and Its Solution,* (New York: Energy Task Force, 1979).

7. John W. Pickering and Harold Bunce, *The Dynamics of Urban Distress: Factors Underlying Troubled Aspects of Urban Life;* Robert P. Boynton, ed., *Occasional Papers in Housing and Community Affairs,* vol. 4 (Washington, D.C.: U.S. Department of Housing and Urban Development, 1979), pp. 50–88.

8. Michael Andreassi, Lorene Yap, and Olson Lee, *The Impact of Residential Conservation Standards on Households* (Washington, D.C.: The Urban Institute, 1980), p. 29.

9. Cf. Hilton, op. cit.; Levine and Raab, op. cit.; Steven Kaye, ed., *Multi-family Energy Conservation: A Reader* (Boston: Coalition of Northeast Municipalities, 1981).

10. Conversation with Pittsburgh Housing Department, July, 1981.

11. Conversation with Betsy Foote, New York City Housing Preservation and Development Office, July, 1981.

12. Richard Fahlander, The Cambridge Energy Strategy: A Combination of Hand Holding and Arm Twisting, in Kaye op. cit., p. 88.

13. Conversation with Mary Brennan, New York City Housing Preservation and Development office, June, 1981.

14. Urban Coalition of Minneapolis, Weatherization of For-Profit Rental Units: Case Study, in Kaye, op. cit., pp. 113–131.

15. Ibid., pp. 43–47.

16. Cf. Compact Cities, op. cit., pp. 43–44

CHAPTER SEVEN

1. Office of Technology Assessment, *Residential Energy Conservation,* two vols. (Washington, D.C.: Office of Technology Assessment, 1979) vol. 2, pp. 469–470.

2. Joe Carter and Robert G. Flower, The Micro Load, *Solar Age,* September, 1980, pp. 22–30.

3. Ibid., pp. 24–25.

4. Cf. OTA, op. cit., vol. 1, p. 17; vol. 2, p. 474.

5. Dennis Moore, Heat Pumps: Hot Water at the Right Price?; *Solar Age,* November, 1981, pp. 37–43.

6. Fedders Corporation, Application for Certification of Fedders Heat Pump Water Heater under Regs. Section 1.44 C-5, Submitted to the Internal Revenue Service, December 6, 1980.

7. Arthur D. Little, Inc., Design, Development, and Demonstration of a Promising Integrated Appliance (Cambridge: 1977), pp. 1–12.

8. OTA, op. cit., vol. 1, p. 253.

9. Carter and Flower, op. cit., p. 30.

10. Kit Mann, Demand Water Heaters, *Home Remedies: Prescriptions and Procedures for Energy Retrofitting* (Philadelphia: Mid-Atlantic Solar Energy Association, 1981), pp. 100–101.

11. OTA, op. cit., vol. 1, pp. 252–253.

12. Mann, op. cit., p. 100.

13. OTA, op. cit., vol. 1, p. 253.

14. Evan Powell, Heat Pump Water Heater, *Popular Science,* April, 1980, pp. 49–52; 57.

15. Cf. Hans Landsberg and Joseph Dukert, *High Energy Costs: Uneven, Unfair, Unavoidable?* (Washington, D.C.: Resources for the Future, 1981), pp. 52–53.

16. Jeff Forker, Get Ready to Replace Thirty Million Water Heaters, *Air Conditioning and Refrigeration Business,* March, 1981, (reprint).

17. Carter and Flower, op. cit.

18. Technically, the determining factor here is the temperature difference between the collector's surface and the bottom of the storage tank. See Peter Hollander, et al., *Installation Guidelines for Solar DHW Systems in One- and Two-Family Dwellings,* (Washington, D.C.: U.S. Department of Housing and Urban Development, 1979), pp. 65–67.

19. Ibid., . 4–9.

20. W. W. Youngblood, et al., *Solar Collector Fluid Parameter Study* (Washington, D.C.: National Bureau of Standards, 1979), p. 4–10; William C. Thomas, Effects of Test Fluid Composition and Flow Rates on the Thermal Efficiency of Solar Collectors (Washington, D.C.: National Bureau of Standards, 1980), p. vi.

21. Ramada Energy Systems, Inc., Technical Bulletin: RES. 4000 Series Transparent Polymer Collector, 1979.

22. Cf. Doug Kelbaugh, Adding Domestic Hot Water to the Kelbaugh Greenhouse, *Home Remedies,* op. cit., pp. 177–181.

23. Bruce Maeda and Bruce Melzer, A Comeback for Breadboxes, *Solar Age,* October, 1980, pp. 56–62.

24. Conversation with Northrop(TM) Distributor, Summer, 1980.

25. Conversation with Appropriate Technology Associates, Davis, California, Summer, 1980.

26. Maeda and Melzer, op. cit., pp. 60–61.

27. Union Electric Company, A Simple Approach to Solar Heating of Domestic Hot Water, St. Louis: Union Electric Company, Undated.

28. New England Electric, Second Year of New England Electric Water Heating Test Program: Summary of Results (Westborough, MA: New England Electric, 1979). Note also that all of the systems installed in this program were ground-mounts. Better performance would be anticipated in roof-mounted systems.

29. David C. Moore, Lessons Learned from HUD's Residential Solar Demonstration Program, Proceedings: AS/ISES Annual Meeting (Newark, Delaware: AS/ISES, 1981), pp. 451–455.

30. In DOE's latest revisions of the methodology for evaluating Residential Conservation Service (RCS) program measures, fixed costs of $1,600 are assumed for active solar energy systems.

31. Robert H. Twiss, et al., *Land Use and Environmental Impacts of Decentralized Solar Energy Use* (Washington, D.C.: U.S. Department of Energy, 1979).

32. Conversation with Edwin Sosa, East 11th Street Project, New York City, Summer, 1980.

33. K. L. Armstrong, Thermal Performance Evaluation of the Aratex Services, Inc., Solar Energy Hot Water System, Ibid., pp. 133–139.

34. Winslow Fuller, Collector Location: No Taboos on East or West. *Solar Age,* December, 1980, pp. 26–27.

35. Michael Shapiro, Boston Solar Retrofits, Discussion Paper E-80-11, Energy and Environmental Policy Center, John F. Kennedy School of Government, Harvard University, p. 21.

36. Conversation with Jerry Engenhauer, Los Angeles Department of Water and Power, Summer, 1980.

CHAPTER EIGHT

1. Compact Cities: Energy Saving Stratgies for the Eighties, Report, Together with Dissenting Views by the Subcommittee on the City of the Committee on Banking, Finance, and Urban Affairs, House of Representatives (Washington, D.C.: Government Printing Office, 1980), p. 15.
2. Calculated on the basis of figures reported by Robert L. Twiss, et al., *Land Use and Environmental Impacts of Decentralized Energy Use* (Washington, D.C.: U.S. Department of Energy, 1980), pp. D-48–D-53.
3. However, estimates do vary about precisely how much of the year 2000's building stock already exists. And figures will differ substantially by region. The Solar Energy Research Institute estimates that up to 40 percent of the housing units existing in 2000 will be built in the next 18 years. *A New Prosperity: Building A Sustainable Energy Future* (Golden, CO: Solar Energy Research Institute, 1981, p. 14.) Others, however, believe that rising interest rates and increasing construction costs will slow growth in the number of units produced. Additionally, many newer units may be in multifamily buildings, meaning that the building stock, as distinct from the number of units, in 2000 will include an even larger number of existing structures.
4. Michael Andreassi, Lorene Yap, and Olson Lee, *The Impact of Residential Energy Conservation Standards on Households* (Washington, D.C.: The Urban Institute, 1980), p. 101.
5. D. T. Harrje, G. S. Dutt, and J. Beyea, *Locating and Eliminating Obscure But Major Energy Losses in Residential Housing* (Princeton, NJ: Center for Environmental Studies, 1979), pp. 5–6.
6. James E. Hill, et al., Performance of the Norris Cotton Federal Office Building for the First Three Years of Operation, *NBS Building Science Series 133* (Washington, D.C.: National Bureau of Standards, 1981), p. 23; National Electrical Contractors Association and National Electrical Manufacturers Association, *Total Energy Management: A Practical Handbook on Energy Conservation and Management* (Washington, D.C.: 1979), pp. 17–18.
7. Gautam S. Dutt, Residential Energy Conservation—The Case for House Doctors, Testimony Before the Subcommittee on Energy and Power of the Committee on Interstate and Foreign Commerce, October 30, 1979, pp. 2–3.
8. Cf. John Rothschild, *Stop Burning Your Money* (New York: Random House, 1981), pp. 217–222; Grant P. Thompson and David Strom, Comments on the Proposed 10 CFR Part 435-Energy Performance Standards for New Buildings (Washington, D.C.: The Conservation Foundation, April 30, 1980), p. 45.
9. William Shurcliffe, *Superinsulated Houses and Double Envelope Houses* (Cambridge, MA, 1980), pp. 5-10–5-15.
10. Andreassi, et al., op. cit., p. 18.
11. Cf. Recent surveys conducted by the National Association of Homebuilder's Research Foundation as referenced in Derivation of an Energy Conservation and Renewable Resource Measures Table for the Proposed RCS Rule, U.S. DOE, Building Conservation Services, November 16, 1981.
12. *A New Prosperity*, op. cit., p. 51; It should be noted that insulation levels in older buildings, often clustered in cities, could be a good deal lower. For example, Dutt and Rowse estimate that little if any insulation was used in buildings built before 1940. Cf. Office of Technology Assessment, *Residential Energy Conservation*, two vols. (Washington, D.C.: Office of Technology Assessment, 1979), 1, p. 99. DOE's Methodology for Evaluating RCS measures (see note 11) assumes R-7 in ceilings and no wall insulation.

13. Cf. National Association of Homebuilders Research Foundation, *Insulation Manual* (Rockville, MD: 1978), p. 10.
14. Rothschild, op. cit., pp. 233–238; Shurcliffe, op. cit., p. 3.01–4.18.
15. Robert L. Twiss, et. al., Land Use and Environmental Impacts of Decentralized Energy Use (Washington, D.C.: U.S. Department of Energy, 1980), p. D-53.
16. Peter Kerr, Debate on Safety of Urea Foam and Plastic Tubing, *New York Times,* Feb. 25, 1982, p. C. 6; Rothschild, op. cit., p. 57; OTA, vol. 1, op. cit., pp. 281–82.
17. Conversation with Chip Tabor, Energy Task Force, New York City, Summer, 1980.
18. Cf. *A New Prosperity,* op. cit., p. 53.
19. OTA, op. cit., p. 231.
20. Andreassi, op. cit., p. 35.
21. Rothschild, op. cit., pp. 76–82.
22. Cf. Michael Shapiro, Boston Solar Retrofits, Discussion Paper E-80-11, Energy and Environmental Policy Center, John F. Kennedy School of Government, Harvard University, 1980, p. 49.
23. Much of this discussion closely follows the material reported by OTA, op. cit., pp. 231–139.
24. Andreassi, op. cit., p. 109.
25. Ibid., pp. 123–125.
26. Ibid., p. 96.
27. Ibid., pp. 27–31; Washington Board of Trade, Energy Conservation Manual for Multifamily Buildings (Washington, D.C.: 1981), pp. 64–70.
28. Washington Board of Trade, op. cit., p. 59. See also: New York State Energy Office, *Multifamily Housing Energy Conservation Workbook* (Albany, undated), pp. 54ff.
29. NECA/NEMA, op. cit., p. 22.
30. Office of Technology Assessment, op. cit., vol. 1, p. 241; vol. 2, p. 408.
31. Washington Board of Trade, op. cit., p. 60.
32. OTA, op. cit., vol. 1, pp. 241–242; vol. 2, pp. 282–403.
33. Dermot McGuigan, *Heat Pumps: An Efficient Heating and Cooling Alternative* (Charlotte, VT: Garden Way, 1981), pp. 7–9.
34. McGuigan, op. cit., pp. 162–165; Edward Kush, Solar Assisted Heat Pumps, *U.S. Department of Energy Active Solar Heating Systems Contractors' Review* (Washington, D.C.: 1978).
35. Cf. McGuigan, op. cit., pp. 15–19.
36. OTA, op. cit., vol. 1, pp. 247–150.
37. Ibid., vol. 2, pp. 437–441.
38. Milton Meckler, *Energy Conservation in Buildings and Industrial Plants* (New York: McGraw-Hill, 1981), pp. 37–40; New York State Energy Office, op. cit., p. 47.
39. Meckler, op. cit., p. 38.
40. *Totem—Total Energy Module* (Brochure), Fiat Auto, 1981.
41. *The Sunpower, Inc.—Total Home Energy System,* (Brochure), Sunpower, Inc., Athens, Ohio, 1981.
42. There is little data on retrofits to support this. There are, however, a number of case studies available on new houses where this has been done. See Shurcliffe, op. cit., Chapters 3 and 4; Rothschild, op. cit., pp. 233–242.
43. NECA/NEMA, op. cit., p. 42; Cf. Also Meckler, op. cit., p. 89.
44. *A New Prosperity,* op. cit., p. 159.
45. NECA/NEMA, op. cit., pp- 22–30; See also New York State Energy Office, op. cit., pp. 54–72.)
46. Ibid. Note, however, that these figures assume a 10 percent real discount rate and a 30-year

loan. This translates into unamortized costs of $30 to $90 per MMBtu saved. Cf. OTA, the Energy Efficiency of Buildings in Cities, op. cit.

47. Ibid., p. 149.
48. Conversation with Chip Tabor, Energy Task Force, New York City.
49. Conversation with Bruce Powerll, Institute for Human Development, Philadelphia, PA.
50. Conversation with Fresno County Economic Opportunity Commission—Energy Project, January, 1981.
51. Conversation with Al Lopez, Urban Coalition of Minneapolis, Summer, 1980.
52. Douglas Sliger, Modular Retrofit Experiment (Princeton Project), June 17, 1980.
53. U.S. Department of Energy, *The Passive and Hybrid Solar Energy Program* (Washington: 1980), pp. 4.28–4.29. Of course, the results of this kind of analysis will differ by climate. In cold, cloudy areas a much larger baseline conservation investment might be justified.
54. T. M. Lechner and W. Quigley, An Electric Utility's View of Passive Solar Backup Energy Costs, in Proceedings: The 5th Passive Solar conference, op. cit., pp. 680–683.
55. In fact, some solar techniques can be used to improve the performance of conventional air-to-air heat pumps. See McGuigan, op. cit., pp. 144–147.
56. U.S. Department of Housing and Urban Development, *New Energy-Conserving Passive Solar Single-Family Homes* (Washington, D.C.: HUD, 1980).
57. Ted L. Kurkowski and Steven E. Ternoey, The Design of Passive Commercial Buildings: Harry T. Gordon and William I. Widdon, Strategies for the Use of Passive Solar Approaches in Commercial Buildings, in *Proceedings: The 5th Passive Conference,* op. cit., pp. 856–863.
58. Named for Felix Trombe, the French inventor who first developed the concept.
59. Cf. Alex Wilson, Trombe Wall Retrofit, *Solar Age,* December 1979, pp. 37–40.
60. For example, *Skysorb, Colored Stainless Steel Solar Collector* (Brochure), available from Ergenics, 681 Lawlins Road, Wyckoff, NJ 07481.
61. The Passive and Hybrid Solar Energy Program, op. cit., pp. 6.2–6.3; 6-24. Fuller Moore and Paul Hemker, Tubewall: A Passive Solar Thermo-siphoning, Field-Fabricated Water Storage Wall System, Proceedings: The 5th National Passive Solar Conference, op. cit., pp. 1119–1122.
62. Shaul Ben-David, A Minimum Benefits Analysis of Residential Passive Solar Sunspace Designs, Proceedings: The 5th National Passive Solar Conference, op. cit., pp. 441–449. Some of the solar energy contributions of a greenhouse are offset, for example, by heating the additional space in the greenhouse itself. Only what is left over would go to the existing building.
63. U.S. Reiniger and C. Morris, Solar Greenhouses in New Mexico: Results of an Extensive Survey, Proceedings, Ibid., pp. 426–420.
64. Conversation with Mary Ann Shepherd, Charlotte (NC) Area Fund, Inc., Summer, 1980.
65. Alex Eldrige, Solar Barnraising in Boston, *New Roots,* Holiday, 1981, pp. 30–33.
66. Much of this discussion relies on material presented by the Underground Space Center, Earth Sheltered Housing: Code, Zoning, and Financing Issues. (Washington, D.C.: U.S. Department of Housing and Urban Development, 1980), pp. 19–47.
67. Cf. Franklin Research Center, Interim Report on Performance Data from the Residential Solar Demonstration Program (Philadelphia: Franklin Research Center, 1988), pp. 2-5, 2-6.
68. Conversation with David Cawley, Anacostia Energy Alliance, Summer, 1980.
69. U.S. Department of Housing and Urban Development, Energy Conserving Passive Solar Multifamily Retrofit Projects (Draft), (Washington, D.C.: U.S. Department of Housing and Urban Development, 1981), pp. 68–69.

70. David C. Moore, Lessons Learned from the HUD Solar Demonstration Program, Proceedings: AS/ISES Annual Meeting (Newark, Delaware: AS/ISES, 1981), pp. 451–455.
71. Joe Kohler and Peter Temple, The Fundamentals of Site-Built Collector Design, *Solar Age,* July, 1980, pp. 12–16.
72. Fedders Solar Products Company, *Application Manual, Carnot II, Solar/Compression Furnace,* (Edison, NJ: 1981). The manufacturer claims that this system avoids one problem that has been ascribed to some solar-assisted heat-pump systems by being able to operate at relatively high temperatures.
73. McGuigan, op. cit., pp. 162–169.
74. OTA, *Application of Solar Energy to Today's Energy Needs,* op. cit., vol. 1, p. 259.
75. Monegon Ltd., Phase Change Thermal Storage: A Comprehensive Look at Developments and Propsects (Rockville, MD: 1980).
76. Moore, op. cit., pp. 17–18; Franklin Research Center, op. cit.
77. Cf. HUD's $23 million demonstration program is paying off, *Builder,* 1979; Spruille Braden and Kathleen Steiner, *Successful Solar Energy Solutions* (New York: Van Nostrand Reinhold, 1980); Donald Best, Measured Output of Two Active Systems, *Solar Age,* July, 1981, pp. 41–44.
78. For economic analysis of heat pumps, see: P. J. Hughes and J. H. Morehouse, *Comparison of Solar Heat Pump Systems to Conventional Methods for Residential Heating, Cooling, and Water Heating,* 3 vols., (Golden, CO: Solar Energy Research Institute, 1980); M. K. Choi, J. H. Morehouse, and P. J. Hughes, *Comparison of Ground-Coupled Heat Pump Systems to Conventional Systems for Residential Heating, Cooling, and Water Hearing* (Draft Final Report) (McLean, LA: Science Applications, Incorporated, 1980); John W. Mitchell, Thomas L. Freeman, and William A. Beckman, Heat Pumps: Do They Make Economic and Performance Sense With Solar? *Solar Age,* July, 1978, pp. 24–18.
79. For example, economic calculations done on prototypical passive systems for both the Building Energy Performance Standards and the Residential Conservation Service assume only variable costs.
80. Real Estate Research Corporation, *Passive Solar Homes in the Marketplace: Final Report of Findings of the 1978–1979 Passive Grant Awards* (Washington, D.C.: U.S. Department of Housing and Urban Development, 1982), pp. 2–3.
81. Travis L. Price, III, *Energy Conservation Guidelines for the Sheffield Block Development,* 3 vols., (Pittsburgh: Department of Architecture, Carnegie-Mellon University, 1981), vol. 1, New Construction, pp. 13–14.
82. Cf. Shapiro, op. cit., (Boston); Twiss, op. cit., Baltimore, Denver, Minneapolis; Philadelphia Solar Planning Project Working Papers—Neighborhoods, Part II, (Philadelphia: 1980), pp. 27–48, Available from Chalres Burnette & Associates, 234 South Third Street, Philadelphia, PA 19106; City of Los Angeles, Solar Envelope Zoning: Application to the City Planning Process (Golden, Colorado: Solar Energy Research Institute, 1980), pp. 59–107.
83. Philadelphia Solar Planning Project, op. cit., p. 28.
84. Cf. Donna Shalala, Testimony, Renewable Energy and the City, Joint Hearings Before the Subcommittee on the City of the Committee on Banking, Finance, and Urban Affairs and the Subcommittee on Oversight and Investigations of the Committee on Interstate and Foreign Commerce, U.S. House of Representatives (Washington, D.C.: Government Printing Office, 1979), pp. 258–259.
85. Greater Roxbury Development Corporation, Solar Typologies Project: Neighborhood-Scaled Solar Energy Development Strategies for Roxbury, MA, November, 1980, Figure 3.
86. Philadelphia Solar Planning Project, op. cit., Audits, p. 30.
87. Samuel A. Cravotta, Jr., The Neglected But Not Forgotten Art of Sun Porch Heating, Proceedings, op. cit., pp. 35–43.

88. White and Converse, op. cit.
89. Philadelphia Solar Planning Project, op. cit., Applications, p. 27.
90. Cf. Vivian Loftness and Volker Hartkepf, Passive Solar Retrofit of All Glass Office Buildings, *Proceedings: The 5th National Passive Solar Conference* (Newark, Delaware: American Section of the International Solar Energy Society, 1980), p. 870.
91. Conversation with Bill Brown, Office of Human Concerns, Summer, 1980.

CHAPTER NINE

1. South Florida Regional Planning Council, South Florida Energy Audit (Draft) (Miami: 1981), p. 76.
2. U.S. Department of Energy, Active Solar Cooling R and D: Background (Draft) (1981), p. 25.
3. Butler and Associates, Energy Audit for Greenville, NC, 1978 (1980), Figure II-2.
4. National Association of Homebuilders Research Foundation, *Insulation Manual* (Rockville, MD: 1979), p. 12.
5. Ibid., p. 13.
6. State of Florida, *Model Energy Efficiency Building Code,* 502.3(b)(3), 1978.
7. Harry B. Zachrickson, Energy Conscious Electrical Design, in Sandra A. Berry, ed., *Research and Innovation in the Building Regulatory Process* (Washington, D.C.: National Bureau of Standards, 1981), pp. 37–56.
8. Los Angeles Energy Management Advisory Board, *The Energy/LA Action Plan,* (1981), p. 55.
9. This depends, of course, on the rate structure. In a time-of-day pricing structure, electricity for heating used in off-peak hours could cost less per kilowatt hour than electricity used for lighting during peak periods.
10. Arkla, *Servel Gas-Fired Chiller* (Brochure), Arkla Industries, Evansville, IN.
11. Argonne National Laboratory, Technology Assessments: Absorption Chillers, Compression Chillers, 1979.
12. Conversation with David Goldstein, Natural Resources Defense Council, San Francisco, Summer, 1981.
13. Argonne, op. cit.
14. Active Solar Cooling R and D: Background, op. cit., p. 9.
15. Argonne, op. cit.
16. Ibid.
17. OTA, op. cit., pp. 44–46.
18. Active Solar Cooling R and D: Background, op. cit., pp. 12; 16–17.
19. For a good discussion of PURPA 210 provisions, see John Plunkett and David Morris, PURPA 210, (Washington, D.C.: Institute for Local Self-Reliance, 1980).
20. National Electrical Contractors Association/National Electrical Manufacturers Association, op. cit., pp. 42–43.
21. Milton Meckler, *Energy Conservation in Buildings and Industrial Plants* (New York: McGraw Hill, 1981), pp. 37–40.
22. Meckler, op. cit., p. 76.
23. State of Florida, op. cit., 503.3.
24. Office of Technology Assessment, *Application of Solar Energy to Today's Energy Needs,* 2 vols. (Washington, D.C.: Office of Technology Assessment, 1979), vol. 1, pp. 524–525; Active Solar Cooling R and D Background, op. cit., pp. 22–23.

25. Cf. Marc Ross and Howard Williams, *Our Energy: Regaining Control* (New York: McGraw Hill, 1981), pp. 112–117; 317.
26. Office of Technology Assessment, *Residential Energy Conservation,* op. cit., vol. 1, p. 250.
27. Ibid., p. 354.
28. Ibid.
29. City of Boulder, *Community Energy Management Plan* (1980), p. 50.
30. Cf. Eunice Grier, *Colder ... Darker: The Energy Crisis and Low-Income Americans* (Washington, D.C.: Community Services Administration, 1977), p. 46.
31. David Morris, Testiomony, Renewable Energy and the City, Joint Hearings Before the Subcommittee on the City of the Committee on Banking, Finance, and Urban Affairs and the Subcommittee on Oversight and Investigations of the Committee on Interstate and Foreign Commerce, U.S. House of Representatives (Washington, D.C.: GPO, 1979), p. 101.
32. U.S. Department of Housing and Urban Development, Energy-Conserving Passive Solar Multifamily Retrofit Projects (Draft) (Washington, D.C.: U.S. Department of Housing and Urban Development, 1981), p. 23.
33. Ibid., p. 43.
34. Charles S. Barnaby, et al., Structural Mass Cooling in a Commercial Building Using Hollowcore Concrete Plank, Proceedings: The 5th National Passive Solar Conference, op. cit., pp. 747–751.
35. The Passive and Hybrid Solar Energy Program, op. cit., p. 5.3.
36. Donald L. Kirkpatrick, Marco Masoero, Robert H. Socolow, Theodore B. Taylor, A Unique Low-Energy Air Conditioning System Using Naturally Frozen Ice, Presented at Solar Rising Conference, International Solar Energy Society Meeting, Philadelphia, PA, May 26–30, p. 4.
37. David C. Moore, Lessons Learned from the HUD Solar Demonstration Program, Proceedings, AS/ISES 1981 Annual Meeting (Newark, Deleware: AS/ISES, 1981), pp. 451–455.
38. Blair Hamilton, et al., Non-Instrumented Performance Evaluation of 335 Passive Solar Homes, the Memphremagog Group, P.O. Box 456, Newport, VT 05855.
39. U.S. Department of Housing and Urban Development, The First Passive Solar Home Awards (Washington, D.C.: U.S. Department of Housing and Urban Development, 1979), pp. 190–192.
40. Active Solar Cooling R and D: Background, op. cit., pp. 10–11.
41. Ibid., pp. 16–17.
42. Application of Solar Technology to Today's Energy Needs, op. cit., pp. 254–255.
43. Emilie Tavel Livesey, Something New Under the Sun, *Christian Science Monitor,* October 27, 1981, pp. B. 12–B.14.
44. Bruce Green, et al., Community Renewable Energy Technologies (Draft) (Golden, CO: Solar Energy Research Institute, 1981), p. 91.
45. Active Solar Cooling R and D: Background, Op. cit., pp. 42–44.
46. Energy-Conserving Passive Solar Retrofit Projects, op. cit., p. 50.

CHAPTER TEN

1. For example, see John Roudususakis, et al., Clark University's Grid-Connected Cogeneration Plant, *District Heating,* vol. 67, no. 1, pp. 7–12.
2. Lennant Lindeberg, District Heating Distribution Systems, *District Heating, Cogeneration,*

Waste Heat Utilization: Avenues to Energy Conservation: Swedish Experiences and Technology (Chicago: Swedish Trade Office, undated), Chapter 3.

3. Much of the information on the St. Paul project comes from James O. Kolb, St. Paul District Heating Demonstration Project: Economic Feasibility and Implementation Strategy, Presentation to Integrated Energy Systems Task Goup, August 11, 1981.

4. Ibid.

5. Office of Technology Assessment, *The Energy Efficiency of Buildings in Cities (Washington, D.C.: OTA, 1982.)*

6. Jack Gleason, District Heating Saves Energy (In Europe), *Public Power,* September–October, 1979, pp. 46–48.

7. Joseph C.Larkin, How to Convert Steam Heat to Hot Water, *The Journal of Property Management,* July/August, 1977, pp. 176–177.

8. Kjell Larson, District Heating Swedish Experience with an Energy Efficient Concept, *Swedish Board of District Heating Catalogue,* 1978, p. 13.

9. Peter J. Robinson, Transmission and Distribution Networks—The Potential for Development, in W. H. R. Orchard and A. F. C. Sherrat, eds., *Combined Heat and Power* (New York: John Wiley and Sons, 1980), pp. 184–186.

10. Cf. Robert W. Timmerman, Utilizing Power Plant Waste Heat, *Heating/Piping, Air Conditioning,* May, 1978, pp. 891–97.

11. W. H. R. Orchard and P. J. Robinson, Combined Heat and Power—The Economic Advantages of Low-Temperature Direct Connection and Their Compatibility With Existing Installation, Orchard and Partners, U.K.

12. Jack Gleason, Efficient Fossil and Solar District Heating Systems, Golden, CO: Solar Energy Research Institute, 1981 (draft), Technical Appendix—Solar District Heating.

13. Robinson, op. cit., p. 201.

14. Kolb, op. cit.

15. Kolb, op. cit.

16. For example, in developing preliminary plans for a district-heating system, the city of Detroit decided on steam distribution because of the large number of projected industrial customers. See Todd Anuskiewicz and Roy Meador, District Heating Backgrounder, Michigan Energy and Resource Research Association, 1980, p. 10.

17. Much of the following material is adapted from Lindeberg, op. cit.

18. Much of the following material is adopted from I. Oliker, Assessment of Existing and Prospective Piping Technology for District Heating Applications, Oak Ridge, TN: Oak Ridge National Laboratory, 1979.

19. Howard S. Geller, *Thermal Distribution Systems and Residential District Heating,* (Princeton: Center for Energy and Environmental Studies, August 1980).

20. Ibid.

21. Mogens Larsen and Hans C. Mortensen, Fundamentals and Economic Principles in District Heating Planning (Odense, Denmark: Danish Board of District Heating), p. 21.

22. Olikeri op. cit., p. 4–6.

23. Conversations with David Oliver (June, 1980) and Theodore Taylor (November, 1980).

24. For example, the financial success of the St. Paul system will require that customers enter into 30-year take or pay contracts. This is rightly identified by Kolb, op. cit. as the key to successful implementation.

25. For a more complete discussion of these technologies, see Norman L. Dean, Jr., *Energy Efficiency in Industry* (Cambridge, MA: Ballinger, 1980).

26. P. F. Bach, Planning District Heating Systems From Danish Generating Stations, Facts About Danish District Heating (Odense, Denmark: Danish Board of District Heating, 1978).

27. Other options are also open to reduce the derating of electrical capacity in power plant cogeneration facilities by, for example, supplementing cogeneration with heat-only peaking boilers. Cf. Margen, op. cit., pp. 43–44.

28. Gleason, Report, op. cit., Technical Appendix, p. 3.

29. Such a policy of industrial attraction has been recommended for consideration by, for example, the team developing policy recommendations for Portland, Oregon. See *Energy Conservation Choices for the City of Portland, Oregon,* 11 vols. (Washington, D.C.: U.S. Department of Housing and Urban Development, 1977), vol. 3, p. 42.

30. Gleason, Report, op. cit., Technical Appendix, pp. 4–5.

31. For a good general criticism of "garbage-to-energy" projects developed without prior recycling, see Tania Lipshutz, *Garbage to Energy: The False Panacea,* Sonoma County Recycling Center, Sept., 1979.

32. OTA, op. cit.

33. U.S. Conference of Mayors, Economic Development and Resource Recovery, Washington, D.C., 1979.

34. Gleason, Report, op. cit., p. 17.

35. Conversation with Milton E. Kirkpatrick, Nashville Thermal Transfer Corporation, Summer, 1980.

36. Lipshutz, op. cit., p. 31.

37. Ibid, pp. 49–51.

38. Resource Recovery Activities, *NCRR Bulletin,* September, 1981.

39. For example, OTA has calculated that the costs of producing steam fall between $10 and $15 per MMBtu. Charging tipping fees of $8 to $10 per ton of garbage can, however, allow the steam to be sold at a competitive price. Conversation with Mary Procter, Office of Technology Assessment, Winter, 1981.

40. Conversation with staff of Institute for Development of the Urban Arts and Sciences, U.S. Conference of Mayors, Summer, 1980.

41. Conversation with Joseph Ferrante, Wheelabrator-Frye, Inc., Summer, 1980.

42. Economic Development and Resource Recovery, op. cit., pp. 10–11.

43. Ibid., p. 27.

44. Lipshutz, op. cit., p. 23.

45. T. J. Marciniak, et al., *Total Energy Technology Alternative Study* (Chicago-Argonne National Laboratory).

46. Gleason, Report, op. cit., Technical Appendix, p. 67.

47. OTA, op. cit.

48. Hampshire College, *Energy Self-Sufficiency in Northampton, Massachusetts* (Washington, D.C.: U.S. Department of Energy, 1979), pp. 126–130.

49. Theodore B. Taylor, Long-Term Storage of Heat and Cold, Princeton/SERI Report 21, Center for Energy and Environmental Studies, Princeton University, June 28, 1981.

50. Allan S. Krass and Roger Laviale III, Solar Ponds as Municipal Energy Systems, Hampshire College, July, 1979.

51. Ibid., Technical Appendix, p. 13.

52. James M. Calm, Heat-Pump-Centered Community Energy Systems: Systems Development Summary, Argonne National Laboratory, February, 1980, p. 7.

53. Communication with Phil Hanson, Idaho Energy Office, Spring, 1981.

54. Communication with Mr. Derrah, City of Klamath Falls (OR), Spring, 1981.

55. For more detailed guidelines, see OTA, op. cit., W. Pferdehirt and N. Kron, *District Heating from Electric Generating Plants and Municipal Incinerators: Local Planner's Assessment Guide* (Argonne, IL: Argonne National Laboratory, 1980); Kolb, op. cit.

CHAPTER ELEVEN

1. David Pomerantz, et al., Energy Management in Municipal Operations: A Framework for Action (Washington, D.C.: International City Managers Association, 1981), pp., 8–9.
2. Conversation with Richard Urbania, Street Lighting Project, City of Oakland, Summer, 1980.
3. Such a planning and oversight function is mandated for the Northeast in the Pacific Northwest Electric Power Planning and Conservation Act: see the summary of the act prepared by the Bonneville Power Administration (BPA), 1981, pp. 5–11.
4. Southern California Edison, Energy Issues . . . And Answers, Congressional Briefing, June, 1981.
5. Ibid., p. 32.
6. Solar Energy Research Institute, *A New Propserity, Building a Sustainable Energy Future* (Andover, MA: Brick House, 1981), p. 74.
7. Milton Meckler, *Energy Conservation in Buildings and Industrial Plants* (New York: McGraw Hill, 1981), p. 53.
8. W. Murgatroyd and B. C. Wilkins, The Efficiency of Electric Motive Power in Industry, *Energy*, vol. 1, 1976, pp. 337–345.
9. City of Portland, *Energy Conservation Choices*, 11 vols. (Washington, D.C.: Department of Urban Housing and Urban Development, 1977), vol. 3c, p. 29.
10. Washington Board of Trade, *Energy Conservation Manual for Multifamily Dwellings* (Washington, D.C.: 1980), p. 74.
11. Meckler, op. cit., p. 62.
12. Ibid, pp. 64–65.
13. John Rothschild, *Stop Burning Your Money* (New York: Random House, 1981), p. 111.
14. Mary Curtis, Stingy Gadgets Delighting Residents, *Save Energy* (Sacramento: California Energy Extension Service, 1980).
15. Alan S. Miller, Application of Least-Cost Principles to the Utility Industry, Testimony of the Energy Conservation Coalition for the House Committee on Energy and Commerce, Subcommittee on Energy Conservation and Power, September 30, 1981, pp. 12–14.
16. Henry C. Kelly and Alan S. Miller, Getting Serious About Utility Regulatory Reform, *Public Utilities Fortnightly*, September 24, 1981, pp. 21–27.
17. David Morris, Testimony, Renewable Energy and the City, Joint Hearings Before the Subcommittee on Banking, Finance, and Urban Affairs and the Subcommittee on Oversight and Investigations of the Committee on Interstate and Foreign Commerce, U.S. House of Representatives (Washington, D.C.: U.S. Government Printing Office, 1979), p. 101. This inefficiency is offset somewhat by the fact that lower-income consumers often use appliances that demand less energy than more sophisticated devices used by higher-income persons (for example, refrigerators that require defrosting rather than frost-free models and black-and-white instead of color televisions). See Eunice Grier, *Colder . . . Darker: The Energy Crisis and Low Income Americans* (Washington, D.C.: Community Services Administration, 1977).
18. Requirements for offsets are far from simple. Cf. Kevin M. O'Brien and Barbara Euser, Solar Energy and the Search for Emissions Offsets, *Solar Law Reporter*, vol. 1, no. 6, March/April, 1980 pp. 1077–1093; Energy Management and Policy Analysis Group, Department of Nuclear and Energy Engineering, University of Arizona, *Assessment of the Potential for Cogeneration in the State of Arizona* (Tucson: Arizona Corporation Commission, 1981), Chapter 7. Arguments calling for the allowance of polluting cogeneration facilities on the grounds that they can offset emissions from electric utilities have some merit.

But offsets can be expected only with very large cogeneration facilities or the widespread use of smaller ones.

19. Cf. University of Arizona, op. cit., pp. 7.4; 7.6–7.8. Existing source clean-up could come from clean-up of facilities owned by the cogeneration system's owner, or that owner could pay for the clean-up of other facilities.

20. Kenneth E. Johnston, Electricity Cogeneration in Urban Areas, Joel T. Werth, ed., *Energy in the Cities Symposium,* PAS Report No. 349 (Chicago: American Planning Association, 1980), pp. 31–36.

21. Richard Stone,Large Residential/Commercial Applications for Cogeneration, Presented at the 8th Energy Technology Conference, 1981.

22. For a detailed description and the analysis of this provision, see Morris and Bardaglio, op. cit., Section 2.

23. Three, according to Morris and Bardaglio (Section 2, p. 12): New Hampshire, California, and North Carolina.

24. Robert L. Bailey, *Solar Electrics: Research and Development* (Ann Arbor: Ann Arbor Science, 1980), p. 247.

25. Jack Park, *The Wind Power Book* (Palo Alto: Cheshire Books, 1981), p. 77.

26. Park (pp. 49–55) gives a good, clear description and assessment of the Rayleigh distribution and its use in wind resource calculations.

27. U.S. Bollmeier, et al., *Small Wind Systems Technology Assessment* (Golden, CO: Rocky Flats Wind Systems Program, 1980), p. 14.

28. Conversations with Sam King, Rick Katzenberg (Natural Power, Inc.), Darryl Dodge (Rocky Flats), Summer, 1980.

29. Energy Task Force, *Windmill Power for City People* (Washington, D.C.: Community Services Administration, 1977), p. 8.

30. For descriptions of various solar thermal systems, see John Thornton, et al., *Final Report: A Comparative Ranking of 0.1–10 (Mw(e) Solar Thermal Electric Power Systems,* 2 vols. (Golden, CO: Solar Energy Research Institute, 1980).

31. Ibid, pp. 6–7.

32. *DOE Guide to Solar Thermal Programs* (Washington, D.C.: 1980), p. 33.

33. Thornton, op. cit., pp. 63–73, indicates that 0.1–1.0 Mw(e) systems can deliver electricity at lifecycle busbar electricity costs under 10 cents/kwh under optimal circumstances. But how soon or if such optimal circumstances can be expected is still questionable.

34. For excellent, clear discussions of the technical workings of photovoltaic technologies, see Paul Maycock and Edward Stirewalt, *Photovoltaics: Sunlight to Electricity in One Step* (Andover, MA: Brick House, 1981), pp. 26–34; Monegon, Ltd., *Solar Electricity* (Rockville, MD: 1981), pp. 69–82.

35. Dan Best, The Advent of Residential Photovoltaics, *Solar Age,* June, 1981, pp. 22–25.

36. Mona Anderson, For Sale: Total Energy, *Solar Age,* September, 1979, pp. 35–37.

37. PRC Energy Analysis Company, Solar Photovoltaics Applications Seminar: Design and Operation of Small Stand-Alone Photovoltaic Power Systems (Washington, D.C.: U.S. Department of Energy, 1980), p. 2–4.

38. Cf. G. L. Angelici, et. al., Urban Photovoltaic Potential: An Inventory and Modeling Study Applied to the San Fernando Valley Region of Los Angeles, Jet Propulsion Laboratory, 1980.

39. PRC, op. cit., pp. 2-13, 2-14.

40. Graham Allison, et al., Governance of Nuclear Power, Submitted to the President's Nuclear Safety Oversight Committee, September, 1981, p. 2.

41. William Snyder, Report on Survey of Utility Investments in Conservation and Renewable Resources, (unpublished paper), Center for Renewable Resources, 1981, p. 60.

42. I. Berman, New Generating Capacity: When, Where, and By Whom,? *Power Engineering,* April, 1980, pp. 70:78.

43. *A New Prosperity,* op. cit., pp. 325–331; General Accounting Office, Electricity Planning— Today's Improvements Can Alter Tomorrow's Investment Decisions (Washington, D.C.: GAO, 1980).

44. This is not to deny that, so far, the safety record of the nuclear industry has been exemplary. But the poor economics of these plants may lead qualified and experienced operating personnel to leave the industry. (Allison, op. cit., p. 16) Moreover, the industry's response to the emergency at Three Mile Island was not reassuring. Finally, long-term problems with reactor structural stability are only now being uncovered. (See Demetrios L. Baskedos, The Risk of a Meltdown, *New York Times,* March 29, 1982.)

45. R. E. Gant, Application of Modular Integrated Utility Systems (MIUS) Concept, as a Strategy for Urban Rehabilitation and Redevelopment (Oak Ridge: Oak Ridge National Laboratory, 1980).

46. Douglas Criner, Peter Steitz, and Barry Flynn, Why the Fuel Cell is Attractive, *Public Power,* January/February, 1980, p. 113.

47. MOD-2, New Source of Windpower in the Northwest, Bonneville Power Administration, 1980.

48. Cf. Solar Thermal Repowering Projects Stay on Track, *Solar Engineering,* March, 1981, pp. 28–32.

49. Ibid. For a description of the Barstow Project, see Central Receiver System Design for Barstow Plant, *Solar Engineering,* Sept., 1979, p. 16.

50. Dave McNary, Political Winds, Other Unknowns Hinder Repowering Projects, *Solar Age,* May, 1981, pp. 48–53.

51. General Accounting Office, Hydropower: An Energy Source Whose Time Has Come Again (Washington, D.C.: General Accounting Office, 1980), p. 6. It should be noted, however, that its regional importance is far greater. It supplied 45 percent of the electrical capacity in the Northwest in 1977.

52. U.S. Army Corps of Engineers, National Hydropower Study Summary Information, July, 1981, pp. 3–4.

53. For a more detailed discussion of different types of turbines useable in hydroelectric systems, see U.S. Army Corps of Engineers, *Feasibility Studies for Small-Scale Hydro Additions,* 6 vols. (Washington, D.C.: U.S. Department of Energy, 1979), vol. 5, pp. 3-1–3-20.

54. Mitre Corporation, Technologies for Commercial and Industrial Heat from Biomass, Working Paper #80 000010 (McLean: VA: 1979), p. 15.

55. Georgia Institute of Technology Wood Energy Systems Branch, *Wood: An Alternative Energy Resource* (Atlanta, 1980).

56. Personal Communication, William I. Toth, Supervisor, Technology Transfer, EG&G, Idaho, Inc., December 11, 1981.

57. R. P. Hartley, Environmental Considerations, in Joseph Krestin, ed., *Sourcebook on the Production of Electricity from Geothermal Energy* (Washington, D.C.: U.S. Department of Energy, 1980), pp. 795–866.

58. Michael Edesess, Solar Ponds Resource Potential, Proceedings: Non-Convecting Solar Pond Workshop (Boulder City, NV: Desert Research Institute, 1980), pp. 2-1–2-14.

59. Karen Anderson, Burlington's Wood and Water, *Public Power,* November–December, 1981, pp. 41–45.

60. Trash Will Fuel New Columbus Plant, *Public Power,* May–June, 1981, pp. 60–61.

61. Fox Butterfield, Massachusetts to Get Biggest Trash-to-Electricity Plant in U.S., *New York Times,* January 7, 1982.

62. *A New Prosperity,* op. cit., p. 419.

63. U.S. Department of Energy, Idaho Operations Office, *A Guide for Small Hydroelectric Development* (1980).
64. Ronald DiPippo, Geothermal Energy as a Source of Electricity (Washington D.C.: U.S. DOE, 1980), pp. 324–325.
65. Lewis Lingell and David Pritchard, O'Shaugnessy and Briggs Dams on the Scioto River, Columbus, Ohio (Burgess and Niple, Ltd., 1979).
66. Glenn Berger, *Why Wood? An Introduction to the Industrial Use of Fuel Wood* (Hanover, NH: Resource Policy Center, Dartmouth College, 1978), p. 8.
67. California Energy Commission, *Electricity Cost by Generation Technology* (September, 1981), p. 6.
68. Rocket Research, Inc., Industrial Electrical Cogeneration Potential in the Bonneville Power Administration Service Area: Phase I, Technical Analysis (Redmond, Washington: Rocket Research Company, 1979), pp. 7-10; 7-14.
69. Conversation with Joseph Ferrante, Wheelabrator Frye, Inc., Summer, 1980.
70. Conversation with Jeffrey Paine, Chief Engineer, City of Idaho Falls Electric Division, Fall, 1980.
71. Anderson, op. cit.
72. Trash Will Fuel New Columbus Plant, op. cit.
73. For example, the City of New York is looking at the feasibility of using the city's own waterways for hydropower generation. Cf. City of New York, op. cit., pp. 30–31.
74. Conversation with John Topalian, City of Paterson, Office of Community Development, Spring, 1981. However, since a large portion of the project would be financed by municipal bonds, the bad bond market may impede project development.

CHAPTER TWELVE

1. U.S. Department of Energy, Energy Information Administration, Annual Report to Congress (Washington, D.C.: 1981) p. 3.
2. *Business Week,* June 1, 1981, p. 64.
3. Greater Bridgeport Regional Planning Agency, op. cit., p. 24.
4. *Business Week,* op. cit.
5. Los Angeles Energy Management Advisory Board, *The Energy/LA Action Plan,* July 1981, p. 38.
6. Cf. *A New Prosperity,* op. cit., p. 225; Norman L. Dean, Jr., *Energy Efficiency in Industry* (Cambridge, MA: Ballinger, 1980); Robert O. Reid and Melvin H. Chiogioji, Technological Options for Improving Energy Efficiency in Industry and Agriculture, in John C. Sawhill, ed., *Energy Conservation and Public Policy* (New York: American Assembly, Columbia University, 1979), pp. 122–141; Marc H. Ross and Robert H. Williams, *Our Energy: Regaining Control* (New York: McGraw Hill, 1980).
7. Dean, op. cit., p. 7.
8. Shirley A. Stadjuhar, *Feasibility Evaluation for Solar Industrial Process Heat Applications* (Golden, CO: Solar Energy Research Institute, 1980), p. 1.
9. *A New Propserity,* op. cit.
10. U.S. Department of Energy, Annual Report to the Congress and the President on the Industrial Energy Efficiency Improvement Program (Washington, D.C.: U.S. Department of Energy, 1980), p. 21.
11. Ibid., p. 23.
12. *The Energy/LA Action Plan,* op. cit., p. 56.

13. U.S. Department of Energy, *The DOE Industrial Conservation Program: A Partnership in Saving Energy* (Washington, D.C.: U.S. DOE. 1979), p. 6.
14. Annual Report to the Congress and the President on the Industrial Energy Efficiency Improvement Program, op. cit., p. 4.
15. Dan Baum, Unconventional Financing Aids Capital—Short Users, *Energy User News,* Sept. 14, 1981, p. 14.
16. City of Seattle, Energy Ltd., Draft Action Plan, October, 1980, p. 85.
17. Ibid., p. 82.
18. City of Portland, *Energy Conservation Choices for the City of Portland, Oregon,* 11 vols.; (Washington, D.C.: U.S. Department of Housing and Urban Development, 1977), vol. 30, pp. 66–67.
19. Pat Lang, *Case Histories of Energy Savings in the Industrial Sector* (Los Angeles: City Energy Office, 1981).
20. Dean, op. cit., p. 59
21. Ibid., p. 52.
22. Ibid.
23. Lang, op. cit.
24. Ibid.
25. Dean, op. cit., p. 50.
26. Marc H. Ross and Robert H. Williams, Energy and Economic Growth, U.S. Congress, Joint Economic Committee, Subcommittee on Energy, 1977, 45.
27. Reid and Chiogioji, op. cit., p. 127.
28. U.S. Department of Energy, Industrial Energy Conservation Strategic Plan (Washington, D.C.: U.S. DOE, 1978), p. 17.
29. Annual Report to Congress and the President on the Industrial Energy Efficiency Improvement Program, op. cit., p. 35.
30. Ibid., p. 53.
31. Lang, op. cit.
32. Douglas Hooker, et al., *Industrial Process Heat Case Studies* (Golden, Colorado: Solar Energy Research Institute, 1980), p. 18.
33. Wieble Alley, Combined Cycle Cogeneration Feasibility Assessment, Presented at the First New Jersey Cogeneration Conference, June 19, 1981.
34. Reid and Chiogioji, op. cit., p. 127.
35. Solar Energy Research Institute, *A New Prosperity: Building a Sustainable Energy Future* (Andover, MA: Brick House, 1981), p. 375.
36. Lang, op. cit.
37. Ibid.
38. Lang, op. cit.
39. Ibid.
40. Ibid.
41. Meckler, op. cit., p. 89.
42. Baum, op. cit., p. 15.
43. Ibid.
44. Meckler, op. cit., p. 162.
45. Dean, op. cit., p. 58.
46. Lang, op. cit.
47. Meckler, op. cit., p. 162.
48. City of Portland, op. cit., p. 31.
49. Dean, op. cit., p. 10.

50. Industrial Energy Conservation Strategic Plan, op. cit., p. 15.
51. Ibid., p. 24.
52. Annual Report to the Congress and the President on the Industrial Energy Efficiency Improvement Program, op. cit., p. 60.
53. Meckler, op. cit., p. 167.
54. Ibid., p. 168.
55. Charles Berg, Process Innovation and Changes in Industrial Use, in Abelson and Rummond, eds., *Energy II: Use, Conservation, and Supply* (Washington, D.C.: American Association for the Advancement of Science 1978), p. 8.
56. Reid and Chiogioji, op. cit., p. 130.
57. The DOE Industrial Energy Conservation Program, op. cit., p. 8.
58. Annual Report to the Congress and the President, op. cit., p. 44.
59. Dean, op. cit., p. 47.
60. Annual Report to Congress and the President, op. cit., p. 53.
61. Ibid.
62. Ibid., p. 53.
63. Personal Communication, National Concrete Masonry Association, January, 1982.
64. E. Kenneth May, *Solar Energy and the Oil Refining Industry* (Golden, CO: Solar Energy Research Institute, 1980), p. 11.
65. Ronald S. Wishart, Industrial Energy in Transition: A Petrochemical Transition, in Abelson and Hammond, op. cit., p. 11.
66. Annual Report to Congress and the President, op. cit., p. 48.
67. Meckler, op. cit.
68. Dean, op. cit., p. 60.
69. Annual Report to Congress and the President, op. cit., p. 34.
70. Hooker, op. cit., p. 29.
71. Lang, op. cit.
72. Hooker, op. cit., p. 108.
73. Ibid., p. 6.
74. Office of Technology Assessment, Application of Solar Technology to Today's Energy Needs (Washington, D.C.: Office of Technology Assessment, 1978), 2 vols., vol. 1.
75. Chalres F. Kutscher and Roger L. Davenport, *Preliminary Operational Results of the Low-Temperature Solar Industrial Process Heat Field Tests* (Golden, CO: Solar Energy Research Institute, 1980). p. 12.
76. Kutscher and Davenport, op. cit., p. 23.
77. Hooker, op. cit., p. 77.
78. Kenneth C. Brown, Solar Industrial Process Heat: Costs and Performance Vary Widely, *Solar Engineering,* June, 1981, p. 23.
79. Ibid., p. 22.
80. *A New Prosperity,* op. cit., p. 267.
81. Kutscher and Davenport, op. cit., p. 15.
82. A. B. Cassomajor, Solar Industrial Process Heat: The Private Sector, Solar Industrial Process Heat Conference Proceedings (Golden, CO: Solar Energy Research Institute, 1979), p. 68.
83. Allen Gatzske and Amy Skewes-Cox, *Regional Comparisons of On-Site Potential in the Residential and Industrial Sectors* (Berkeley: Lawrence Berkeley Laboratory, 1980), p. 12.
84. Ibid., p. 15.
85. Ibid., p. 86.
86. Valerie M. Greaver, et. al., Applications of Solar Energy in Industrial Parks (Golden, CO: Solar Energy Research Institute, 1980), p. 12.

87. Hooker, op. cit., p. 21.
88. Kevin M. O'Brian and Barbara Euser, Solar Energy and the Search for Emissions Offsets, *Solar Law Reporter,* March/April, 1980, p. 1093.
89. *A New Prosperity,* op. cit., p. 257.
90. Ibid.
91. Mitre Corporation, Technologies for Commercial and Industrial Heat from Biomass, Working Paper #80 000010 (McLean, VA: 1979).
92. Office of Technology Assessment, Energy from Biological Processes (Washington, D.C.: Office of Technology Assessment, 1980), p. 82.
93. For a discussion of the complex offset procedures that can face industrial cogenerators or biofuels users, see: Kenneth E. Johnston, Electricity Cogeneration in Urban Areas, in Joel T. Werth, ed., *Energy in the Cities Symposium,* PAS Report No. 349 (Chicago: American Planning Association, 1980), pp. 31–36; Glenn Berger and John Townes, *A User's Guide to Legal and Institutional Factors Affecting the Use of Industrial Wood Boilers in New England* (Boston: New England Regional Commission, 1979).
94. Energy From Biological Processes, op. cit., p. 82.

CHAPTER THIRTEEN

1. F. J. Collins, et al., *Fuel Alcohol: An Energy Alternative for the 1980s* (Washington, D.C.: National Fuel Alcohols Commission, 1981), appendix, p. 173.
2. *Fuel Alcohol,* p. 48.
3. *Fuel Alcohol,* p. 49.
4. Estimate of Bill Holmberg, Department of Energy, April, 1982.
5. Donald Hertzmark, *The Agricultural Sector Impacts of Making Ethanol from Grain,* The Solar Energy Research Institute, 1980.
6. This figure was derived by averaging the transportation energy bills of nine CCEMP cities: Boulder (28.2 percent), Knoxville (31.5 percent), Portland (27.0 percent), Bridgeport (26.9 percent), Wayne, CO (22.0 percent), Toledo (18.1 percent), Allegheny, CO (21.0 percent), Philadelphia (19.0 percent), and Greenville (23.0 percent).
7. Telephone estimate, The Renewable Fuels Association, 499 South Capital Street, S.W., Suite 420, Washington, D.C. 20003.
8. Office of Technology Assessment, *Gasohol—A Technical Memorandum,* September 1979, p. iii.
9. Steve Meyers, "Can Methanol Be Managed?" *Soft Energy Notes,* v.3, n.4, (August/September, 1980).
10. National Alcohol Fuel Commission, *Alcohol Fuels From Biomass: Production Technology Overview,* Aerospace Corporation, July, 1980.
11. NAFC, *Alcohol Fuels From Biomass,* op. cit.
12. NAFC, *Alcohol Fuels From Biomass,* op. cit.
13. V. Daniel Hunt, *The Gasohol Handbook,* Industrial Press, Inc., 1981, pp. 247–272.
14. Op. cit., pp. 31–34.
15. For a general summary of this issue, see *The Gasohol Handbook* (pp. 39–46); Dan Jantzen and Tom McKinnon, *Preliminary Energy Balance and Economics of a Farm-Scale Ethanol Plant,* Solar Energy Research Institute, 1980.
16. Donald Hertzmark, *The Economics of Ethanol from Grain,* Solar Energy Research Institute, 1980.
17. D. Medville, *Comparative Economic Assessment of Ethanol from Biomass,* Mitre Corporation, 1978.

18. Solar Energy Research Institute, Strategic Planning Branch, Cost of Energy from Some Renewable and Conventional Technologies, 1981.
19. Margaret Hilton, Memo: Economics of Liquid Fuels from Biomass, Center for Renewable Resources, 1981, p. 6.
20. Steven R. Beck, Methanol from Wood Through Gasification in A Dual Fluidized-Bed Gasifier, in Schnorr et. al., *Biomass Feasibility Study for the State of Arizona*, 1981.
21. Solar Energy Research Institute, Strategic Planning Branch, op. cit.
22. F. J. Collins et al., Alcohol Fuels from Biomass: Production Technology Overview, in *Fuel Alcohol: An Energy Alternative for the 1980s*, Appendix, 1981.
23. U.S. National Alcohol Fuels Commission, *Fuel Alcohol: An Energy Alternative for the 1980s*, 1981, p. 64.
24. For a thorough discussion of ethanol's use as a substitute for oil, see *Fuel Alcohol: An Energy Alternative for the 1980s*, U.S. National Alcohol Fuels Commission, 1981.
25. Donald Hertzmark, et al., *The Agricultural Sector Impacts of Making Ethanol from Grain*, the Solar Energy Research Institute, 1981.
26. Office of Technology Assessment, *Gasohol—A Technical Memorandum*, September, 1979, p. 11.
27. Lester R. Brown, Food or Fuel: New Competition for the World's Cropland, Worldwatch Paper #35, March 1980.
28. For a readable discussion of cropland degradation, see *Building a Sustainable Society*, Lester R. Brown (New York. Norton, 1981.)
29. V. Daniel Hunt ,*Gasohol Handbook*, Industrial Press, Inc., 1981.
30. U.S. Congressional Committee on Banking and Urban Affairs, Compact Cities: Energy Saving Strategies for the Eighties, Committee Print 96-15, July, 1980.
31. David Zwick, Our Inland Waters, in *Life After '80: Environmental Choices We Can Live With* (Kathleen Courrier, editor), Brick House, 1980, p. 16; David Sheridan, The Underwatered West, *Environment*, March, 1981.
32. U.S. Department of Energy, Office of the Assistant Secretary for Environment, Comparing Energy Technology Alternatives from an Environmental Perspective, February, 1981, pp. 35–39.
33. Office of Technology Assessment, *Energy from Biological Processes*, July, 1980, pp. 108, 109.
34. Office of Technology Assessment, *Gasohol—A Technical Memorandum*, September 1979, p. 48.
35. V. Daniel Hunt, *The Gasohol Handbook*, Industrial Press, Inc., 1981, p. 388.
36. Robert Stobaugh and Daniel Yergin, *Energy Future* (New York: Ballantine paperback edition, 1979), p. 300. Based on 1979 Department of Energy figures.
37. Stobaugh and Yergin, p. 97–99.
38. Office of Technology Assessment (OTA), *Energy from Biological Processes*, vol. 1, Washington, D.C., 1980, p. 61.
39. OTA, p. 59.
40. Solar Energy Research Institute (SERI), *A New Prosperity*, Andover, MA, 1981, p. 250.
41. OTA, p. 61.
42. OTA, p. 68.
43. Timothy S. Cronin, "New Wood-Fired Generating Unit Brings Savings," *Public Power*, January/February, 1978, pp. 32–33.
44. Charles E. Hewett and Thomas E. Hamilton, *Forests in Demand* (Boston: Auburn House, 1982), p. 45.
45. OTA, p. 71.

46. Charles Hewett, Institutional Constraints on the Expanded Use of Wood Energy Systems, paper presented to the Third Annual National Biomass Systems Conference, June 6, 1979, p. 8.
47. OTA, pp. 75–81.
48. OTA, p. 59.
49. OTA, pp. 81–82.
50. U.S. General Accounting Office (GAO), Conversion of Urban Waste to Energy: Developing and Introducing Alternate Fuels from Municipal Solid Waste, February 1979, p. 1–2.
51. Robert Stobaugh, and Daniel Yergin, *Energy Future* (New York: Ballantine paperback edition 1979, updated), p. 99. Based on figures of National Electric Reliability Council.
52. Staff phone calls. Cf. James Ridgeway, *Energy Efficient Community Planning* (Emmaus, PA: J.G. Press, 1979), pp. 163–168.
53. Tania Lipshutz, *Garbage to Energy: The False Panacea,* Sonoma County Community Recycling Center, Santa Rosa, CA, September, 1979, pp. 14–17. Environmental Protection Agency, *Resource Recovery Plant Implementation: Guide for Municipal Officials: Technologies,* Washington, D.C., 1976, pp. 29–41.
54. Lipshutz, pp. 19–23.
55. Lipshutz, pp. 47–48.
56. Staff phone calls. Personal communications with Kenneth Cramer, Teledyne National Padonia Center (Baltimore); Bob Bartalotto, Refuse Recovery Plant (Ames); Robert Letter, Ted Jagelski, Division of Engineering (Madison).
57. EPA, p. 5.
58. Lipshutz, p. 44.
59. Lipshutz, p. 23.
60. Office of Technology Assessment, *Energy from Biological Processes,* 2 vols. (Washington, D.C.: OTA, 1981), vol. 1.
61. Ibid., vol. 2, p. 137.
62. Booz-Allen-Hamilton, *Assessment of Industrial Activity in the Utilization of Biomass for Energy* (Washington, D.C.: U.S. DOE, 1980).
63. General Accounting Office, *Conversion of Urban Waste to Energy: Developing and Introducing Alternative Fuels from Municipal Solid Waste* (Washington, D.C., GAO, 1979), p. 2–10.
64. Environmental Protection Agency, Resource Recovery Plant Implementation Guides for Municipal Offices, 8 vols. (Washington, D.C.: U.S. EPA, 1976), vol. 2, pp. 4–8.
65. General Accounting Office, op. cit., p. 2–9.
66. Ibid., pp. 2-9, 2-10.
67. Tania Lipshutz, *Garbage-to-Energy—The False Panacea,* Sonoma County Recycling Center, 1979, pp. 30–31.
68. Booz-Allen-Hamilton, op. cit., p. III-23.
69. OTA, op. cit., p. 73.
70. OTA, op. cit., p. 127.
71. Ibid.
72. F. J. Collins, et al., *Fuel Alcohol: An Energy Alternative for the 1980s* (Washington, D.C.: National Fuel Alcohols Commission, 1981), Appendix, p. 146.
73. Bruce Green, et al., *A Guidebook to Renewable Energy Technologies* (Golden, CO: Solar Energy Research Instiute, 1981).
74. Conversation with Joseph Mattera, Cranston, Rhode Island Methane Facility, Fall, 1981.
75. Collins, op. cit., p. 173.
76. *Technology Tomorrow,* June, 1980.

77. Conversation with Ted Lander, October, 1981.
78. Collins, op. cit., p. 173.
79. Ibid., p. 177.
80. Bureau of Sanitation, Baltimore, *Evaluation of Landfill Gas as an Energy Source* (Washington, D.C.: Energy Task Force of the Urban Consortium, 1980).
81. *Resource Recovery Activities,* National Center for Resource Recovery, September, 1981.
82. Blacket, Pacific Gas and Electric, private communication.
83. Conversation with Mr. Aliwalos, City of Los Angeles, Sanitation Department, Fall, 1981.
84. Conversation with Petr Mooij, REFCON, October, 1981.
85. Blancket, op. cit., p. 5.
86. OTA, op. cit., vol. 2, pp. 148:149.
87. Lipshutz, op. cit., p. 26.

References for Further Reading

I. COMMUNITY ENERGY PLANNING

A. General

John H. Alschuler, Jr., Community Energy Strategies: A Preliminary Review, May 1, 1980 (Draft).

J. L. Moore, et al., *The Comprehensive Community Energy Management Program: An Evaluation,* Argonne National Laboratories, 1982. Available from National Technical Information Service, 5285 Port Royal Road, Springfield, VA 22161 (NSV-tm-89).

Alan Okagaki, *County Energy Plan Guidebook,* Institute for Ecological Policies, 1979.

James Ridgeway, *Energy Efficient Community Planning,* J. G. Press, 1979.

David Pomerantz, et al., *Energy Mnagement in Municipal Operations—A Framework for Action,* International City Managers Association, 1981.

Annette Woolson, *The County Energy Production Handbook,* National Association of Counties Research Foundation, 1981.

B. Specific Communities

Greater Bridgeport Regional Planning Agency, Greater Bridgeport Region—Energy Audit, 1980 (CCEMP CITY).

City of Boulder, Community Energy Management Plan, May, 1980.

Knoxville/Knox County Metropolitan Planning Commission, Community Energy Use Profile, Sept. 1980 (CCEMP CITY).

Los Angeles Energy Management Advisory Board, *The Energy/LA Action Plan,* July, 1980 (CCEMP CITY).

Hampshire College, Energy Self-Sufficiency in Northampton, Massachusetts, U.S. Department of Energy, 1979.

David Morris, et al., *Planning for Local Self-Reliance: A Case Study for the District of Columbia,* Institute for Local Self-Reliance.

Philadelphia Solar Planning Project, Working Papers, 1980–81.

City of Portland, *Energy Conservation Choices for the City of Portland,* Oregon, 11 vols., U.S. Department of Housing and Urban Development, 1977.

St. Paul Energy Committee of 100, Reports, 1980.

Seattle Energy Office, Energy Ltd. (Draft Action Plan), October, 1980 (CCEMP CITY).

South Florida Regional Planning Council, South Florida Energy Audit, 1981, (CCEMP CITY).

Southern Tier Central Region, *Renewable Energy Resource and Technology Assessment,* U.S. Department of Energy, 1978.

II. TECHNOLOGIES

A. Energy Conservation in Buildings

Dermot McGuigan, *Heat Pumps: An Efficient Heating and Cooling Alternative,* Garden Way, 1981.

National Association of Homebuilders Research Foundation, *Insulation Manual,* 1979.

National Electrical Contractors Association and National Electrical Manufacturers Association, Total Energy Management Association, *Total Energy Management: A Practical Handbook of Energy Conservation and Management,* 1979.

New York State Energy Office, *Multifamily Housing Energy Conservation Workbook* (Undated).

Office of Technology Assessment, *Residential Energy Conservation,* 1979.

Travis L. Price, III, et al., *Energy Conservation Guidelines for the Sheffield Block Development,* Department of Architecture, Carnegie-Mellon University, 1981.

John Rothschild, *Stop Burning Your Energy,* Random House, 1981.

Tom Wilson, ed., *Home Remedies,* Mid-Atlantic Solar Energy Association, 1981.

B. Solar Energy for On-Site Space and Water Heating Applications

Bruce Anderson and Malcolm Wells, *Passive Solar Energy,* Brick House, 1980.

David Bainbridge, *The Integral Passive Solar Water Heater Book,* Passive Solar Institute, 1981.

Barbara Glenn and Gregory Franta, eds., *AS/ISDES 1981: Proceedings of the 1981 Annual Meeting,* 1981 (Past Proceedings also available).

Bruce Green, et al., *A Guidebook to Renewable Energy Technologies,* Solar Energy Research Institute, 1981.

John Hayes and Rachel Snyder, *The 5th National Passive Solar Conference,* 1980 (Past Proceedings also available).

U.S. Department of Housing and Urban Development, *Guidelines for Solar DHW in One- and Two-Family Dwellings,* 1979.

U.S. Department of Housing and Urban Development, *Energy Conserving Passive Solar Multifamily Retrofit Projects,* 1982.

Elizabeth McPherson, *Hot Water from the Sun,* U.S. Department of Housing and Urban Development, 1981.

Office of Technology Assessment, *Application of Solar Energy to Today's Energy Needs,* 2 vols., 1978.

C. Solar Energy for Electrical Applications

Wind

Energy Task Force, *Windmill Power for City People,* Community Services Administration, 1977.

V. Daniel Hunt, *Windpower,* Van Nostrand Reinhold, Inc., 1981.

Jack Park, *The Wind Power Book,* Cheshire Books, 1981.

Harry L. Wegley, et al., *A Siting Handbook for Small Wind Energy Conversion Systems,* Battelle Memorial Institute, 1980.

Photovoltaics
Paul Maycock and Edward Stirewalt, *Photovoltaics: Sunlight to Electricity in One Step,* Brick House, 1981.
Monegon, Ltd., *Solar Electricity: Making the Sun Work for You,* 1981.
Office of Technology Assessment, *Application of Solar Energy to Today's Energy Needs,* 1978.
Planning Research Corporation, *Solar Photovoltaic Applications Seminar,* U.S. DOE, 1980.

Hydropower
U.S. Army Corps of Engineers, *Feasibility Studies for Small-Scale Hydropower Additions,* A Guide Manual, 1979.
Energy Law Institute, Franklin Pierce Law Institute Center, *A Manual for Development of Small-Scale Hydroelectric Projects by Public Entities,* U.S. DOE, 1981.
U.S. DOE, *The Financing of Private Small-Scale Hydroelectric Projects,* 1981.
General Accounting Office, *Hydropower: An Energy Source Whose Time Has Come Again,* 1980.
National Center for Appropriate Technology, *Micro-Hydro Power: Reviewing an Old Concept,* 1979.

Wood
Georgia Institute of Technology, Wood Energy Systems Branch, *Wood: An Alternative Energy Resource,* 1980.
Charles Hewett and Colin High, *Construction and Operation of Small, Dispersed Wood-Fired Power Plants,* Resource Policy Center, Dartmouth College, 1978.
Georgia Institute of Technology, *The Industrial Wood Energy Handbook,* Van Nostrand Reinhold, Inc., 1983.

Geothermal
Ronald DiPippo, *Geothermal Energy as a Source of Electricity,* U.S. DOE, 1980.

Solar Ponds
Desert Research Institute, *Non-Convecting Solar Pond Workshops,* 1980.

D. Community Heating Systems

District Heating
District Heating, *Cogeneration Waste Heat Utilization, Avenues to Energy Conservation: Swedish Experiences and Technology,* Swedish Trade Office (Undated).
Jack Gleason, Efficient Fossil and Solar District Heating Systems: Preliminary Report, Solar Energy Research Institute, 1981.
Peter Margen, et al., *District Heating/Cogeneration Applications Studies for the Minneapolis/ St. Paul Area,* Oak Ridge National Laboratory, 1979.

Municipal Solid Waste Systems
U.S. Conference of Mayors, *Economic Development and Resource Recovery,* 1979. (See also the series of case studies produced by the Conference on Existing Resource Recovery Projects.)
Environmental Protection Agency, *Resource Recovery Plant Implementation: Guides for Municipal Officials,* 8 vols., 1976.
Tania Lipshutz, *Garbage-to-Energy, The False Panacea,* Sonoma County Recycling Center, Sept., 1979.

Geothermal

U.S. DOE, *Geothermal Direct Heat Applications: Program Summary,* 1980.
Geothermal Resources Council, *Direct Utilization of Geothermal Energy: A Technical Handbook,* 1979.

E. Energy Conservation and Renewable-Energy Opportunities in Industry

Glenn Berger and John Lounes, *A User's Guide to Legal and Institutional Factors Affecting the Use of Industrial Wood Boilers in New England,* New England Regional Commission, 1979.
Norman L. Dean, *Energy Efficiency in Industry,* Ballinger, 1980.
Douglas Hooker, et al., *Industrial Process Heat Case Studies,* 1980.
Meg Schacter, *Creating Jobs Through Energy Policy,* U.S. DOE, 1979.

III. URBAN ENERGY CONTEXT

A. Land Use

Robert U. Burchell and David Listokin, eds., *Land Use and Energy,* Center for Urban Policy Research, 1982.
Compact Cities: Energy Saving Strategies for the Eighties, Report, Together with Dissenting Views by the Subcommittee on the City of the Committee on Banking, Finance, and Urban Affairs, 1980.
Gail Boyer Hayes, *Solar Access Law,* Ballinger, 1979.
Martin Jaffe and Duncan Erley, *Protecting Solar Access for Residential Development: A Guidebook for Planning Officials,* U.S. Department of Housing and Urban Development, 1973.
Berg, *Solar Envelope Concepts,* Solar Energy Research Institute, 1980.
City of Los Angeles, *Solar Envelope Zoning: Application to the Solar Envelope Concept,* Solar Energy Research Institute, 1980.
Philadelphia Solar Planning Project, Working Papers—Neighborhoods, 1980.
Michael Shapiro, Boston Solar Retrofits, John F. Kennedy School of Government, 1980.
Rober Twiss, et al., *Land Use and Environmental Impacts of Decentralized Energy Use,* U.S. DOE, 1979.

Utilities

Glenn J. Berger and Raymond Royce, *A Handbook on the Sale of Excess Electricity by Industrial and Individual Power Producers Under the Public Utility Regulatory Policies Act,* Solar Energy Research Institute, 1980.
David Morris and David Bardaglio, Municipalities and PURPA 210: Opportunities for Local Revenue Generation (Draft), Institute for Local Self-Reliance, 1981.

Socio-economic Factors

Steven Ferrey, Solar Banking: Constructing New Solutions to the Urban Energy Crisis, *Harvard Journal of Legislation,* 1981.

Eunice Grier, *Colder . . . Darker,* Community Services Administration, 1977.

Steven Kaye, ed., *Multifamily Energy Conservation: A Reader,* Coalition of Northeast Municipalities, 1981.

Hans Landsberg and Joseph Dukert, *High Energy Costs: Uneven, Unfair, Unavoidable?* Resources for the Future, 1981.

Leonard Rodberg, Employment Impact of the Solar Transition, Joint Economic Committee, 1979.

Fuels from Renewable Resources

Office of Technology Assessment, *Energy from Biological Processes,* July, 1980.

U.S. National Alcohol Fuels Commission, *Fuel Alcohol: An Energy Alternative for the 1980s,* Appendix, 1981.

V. Daniel Hunt, *The Gasohol Handbook,* Industrial Press, Inc., 1981.

National Center for Appropriate Technology, *Alcohol Fuels: An Annotated Bibliography,* n.d.

Solar Energy Research Institute, *A Guide to Commercial-Scale Ethanol Production and Financing,* 1980.

Index